High Frequency Sources of Coherent Radiation for Fusion Plasmas

IOP Series in Plasma Physics

Series Editors

About the series

The IOP Plasma Physics ebook series aims at comprehensive coverage of the physics and applications of natural and laboratory plasmas, across all temperature regimes. Books in the series range from graduate and upper-level undergraduate textbooks, research monographs and reviews.

The conceptual areas of plasma physics addressed in the series include:

- Equilibrium, stability and control
- Waves: fundamental properties, emission, and absorption
- Nonlinear phenomena and turbulence
- Transport theory and phenomenology
- Laser-plasma interactions
- Non-thermal and suprathermal particle populations
- Beams and non-neutral plasmas
- High energy density physics
- Plasma-solid interactions, dusty, complex and non-ideal plasmas
- Diagnostic measurements and techniques for data analysis

The fields of application include:

- Nuclear fusion through magnetic and inertial confinement
- Solar-terrestrial and astrophysical plasma environments and phenomena
- Advanced radiation sources
- Materials processing and functionalisation
- Propulsion, combustion and bulk materials management
- Interaction of plasma with living matter and liquids
- Biological, medical and environmental systems
- Low temperature plasmas, glow discharges and vacuum arcs
- Plasma chemistry and reaction mechanisms
- Plasma production by novel means

High Frequency Sources of Coherent Radiation for Fusion Plasmas

G Dattoli
ENEA Frascati Research Center, Frascati, Rome, Italy

E Di Palma
ENEA Frascati Research Center, Frascati, Rome, Italy

S P Sabchevski
Institute of Electronics of the Bulgarian Academy of Sciences, Sofia, Bulgaria

I P Spassovsky
ENEA Frascati Research Center, Frascati, Rome, Italy

IOP Publishing, Bristol, UK

ISBN 978-0-7503-2464-9 (ebook)
ISBN 978-0-7503-2462-5 (print)
ISBN 978-0-7503-2465-6 (myPrint)
ISBN 978-0-7503-2463-2 (mobi)

DOI 10.1088/978-0-7503-2464-9

Version: 20210801

IOP ebooks

British Library Cataloguing-in-Publication Data: A catalogue record for this book is available from the British Library.

Published by IOP Publishing, wholly owned by The Institute of Physics, London

IOP Publishing, Temple Circus, Temple Way, Bristol, BS1 6HG, UK

US Office: IOP Publishing, Inc., 190 North Independence Mall West, Suite 601, Philadelphia, PA 19106, USA

Contents

Preface

This book deals with the scientific and technological aspects associated with those devices currently exploited for the heating of magnetically confined fusion plasma with intense electromagnetic waves.

The book consists of two parts; the first dealing with the basics elements of plasma physics, the second accounts for electromagnetic sources designed as external plasma heaters. We will discuss particular items, like free electron laser (FEL) type generators of coherent electromagnetic radiation, designed to provide the supplementary power necessary to ignite, heat and sustain the plasma. They cannot, therefore, be separated from the physical aspects of magnetic fusion and of magnetized plasma, which are initially sketched in the first three chapters.

Before getting into the book plan, we would like to mention that this effort merges four different expertises, which encompass the physics of undulator FELs, the physics of gyrotrons, that of cyclotron auto-resonance maser (CARM), and more generally the physics of charged particles–radiation interaction. The skills of the authors are equally distributed between theoretical, experimental and numerical competencies, but do not include any long-term experience of direct work on plasma.

Even though this created a major problem in writing the book, it turned out to be a decisive advantage because we have been obliged to study deeply what we had learned in the past in amateur terms.

We filtered the physics and Tokamak engineering through our own scientific experiences and the results are those summarized in the first three chapters, which deal with the physics of fusion plasma, the physics of Tokamaks, the associated technical issues and some aspects of the additional heating.

The subsequent chapters cover a general introduction to the FEL-like coherent sources and more specific descriptions of the physics of undulator FELs, gyrotrons and CARM.

We have placed particular emphasis on the description of the latter device, for different reasons, including the fact that it was the main topic of our research in recent years. It is an extremely challenging device, because it requires significant technological efforts, which go from an extremely demanding power supply, to a high performing electron gun capable of providing a high quality beam and a corresponding beam transport system.

The compilation of the book has been conceived in a painful period for the authors' life. They have lost beloved people who had played a central role in terms of affection, support, and presence in their growth, not just professional. Therefore Giuseppe Dattoli dedicates the book to his mother Ada, Emanuele Di Palma to his father Andrea, Svilen Sabchevski to his wife Petia and Ivan Spassovsky to his parents Rumiana and Pano.

The book comes after years of intensive research work in the field. We had been assisted by more experienced colleagues.

We express our gratitude to Dr A Cardinali and Ing. F Mirizzi for reading and correcting the first two chapters. Discussions on weak interactions and solar 'burning' with Dr F Alladio have been useful to clear our minds of many dangerous (wrong) commonplaces.

The understanding of the physics of gyrotrons and CARM has benefitted from discussions with Professor G Nusinovich who shared with us years of experience in these fields.

The generous effort in correcting our misconceptions by Professors N Ginzburg, M Glyavin, N Peskov and A Savilov from IPA-RAS Nizhny Novgorod has been greatly appreciated.

It is finally a great pleasure to thank the colleagues of the CARM task force at ENEA Frascati for sharing with us many enlightening discussions on the topics treated in this book.

Giuseppe Dattoli
Emanuele Di Palma
Svilen Petrov Sabchevski
Ivan Panov Spassovsky

Author biographies

Giuseppe Datoli

Giuseppe Datoli was born in Lagonegro, Italy, in 1953. He received a PhD degree in physics from La Sapienza University of Rome Italy, in 1976. He is an ENEA Researcher and has been involved in different research projects, including high energy accelerators, free electron lasers, and applied mathematics networks since 1979. Dr Dattoli has taught in Italian and Foreign universities, and has received the FEL Prize Award for his outstanding achievements in the field.

Emanuele Di Palma

Emanuele Di Palma received the Laurea degree in mathematics from La Sapienza University of Rome Italy, in 1996. He started his research activity on development of advanced FDTD numerical modeling applied to the electromagnetic simulation with the Italian Institute of Mathematics (INdAM). He joined ENEA laboratories in 2003, where he started his research activities on free electron laser application and microwave tube. He received a master degree in 'Fusion Energy: Science and Engineering' from Tor Vergata University of Rome Italy, in 2013 and a PhD degree in 'Fusion Science and Engineering' from the University of Padova Italy in cotutelle with the Universidade de Lisboa—Instituto Superior Tecnico-IST (Portugal) in 2018. Dr Di Palma is presently serving as Task Force Leader for the design and construction of a CARM device at the ENEA Frascati Laboratories.

Svilen Petrov Sabchevski

Svilen Petrov Sabchevski graduated from St. Petersburg State Electrotechnical University (Russia) in 1984 as an MSc in Electron Devices (diploma with distinction summa cum laude). In 1991 he received a PhD degree in physics from the Institute of Electronics of the Bulgarian Academy of Sciences (IE-BAS). Currently, he is head of Laboratory Plasma Physics and Engineering at IE-BAS. His research interests are in the fields of physics and applications of intense electron beams, computer-aided design and development of gyrotrons for various novel applications in the fundamental physical research and high-power THz science and technologies.

Ivan Panov Spassovsky

Ivan Panov Spassovsky received a PhD degree in physics from Sofia University, Bulgaria and started his research activity at the laboratory of Plasma Electronics. In 1992 he moved to the Brazilian Institute of Space Research (INPE) to work on the gyrotron. In 1994–1995 he joined the ENEA, Frascati FEL team developing at that time a compact microtron driven FEL. He spent two years with the Korean Atomic Energy Institute (KAERI) collaborating in the realization of far-infrared FEL. In 1999 he moved to the University of Maryland, where he participated in the development of harmonic Gyroklystron. In 2002 he returned to the ENEA to work on the compact IR FEL and SPARC FEL project. Currently he is responsible for the realization of a 250 GHz CARM oscillator.

Operators properties

	Cartesian coordinates (x, y, z)	Cylindrical coordinates (ρ, ϕ, z)
$\vec{A}$	$A_x\hat{x} + A_y\hat{y} + A_z\hat{z}$	$A_\rho\hat{\rho} + A_\phi\hat{\phi} + A_z\hat{z}$
$\vec{\nabla} \cdot \vec{A}$	$\frac{\partial A_x}{\partial x} + \frac{\partial A_y}{\partial y} + \frac{\partial A_z}{\partial z}$	$\frac{1}{\rho}\frac{\partial(\rho A_\rho)}{\partial \rho} + \frac{1}{\rho}\frac{\partial A_\phi}{\partial \phi} + \frac{\partial A_z}{\partial z}$
$\vec{\nabla} \times \vec{A}$	$\hat{x}\left(\frac{\partial A_z}{\partial y} - \frac{\partial A_y}{\partial z}\right) + \hat{y}\left(\frac{\partial A_x}{\partial z} - \frac{\partial A_z}{\partial x}\right) +$ $+ \hat{z}\left(\frac{\partial A_y}{\partial x} - \frac{\partial A_x}{\partial y}\right)$	$\hat{\rho}\left(\frac{1}{\rho}\frac{\partial A_z}{\partial \phi} - \frac{\partial A_\phi}{\partial z}\right) + \hat{\phi}\left(\frac{\partial A_\rho}{\partial z} - \frac{\partial A_z}{\partial \rho}\right) +$ $+ \hat{z}\left(\frac{\partial(\rho A_\phi)}{\partial \rho} - \frac{\partial A_\rho}{\partial \phi}\right)\frac{1}{\rho}$
$(\vec{A} \cdot \vec{\nabla})\vec{B}$	$\vec{A} \cdot \nabla B_x\ \hat{x} + \vec{A} \cdot \nabla B_y\ \hat{y} + \vec{A} \cdot \nabla B_z\ \hat{z}$	$\left(A_\rho\frac{\partial B_\rho}{\partial \rho} + \frac{A_\phi}{\rho}\frac{\partial B_\rho}{\partial \phi} + A_z\frac{\partial B_\rho}{\partial z} - \frac{A_\phi B_\phi}{\rho}\right)\hat{\rho} +$ $+ \left(A_\rho\frac{\partial B_\phi}{\partial \rho} + \frac{A_\phi}{\rho}\frac{\partial B_\phi}{\partial \phi} + A_z\frac{\partial B_\phi}{\partial z} - \frac{A_\phi B_\rho}{\rho}\right)\hat{\phi} +$ $+ \left(A_\rho\frac{\partial B_z}{\partial \rho} + \frac{A_\phi}{\rho}\frac{\partial B_z}{\partial \phi} + A_z\frac{\partial B_z}{\partial z}\right)\hat{z}$
ψ	$\psi: \mathbb{R}^3(x, y, z) \to \mathbb{R}$	$\psi: \mathbb{R}^3(\rho, \phi, z) \to \mathbb{R}$
$\vec{\nabla}\psi$	$\hat{x}\frac{\partial \psi}{\partial x} + \hat{y}\frac{\partial \psi}{\partial y} + \hat{z}\frac{\partial \psi}{\partial z}$	$\hat{\rho}\frac{\partial \psi}{\partial \rho} + \hat{\phi}\frac{\partial \psi}{\partial \phi} + \hat{z}\frac{\partial \psi}{\partial z}$
$\nabla^2\psi$	$\frac{\partial^2 \psi}{\partial x^2} + \frac{\partial^2 \psi}{\partial y^2} + \frac{\partial^2 \psi}{\partial z^2}$	$\frac{1}{\rho}\frac{\partial}{\partial \rho}\left(\rho\frac{\partial \psi}{\partial \rho}\right) + \frac{1}{\rho^2}\frac{\partial^2 \psi}{\partial \phi^2} + \frac{\partial^2 \psi}{\partial z^2}$

Part I

Fusion plasma generalities

IOP Publishing

High Frequency Sources of Coherent Radiation for Fusion Plasmas

G Dattoli, E Di Palma, S P Sabchevski and I P Spassovsky

Chapter 1

Magnetically confined plasma for fusion energy

1.1 Worldwide energy needs and fusion plants

Whatever energy plant one has in mind, the final goal is a device accomplishing the paradigmatic cycle associated with a generator of heat and a heat exchanger powering an electrical generator (see figure 1.1).

The black-box (dashed cylinder in figure 1.1) may be provided by any device producing energy, based on any available mechanism of mechanical, chemical or nuclear nature [1].

In order to understand what is the amount of energy needed to support a reasonable model of development for the next years, we have reported in figure 1.2 the world population growth, the associated energy requirements and the per-capita energy consumption.

The distinction between rich and poor countries can be measured in terms of the energy usage, we start therefore with a naive (but effective) argument, allowing the gross understanding of the numbers involved in the management of energy demand [2]. A 'democratic' figure of merit is the basal metabolic rate (BMR), namely the amount of power necessary to sustain the human body in conditions of absolute resting. In normal conditions, the BMR power is on the order of 100 W. Therefore, in one year a population of $7 \cdot 10^9$ individuals demands a survival energy expenditure amounting to

$$E[J] \cong 10^2 \times 7 \times 10^9 \times 8.64 \times 10^4 \times 3.65 \times 10^2 \cong 2.2 \times 10^{19} = 22 \text{ EW (Exa–Watt)}$$

where the suffix EW stands for 10^{18}. Such an amount of energy is equivalent to about 4% of the total value, which includes not only the mere survival contingency, but all the rest (production costs, housing, transport, education, defense etc) [3].

doi:10.1088/978-0-7503-2464-9ch1

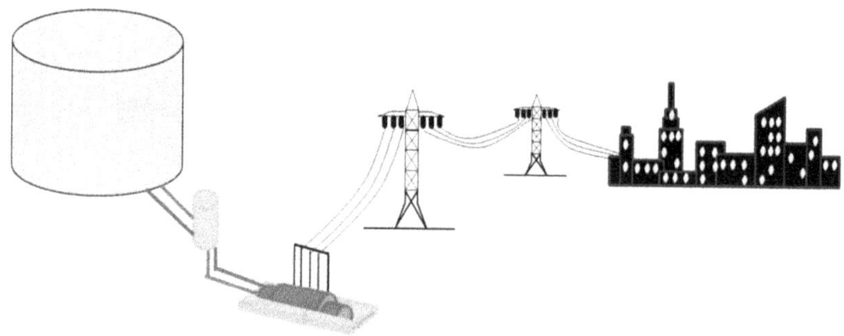

Figure 1.1. Heat generator, heat exchanger, turbine and electric power generator.

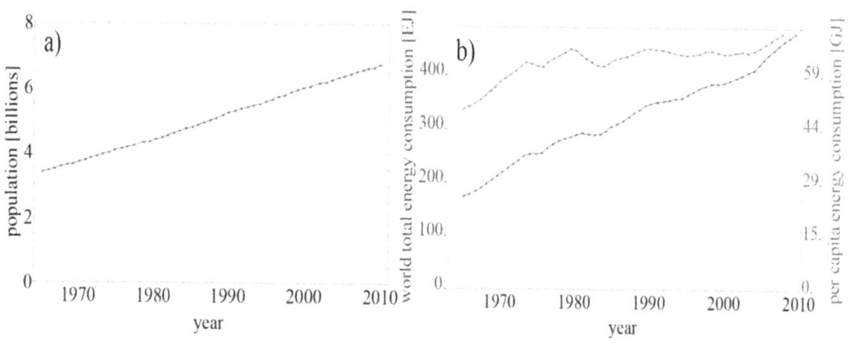

Figure 1.2. (a) Population growth in the last 50 years; (b) associated energy demand (blue) and the per capita energy request (in red with the y-axes scale on the right).

Just to have some reference numbers, we note that the energy released in a chemical reaction is about 1 eV per nucleon, 1 MeV in fission reactions and 3.5 MeV in fusion reactions[1].

Stated in more macroscopic terms we underscore that, to sustain 100 W of basal metabolism for one day, 3 kg of coal should be burnt. On the other hand, $1.2 \cdot 10^{-4}$ g of deuterium–tritium (D–T) mixture ensures the same amount of energy.

Stated in these terms, the choice of power plants based on fusion processes is an almost obligatory choice; let us assume[2] that this is true and that we can provide a significant amount of the world-wide energy consumption by burning D–T mixtures in fusion reactors.

The accomplishment of such a task requires:

(a) The availability of deuterium; this is not a great problem, about 10^{16} kg are in the surface water on Earth.
(b) Tritium does not occur naturally, it is radioactive (β-decay) with a half-life of 12 years. It must be produced inside the fusion reactor through the

[1] In MKSA units 1 eV = $1.6 \cdot 10^{-19}$ J.
[2] We discuss cons at the end of this section.

following reactions involving lithium (another abundant element on Earth, sufficient for thousands of years):

$$^{7}Li + \text{neutron} \Rightarrow {}^{4}He + T + \text{neutron} - 2.49\ \text{MeV}$$
$$^{6}Li + \text{neutron} \Rightarrow {}^{4}He + T + \text{neutron} + 4.8\ \text{MeV}$$

(c) Once the D–T mixture is available the conditions to achieve fusion reaction

$$D + T \Rightarrow^{4} \text{He}(3.5\ \text{MeV}) + n(14.1\ \text{MeV}),$$

must be ensured. Both D and T ions are positively charged and therefore experience Coulomb repulsion, which prevents them from coming sufficiently close to fuse. The problem can be avoided by embedding the fusion elements in a hot plasma environment, where the high temperature yields sufficient energy to overcome the Coulomb barrier [4, 5].

The architecture of a fusion plant can be that shown in figure 1.3. The D–T mixture is injected inside a 'confining' chamber and tritium is bred through an external blanket, containing lithium. The heat produced inside the chamber is then exchanged, via ordinary components, to electric power.

The quantities to be controlled in order to achieve the conditions to produce sufficient power in the plasma chamber are listed below:

(i) The density of the injected particles (n);
(ii) The energy acquired by the particles to ensure the reaction, specified in terms of the plasma temperature (T [keV])[3]
(iii) The confining time (τ_E) that guarantees a sufficient number of reactions.

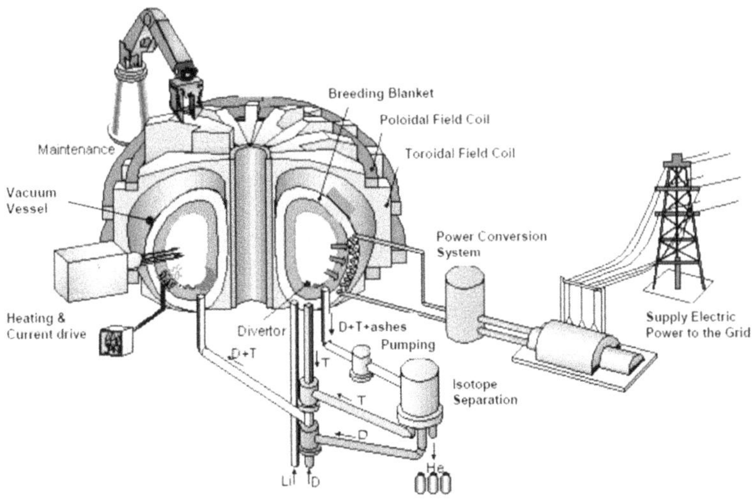

Figure 1.3. The fusion plant architecture (credit: EUROfusion; see reference [6]).

[3] The conversion from temperature to energy is realized through the Boltzmann constant k_B, as discussed in the following.

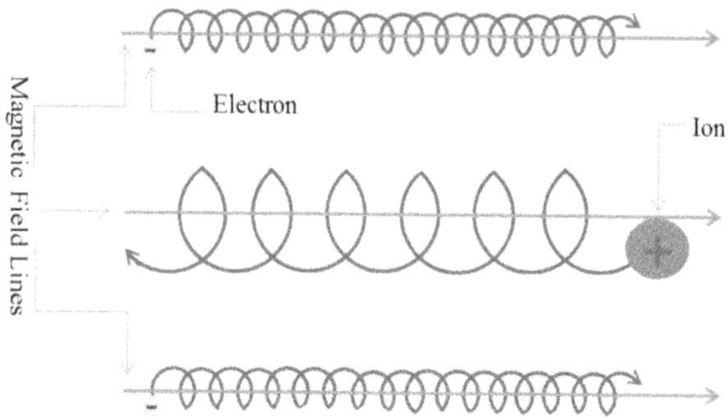

Figure 1.4. Trajectories of charged particles along the magnetic field lines, for light (electrons) and heavy particles (ions).

In conclusion, the product $nT\tau_E$ should be kept sufficiently high to sustain a large number of processes, allowing the production of net energy, namely larger than that employed to bring the device to the energy break-even level [7].

So far, we have mentioned the necessity of confining the D–T mixture in rather vague terms The most effective tool to confine charged particles is a magnetic field, along whose lines the particles execute helical trajectories (see figure 1.4).

It is evident that a magnetic field with the straight intensity lines in figure 1.4 is not a confining tool since the particles flow outside at the end points and are therefore lost.

The solution to achieving confinement is that of closing the magnetic field line, by exploiting the arrangement shown in figure 1.5, where the essential elements of a Tokamak[4] device are reported. The chamber containing the fusion materials has the geometrical shape of a torus and the guiding field inside is ensured by the toroidal coils. The other magnetic fields associated with the solenoid, namely the poloidal coils (with a current, circulating along the torus axis) guarantee the plasma stability, and the relevant technical details, along with the associated physical issues, will be described later in this chapter.

Let us summarize and underscore the main points of the discussion developed so far:

1. A mixture of D–T is injected inside a toroidal chamber where it circulates and the components eventually fuse.
2. Both deuterium and tritium are ionized, this occurs because at the energies required to ensure the fusion conditions (>10 keV) the electrons are stripped during the collisions, and in these conditions the mixture becomes a plasma.
3. The electrons do not play any direct role in the fusion reactions, but their presence guarantees the neutrality of the plasma and their higher mobility, with respect to the heavier particles, strongly affects the plasma properties.

[4] This is an acronym from Russian: toroidal'naya kamera s magnitnymi katushkami (toroidal chamber with magnetic coils).

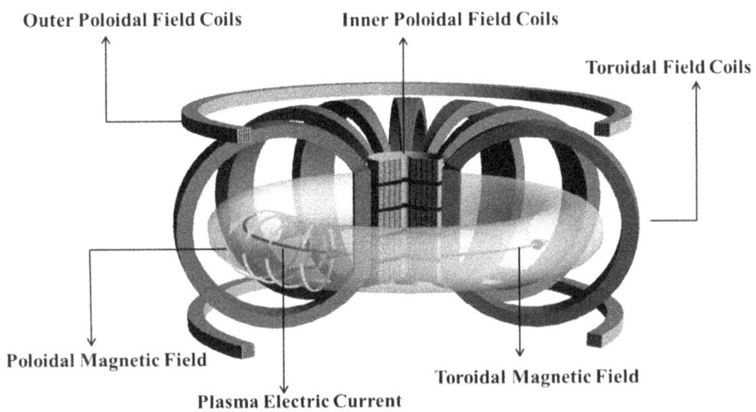

Figure 1.5. Schematic of the Tokamak chamber and surrounding coils.

In the following section, we will expand this qualitative discussion by introducing further elements clarifying the physical quantities that are of primary importance.

Before closing this introductory section, we would like to dampen the enthusiasm we may have raised on the feasibility of the fusion devices and on their short-term use.

Despite the benefits which are commonly mentioned (almost inexhaustible availability of the fusion elements, no waste products, no CO_2 emission, no radioactivity, no proliferation risks, intrinsic safety, etc) the cons against fusion are many and not trivial.

In many years of active studies and after significant capital investment, no plants providing electricity by fusion have been realized. There are intrinsic problems which make such devices significantly more complicated than fission reactors. Extremely dense and hot plasma is needed to overcome the Coulomb repulsion between deuterium and tritium and make them sufficiently close to fuse. A temperature near that of the Sun's core is therefore required. When the fusion takes place, it ranges around hundreds of millions of Kelvin, which poses severe problems on the choice of the materials necessary to keep the hot 'particle soup'.

Apart from the technical discussions (touched on in the forthcoming sections) a few points based on common sense can be listed[5]. For example, one could argue the following for fusion currently:

- it is unproven at anything resembling a commercial scale;
- it is not expected for full-scale production until at least 2050;
- commercial power plants would be extremely expensive to build;
- the billions in research funding could be spent on renewables instead.

We have just mentioned a number of specific issues, which are part of an ongoing discussion to be carefully accounted for, when dealing with future perspectives of the

[5] See http://energy-pros-and-cons.com/2018/03/22/nuclear-fusion-pros-and-cons/.

commercial use of fusion. We will not address any further comments in this direction, because they are not strictly within the scope of this book.

1.2 Fusion products and energy balance

We have mentioned that the key fusion reaction D–T is characterized by the production of an α particle ^4H_e and a neutron, together with an energy release of 17.6 MeV. This last statement sounds obscure and needs to be clarified. The energy accompanying the fusion mechanism is just due to the mass defect of the nuclear components involved in the reaction, which is evaluated as [8]

$$\begin{aligned} m_D &= (2 - 9.9400 \cdot 10^{-4})m_H \qquad & m_T &= (3 - 6.284 \cdot 10^{-3})m_H, \\ m_{H_e} &= (4 - 2.7404 \cdot 10^{-2})m_H \qquad & m_n &= (1 + 1.378 \cdot 10^{-3})m_H, \end{aligned} \tag{1.1}$$

where m_H, m_D, m_T, m_n are the hydrogen, deuterium, tritium, and neutron mass, respectively. The mass deficit is therefore

$$\Delta m = (m_D + m_T) - (m_\alpha + m_n) = 1.87 \cdot 10^{-2} m_H, \tag{1.2}$$

where $m_H = 1.627 \cdot 10^{-27}$ kg and $m_\alpha = 4.0015\, m_H$.

According to the Einstein relation we find, for the released energy [9–11],

$$E_F = \Delta m\, c^2 = 1.87 \cdot 10^{-2} m_H c^2 = 2.8184 \cdot 10^{-12}\ \mathrm{J} \cong 17.591\ \mathrm{MeV}. \tag{1.3}$$

This amount of energy is carried out by the fusion products as kinetic energy. It is a matter of a simple exercise to understand how this amount of energy is distributed among H_e and the neutron. From the energy and momentum conservation we find

$$\begin{aligned} \frac{1}{2} m_{He} v_{He}^2 + \frac{1}{2} m_n v_n^2 &= E_{fus}, \\ m_{He} v_{He} + m_n v_n &= 0. \end{aligned} \tag{1.4}$$

Thus yielding

$$\begin{aligned} \frac{1}{2} m_{He} v_{He}^2 &= \frac{m_n}{m_{He} + m_n} E_{fus}, \\ \frac{1}{2} m_n v_n^2 &= \frac{m_{He}}{m_{He} + m_n} E_{fus}. \end{aligned} \tag{1.5}$$

It is accordingly evident that, as $m_n/m_{He} \cong 1/4$, almost 80% of the fusion energy is carried by the neutrons.

The energy can be expressed in Kelvin temperature by using the relation

$$\begin{aligned} E &= k_B T[\mathrm{K}], \\ k_B &\cong 1.38 \cdot 10^{-23}\ \mathrm{J\,K^{-1}}. \end{aligned} \tag{1.6}$$

We argue therefore that 1 eV $\cong 1.16 \cdot 10^4$ K, an energy of 15 keV amounts to $1.8 \cdot 10^8$ K, corresponding to the Sun's core temperature, we have quoted before.

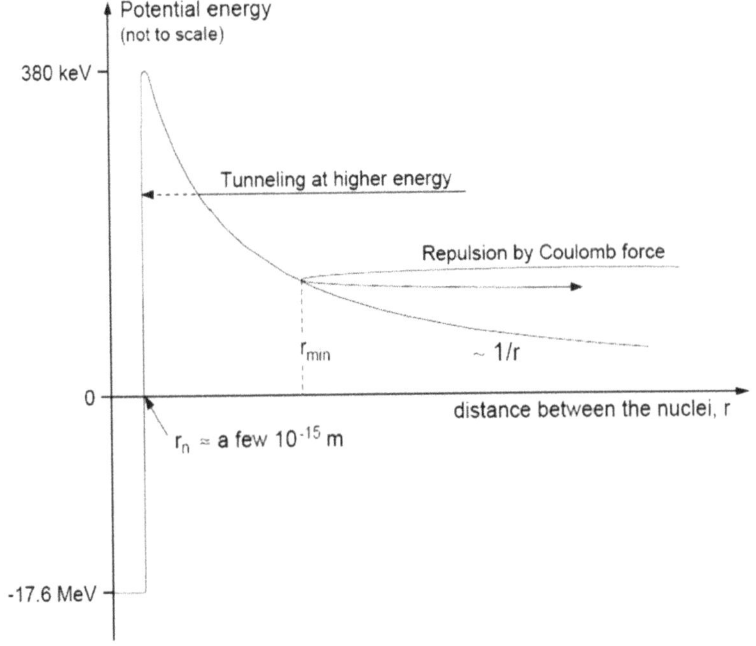

Figure 1.6. Coulomb potential U of two nuclei versus their mutual distance. Nuclei (in the present case D, T) fuse when they overcome the Coulomb barrier. The energy of 380 keV is a 'classical' reference value, whereas for the process ruled by quantum mechanical effects, the fusion occurs at significantly lower values. Reproduced from [1]. Copyright The Authors 2015. CC BY 4.0.

We furthermore recall that the kinetic average energy and velocity in terms of the temperature reads [12]

$$\begin{aligned} \langle T \rangle &= \frac{3}{2} k_{\mathrm{B}} T[\mathrm{K}], \\ v_{\mathrm{th}} &= \sqrt{\frac{2\, k_{\mathrm{B}} T}{m}}. \end{aligned} \tag{1.7}$$

The plasma must be heated to a temperature of the order of tens of keV for the simple reason that the fusion takes place when the corresponding energy is sufficient to overcome the Coulomb barrier. Particles with the same charge are subject to a potential of the type shown in figure 1.6. The Coulomb repulsion increases till the nuclear binding forces become larger. Loosely speaking, two nuclei a, b with charges and radii ($Z_{a,b}$, $R_{a,b}$) undergo a fusion process if their energies satisfy the condition

$$k_{\mathrm{B}} T > \frac{Z_a Z_b}{R_a + R_b} \frac{e^2}{4\pi\varepsilon_0}, \tag{1.8}$$

which in the case of D and T yields[6]

[6] The numerical values for the T–D system are $Z_{\mathrm{T}} = 1$, $R_{\mathrm{T}} = 1.7 fm$, $Z_{\mathrm{D}} = 1$, $R_{\mathrm{D}} = 1.5$ being $fm = 10^{-15}$ m.

$$k_{\mathrm{B}}T \approx 400\ \mathrm{keV}, \tag{1.9}$$

corresponding to a temperature of 10^9 K. It must be stressed that actually the quoted values refer to the total energy, therefore they can be halved.

We have noted that the quantity to be kept under control and maximized is the triple product involving energy, D and T number densities, and the confinement time. Furthermore, we are in the position of providing a more quantitative statement.

The Lawson criterion is the key argument to get a fairly good understanding of the physical reasons underlying the previous statement.

The fusion power available in a fusion process is roughly given by

$$P_{\mathrm{fus}} \cong V \cdot n_{\mathrm{D}} \cdot n_{\mathrm{T}} \cdot E_{\mathrm{fus}} \cdot \Sigma. \tag{1.10}$$

The first term V is the volume containing $n_{\mathrm{D,T}}$ number densities of deuterium and tritium, E_{fus} is the energy released per fusion process with the assumption that it will all contribute to the plasma heating (we will specify more precisely its specific value in the following; for the moment, however, we assume that it is of the order of E_{fus} that we have just mentioned. By a dimensional analysis, we can easily find that Σ is a quantity with the dimensions of a volume divided by a time. It is evident that P_{fus} depends also on the probability rate at which the fusion event occurs. We argue therefore that

$$\Sigma \cong \frac{\sigma(v)\,\Delta s}{\Delta t} = \sigma(v)\,v, \tag{1.11}$$

where $\sigma(v)$ is the cross section of the process at a given velocity v [13].

In less uneducated terms, we can say that Σ is the reactivity, provided by the following average

$$\Sigma = \langle \sigma(v)\,v \rangle = \int_0^{\infty} \sigma(v)\,v\,f(v)dv, \tag{1.12}$$

where $f(v)$ is the velocity distribution (which we will comment on later in this chapter).

In figure 1.7, which shows the reactivity versus the particle energy, the maximum is located around a few tens of keV.

The power due to the fusion process should now be compared to the power loss P_{loss}, which is the energy flow to the environment during the confinement time τ_E

$$\begin{aligned} P_{\mathrm{loss}} &= \frac{W}{\tau_E}, \\ W &= \frac{3}{2}k_{\mathrm{B}} \int [n_e T_e + (n_{\mathrm{D}} + n_{\mathrm{T}})T_i]dV, \end{aligned} \tag{1.13}$$

where n_e is the electron density, $T_{e,i}$ are the temperatures of the electrons and ions, respectively

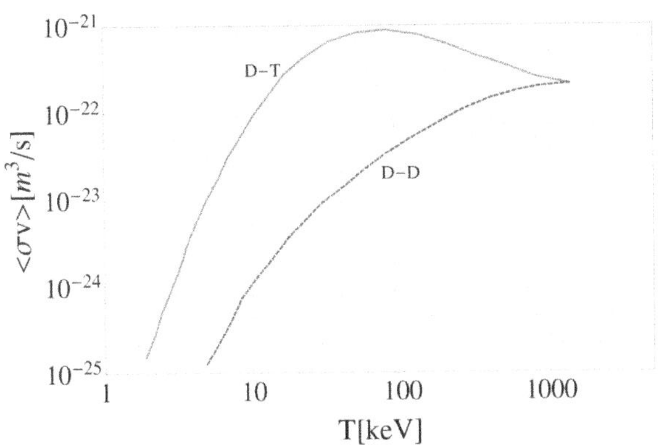

Figure 1.7. Reactivity versus energy for deuterium–tritium (D–T) and deuterium–deuterium (D–D) reactions.

The break-even is reached when $P_{\text{fus}} = P_{\text{loss}}$ and a self-sustained operation takes place when

$$P_{\text{fus}} > P_{\text{loss}}. \tag{1.14}$$

We assume thermal equilibrium ($T_e = T_i$) and neutrality ($n_{\text{D}} + n_{\text{T}} = n_e = n$) thus finding for the power loss

$$\frac{W}{V} = 3k_{\text{B}}n\, T. \tag{1.15}$$

Assuming furthermore $n_{\text{D}} = n_{\text{T}} = \frac{1}{2}n$ we find from equation (1.14)

$$n\, \tau_E > L, \tag{1.16}$$

where

$$L = \frac{12k_{\text{B}}\, T}{E_f\langle\sigma(v)\, v\rangle}. \tag{1.17}$$

The minimum value of the quantity L can be inferred from figure 1.7, which shows that for the D–T fusion it is around 25 keV[7] and therefore

$$n\, \tau_E > 1.5 \cdot 10^{20}\text{s m}^{-3}. \tag{1.18}$$

This inequality is the well-known Lawson criterion, which must be satisfied by the design of any fusion device, as discussed below.

1.3 Magnetic field and confinement

We have underscored that the geometrical arrangement of the magnetic fields, around and inside the fusion chamber, are the key tools determining the confinement but not the only ones.

---

[7] This value (25 keV) may sound lower than the values necessary to overcome the Coulomb barrier. It is just a reference value based on the classical physics. The presence of the quantum tunneling significantly lowers this value thus allowing the reactivity to peak around 20 keV.

The inspection of figure 1.5 yields an idea of the complexity of the topology of coils leading to the field suitable for the confinement.

Before explaining how plasma reacts as a whole to an external field, it is worth starting from a microscopic picture aiming at explaining the motion of charged particles inside the torus. Within this context, the most natural tool is the use of the Lorentz force, which allows the understanding of a significant amount of the phenomenology of the confinement. In figure 1.4 we have reported the motion of particles (electrons and ions), moving along a magnetic field line. The particle trajectory is a helix, characterized by a radius and a gyration frequency. The most natural solution to confine the particles is the torus geometry, in which the charged particles move around the inner magnetic field lines (see figure 1.8) of the toroidal coils.

In MKSA units the (non-relativistic) Lorentz equation reads (we have included here the electric field $\vec{E}$ too) [14]

$$m\frac{d}{dt}\vec{v} = q(\vec{E} + \vec{v} \times \vec{B}). \tag{1.19}$$

The vector structure of the equation confirms that the trajectories wrapping around the magnetic field lines, are simple helices, if the fields are constant in time and independent of the spatial coordinates (homogeneous fields). In the case of realistic configurations, involving not only time and coordinate dependent fields but elaborate magnet geometries too, exact solutions are hardly achievable. The information which can be obtained using appropriate approximations, however, is of noticeable importance.

A straightforward solution of equation (1.19) is available for time dependent and homogeneous fields and after setting it in the form

Figure 1.8. Ions (blue trajectories) and electrons (red trajectories) moving along the toroidal field lines.

$$\frac{d\vec{v}}{dt} = q\,\frac{\vec{E}}{m} - \vec{\Omega}_c \times \vec{v},$$
$$\vec{\Omega}_c = \frac{q\,\vec{B}}{m}, \tag{1.20}$$

the use of the evolution operator formalism yields the solution [15]

$$\vec{v}(t) = \hat{U}(t)\left[\vec{v}_0 - \int_0^t \hat{U}(-t')q\,\frac{\vec{E}}{m}dt'\right],$$
$$\hat{U}(t) = e^{-[\vec{\Omega}_c]t},\ [\vec{\Omega}_c] = \vec{\Omega}_c\times. \tag{1.21}$$

The action of the evolution operator on the vector, specifying the initial velocity and electric field is simply obtained by expanding the exponential and then by exploiting the cyclic properties of the vector product, namely

$$\hat{U}(t) = \sum_{n=0}^{\infty}\frac{(-t)^n}{n!}[\vec{\Omega}_c]^n\vec{v}_0,$$
$$[\vec{\Omega}_c]^1\vec{v}_0 = \vec{\Omega}_c \times \vec{v}_0, \tag{1.22}$$
$$[\vec{\Omega}_c]^2\vec{v}_0 = \vec{\Omega}_c \times (\vec{\Omega}_c, \times\vec{v}_0),\ \ldots$$

Leaving the technicalities of the general solution to the specialized literature [16], let us deal with a more elementary treatment.

In the absence of electric field, the Lorentz force equation is ruled by the torque vector $\vec{\Omega}_c$, whose modulus defines a quantity of pivotal importance, namely the cyclotron frequency. The numerical values in the case of electron and protons, respectively, and corresponding to $B = 1T$, are

$$\Omega_{c,e} \cong 28\ \text{GHz},$$
$$\Omega_{c,p} \cong 15.2\ \text{MHz}.$$

To better visualize the motion, it is convenient to make reference to the vertical ($v_\perp$) and parallel ($v_{||}$) velocity components (with respect to the magnetic field). The latter remains unaffected by the interaction with the field and contributes to the axial drift of the helix, while the former component contributes to the circular motion and defines the gyration (Larmor) radius, which is obtained by balancing the Lorentz and centrifugal forces (see figure 1.9), namely

$$r_L = \frac{m\,v_\perp}{|q|\;B} = \frac{v_\perp}{\Omega_c}. \tag{1.23}$$

It is therefore evident that, in the reference frame moving with velocity $v_{||}$, the particle trajectory is just a circle with radius r_L and the circular motion is characterized by a period $2\pi/\Omega_c$.

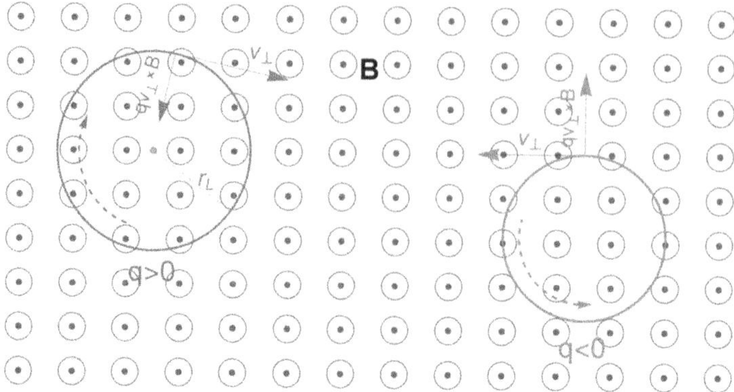

Figure 1.9. Larmor radii for electrons and ions.

The inclusion of the electric field complicates the motion dynamics, which depends on the relative orientation of the $\vec{E}$, $\vec{B}$ vectors. A particularly important example is the case in which the electric and magnetic fields are orthogonal.

We consider equation (1.20) assuming that the velocity can be written as

$$\vec{v} = \vec{w} + \vec{v}_E,$$
$$\vec{v}_E = \frac{\vec{E} \times \vec{B}}{|\vec{B}|^2}. \tag{1.24}$$

If we substitute equation (1.24) in equation (1.19) we find

$$m\frac{d}{dt}\left(\vec{w} + \frac{\vec{E} \times \vec{B}}{|\vec{B}|^2}\right) = q\left(\vec{E} + \left(\vec{w} + \frac{\vec{E} \times \vec{B}}{|\vec{B}|^2}\right) \times \vec{B}\right), \tag{1.25}$$

and therefore both the electric and magnetic fields time independent are

$$\frac{d\vec{w}}{dt} = -\vec{\Omega}_c \times \vec{w} + \frac{q}{m}\left(\vec{E} + \frac{(\vec{E} \times \vec{B}) \times \vec{B}}{|\vec{B}|^2}\right). \tag{1.26}$$

According to the standard vector calculus rules we find (for $\vec{E}$ and $\vec{B}$ perpendicular to each other), $(\vec{E} \times \vec{B}) \times \vec{B} = -|\vec{B}|^2\vec{E}$, the term in parenthesis on the rhs, therefore, vanishes, so that the vector velocity $\vec{w}$ is governed by the equation

$$\frac{d\vec{w}}{dt} = -\vec{\Omega}_c \times \vec{w}. \tag{1.27}$$

The reference velocity is, accordingly, ruled by the Lorentz equation in the absence of an electric field, we can, therefore, visualize the particle trajectory as shown in figure 1.9. The conclusion we have drawn, holds for any force acting on the particle and orthogonal to the magnetic field. Such a statement is implicitly contained in the definition of $\vec{v}_E$

$$\vec{v}_E = \frac{q\vec{E} \times \vec{B}}{q\,|\vec{B}|^2} = \frac{\vec{F} \times \vec{B}}{q\,|\vec{B}|^2}, \tag{1.28}$$

where we have replaced the electric force $q\vec{E}$ with a generic force $\vec{F}$.

An example of the velocity $\vec{v}_E$ induced by a force of non-electric nature acting on charged particles moving in a magnetic field is provided by the drift velocity with modulus $v_g = mg/(qB_T)$, induced by the gravitational force perpendicular to the Earth's magnetic field (B_T), on the ion motion along the terrestrial field lines and manifests itself through a current (see figure 1.10)[8].

We have argued that the charged particle, moving along the line forces of $\vec{B}$ field, executes a circular trajectory of radius r_L. If the field is not homogenous and increases along the y-direction (see figure 1.11), with a variation appreciable for a length of the order of $r_L \propto 1/B$, it happens that the Larmor radius is larger at the lower portion of the orbits. The magnetic dipole moment associated with the

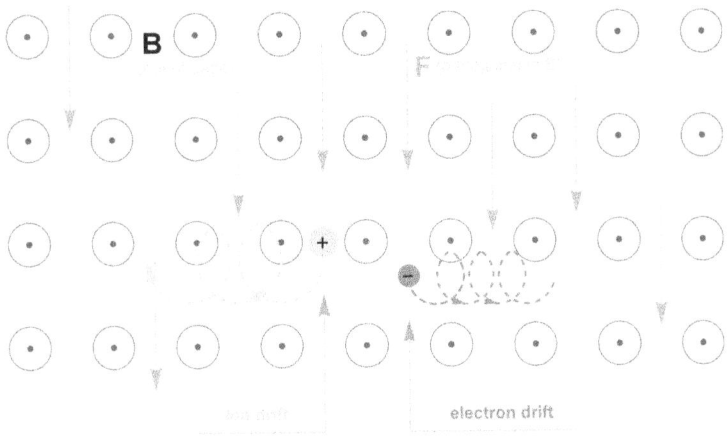

Figure 1.10. Drift velocity induced on ions or electrons by a force orthogonal to the magnetic field (note the dependence on the sign of the charge).

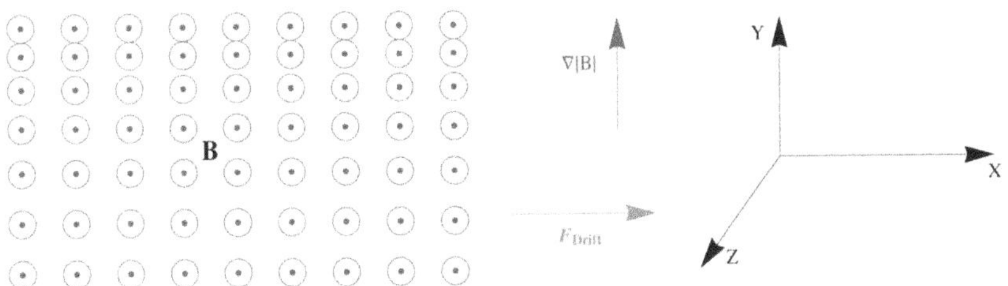

Figure 1.11. Drift induced by an inhomogeneous magnetic field.

[8] In laboratory plasma the effect of gravity is too small to be accounted for.

circulating charge and the magnetic field variation determine a force, which can eventually lead to a drift velocity affecting the particle's confinement.

The field inhomogeneities can be induced by the topology in which they are embedded. The following discussion considers the toroidal geometry for which in a more quantitative term we can state that [17]:

(a) The toroidal field, due to the coils wrapped around the torus, exhibits a $1/R$ dependence which determines a non-uniformity[9], due to the fact that it is larger at smaller R. If these variations are experienced by the circulating particles within a Larmor radius, we can express the extra force acting on the particle as

$$F = qv_{\perp}[B(R - r_L) - B(R + r_L)] \approx -2qv_{\perp}r_L\frac{\partial B}{\partial R} \tag{1.29}$$

whose direction is shown in figure 1.12. It is responsible for the drift velocity determining electron–ion charge separation, responsible for the on-set of an electric field, which coupled to the magnetic field yields a subsequent force pulling the particles towards the external wall.

It is worth noting that equation (1.29) can also be cast in the form

$$F = -\mu\frac{\partial B}{\partial R}, \tag{1.30}$$

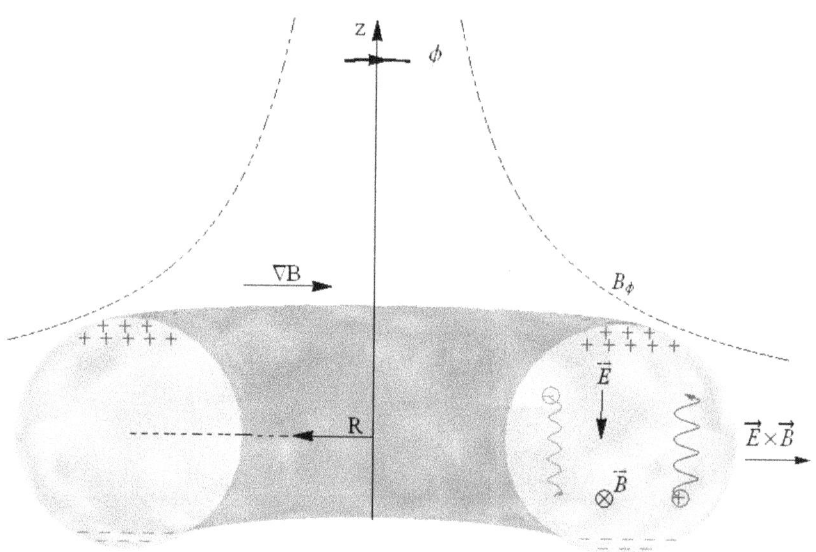

Figure 1.12. Drift motion induced by the gradient of the magnetic field, charge separation, induced electric field, and the associated particle loss.

[9] If I is the total current flowing inside the toroidal coils, for any circle inside the torus around the Z-axis with a major radius R from the Ampère law we get $\oint \vec{B} \cdot dl = 2\pi RB = \mu_0 I$ which leads to $B = \text{const}\frac{1}{R}$.

where

$$\mu = I\ \pi\ r_L^2,\ I = |q|\ \frac{\Omega_c}{2\ \pi}. \tag{1.31}$$

with μ being the magnetic dipole moment associated with the electron executing a circular orbit in a magnetic field. According to this, and using the constitutive relation between the magnetic field strength H and the magnetic induction (flux density) $\vec{B}$

$$H = \vec{\mu} \cdot \vec{B} \tag{1.32}$$

the force acting on the dipole is

$$F = -\frac{\partial H}{\partial R}. \tag{1.33}$$

The last derivation is more rigorous than the first one, which is heuristic but insightful.

This implies that the drift velocity is given by

$$\vec{v}_E^{(\nabla B_\phi)} = \frac{|q|}{2} \pi v_\perp r_L \frac{\vec{\nabla} B \times \vec{B}}{|\vec{B}|^2} \tag{1.34}$$

The physics characterizing the radial dependence of the toroidal field is summarized in figure 1.12.

(b) The dependence of the drift velocity on q leads to a charge separation and the resulting electric field induces a further drift pushing the particles towards the walls of the toroidal chamber.

(c) An analogous effect is produced by the centrifugal force $F_c = mv_{||}^2/R_c$, experienced by the particles moving along the axes of the toroidal chamber (see figure 1.13), where the associated motion is ruled by

$$\vec{v}_E^{(c)} = \frac{mv_{||}^2}{q} \frac{\vec{R}_c \times \vec{B}}{|\vec{R}_c|^2\ |\vec{B}|^2}. \tag{1.35}$$

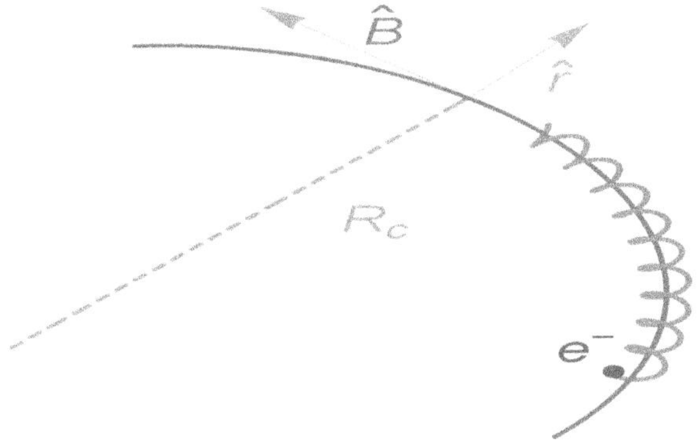

Figure 1.13. Helical motion of a charge particle along the bent field lines.

The physical origin of this drift is associated with the field non-homogeneity due to the curvature

All these drifts combine and the total effect is that described in figure 1.13. The lesson we gain from this discussion is that the only toroidal field does not ensure an efficient confinement.

The solution is provided by imposing additional magnetic fields, with lines orthogonal to those of the toroidal component. The (poloidal) equilibrium field coils in figure 1.5, for example, stabilize the particle drift. However, the final solution requires additional coils inducing a field B_θ orthogonal to the toroidal plane, which, once combined with B_ϕ, yields the helical shape displayed in figure 1.14(a). This field can be generated by a plasma current I_p circulating along the torus axis (figure 1.14(b)). The particles' orbits in the resulting combined field are shown in figure 1.14(a).

The obvious solution is inducing such a current by the use of a transformer in which the secondary winding is the plasma itself, as shown in figure 1.15. The induced plasma current is due to the variation of the magnetic flux, inside the

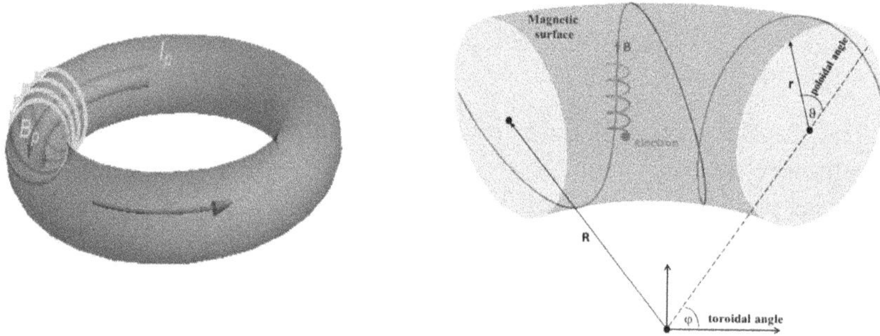

Figure 1.14. Plasma current and associated magnetic field (left). Composition of the toroidal and poloidal fields (right).

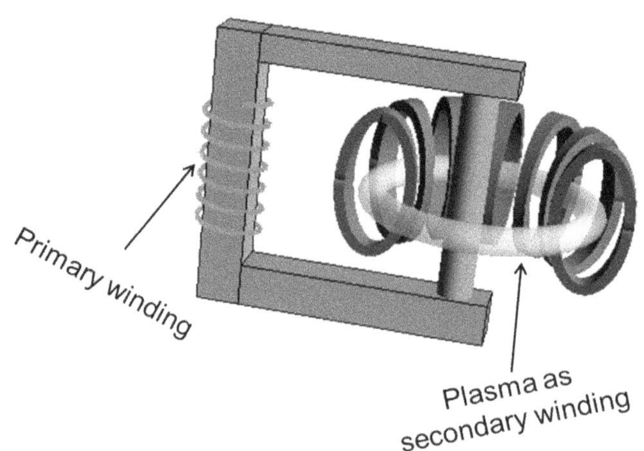

Figure 1.15. Tokamak solenoid transformer.

solenoid, and is related to the corresponding electromotive potential V_I associated with it as follows:

$$I_p = \frac{V_I^2}{R_p}, \tag{1.36}$$

where R_p is the plasma resistance. The flowing current induces both ohmic heating power

$$P_H = R_p I_p^2, \tag{1.37}$$

and a magnetic field

$$B_p = \mu_0 \frac{I_p}{2\,\pi\,a}. \tag{1.38}$$

This heating contribution is only a part of the overall budget, to be provided to the plasma, in order to achieve the fusion conditions. We will see in the following that the plasma current heating is not sufficient to reach temperatures (>10 keV) necessary to provide a sufficient alpha particle heating, since the plasma resistance drops dramatically at 3 keV. To understand these points we need to break from Tokamaks architecture and consider some fundamental aspects of plasma physics.

1.4 Magnetic mirror and confinement

Before discussing the problems of Tokamak-assisted fusion from the point of view of plasma physics, we consider it worth adding a few comments on a confining mechanism known as a magnet mirror. It was a solution foreseen in the early fusion devices [18].

The so-called axisymmetric magnetic mirror shown in figure 1.16, realizes one of the most natural confining schemes.

In such a magnetic field configuration, the strength increases in the region where the force lines become closer, the conservation of the particle magnetic moment and of the energy determine (in principle) the bouncing back and forth of the particles with the appropriate kinetic conditions. Something similar occurs in the trapping of the charged particles in the Earth's magnetic field.

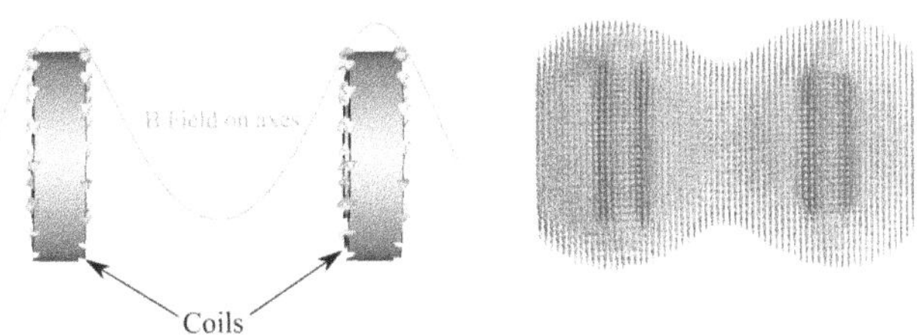

Figure 1.16. Left: magnetic mirror configuration and on-axes profile. Right: field distribution; at the center of the magnet, the parallel field component is maximum, moving towards the external coils a radial component develops, which is responsible for the appearance of a parallel force component (opposed to the parallel motion when the field decreases and the other way around when it increases).

The key concept, to understand the mechanism, is that of the adiabatic invariant. It is not an exactly conserved quantity, like the energy or angular momentum; rather, it is a physical quantity which changes very slowly, with respect to some characteristic time of the system under study.

The motion of a charged particle in a slowly changing magnetic field in the parallel direction is characterized by the adiabatic invariant [19]

$$\mu = \frac{K_\perp}{B}, \tag{1.39}$$

where

$$K_\perp = \frac{1}{2} m v_\perp^2. \tag{1.40}$$

The proof goes as follows:

(a) The magnetic field is assumed to be specified by the radial and axial components in cylindrical coordinates

$$\vec{B} = B_r \hat{r} + B_z \hat{z}. \tag{1.41}$$

The condition $\vec{\nabla} \cdot \vec{B} = 0$ yields[10] the relation between the radial and axial components given below (in a paraxial approximation)

$$B_r = -\frac{1}{2} r \frac{\partial B_z}{\partial z}|_{r=0}. \tag{1.42}$$

Regarding the particle motion, the presence of the radial field component induces[11] the parallel force

$$F_\| = \frac{q}{2} \frac{v_\vartheta r}{} \frac{\partial B_z}{\partial z}|_{r=0}. \tag{1.43}$$

After averaging on the gyro orbit, we make the replacements

$$\begin{aligned} v_\vartheta &\to v_\perp, \\ r &\to r_L, \end{aligned} \tag{1.44}$$

and rewrite equation (1.43) as

$$F_\| = \frac{q}{2} \frac{v_\perp r_L}{} \frac{\partial B_z}{\partial z} = -\frac{1}{2} \frac{m v_\perp^2}{B} \left| \frac{\partial B_z}{\partial z} \right|, \tag{1.45}$$

and if we set $\mu = \frac{1}{2} \frac{m v_\perp^2}{B}$, we find

$$F_\| = -\mu \left| \frac{\partial B_z}{\partial z} \right|. \tag{1.46}$$

[10] For the $\vec{\nabla}\cdot$ operator in cylindrical coordinates see the introductory table.

[11] For the vector product in cylindrical coordinates see the introductory table.

Furthermore, noting that

$$\left|\frac{\partial B_z}{\partial z}\right| \cong \frac{1}{v_{\|}}\left|\frac{\partial B_z}{\partial t}\right|, \tag{1.47}$$

we infer that the averaging condition holds if the time associated with the field change is much longer than the gyration period, namely

$$\left|\frac{\partial B_z}{\partial t}\right|^{-1} B_z \gg \frac{1}{\Omega_c}. \tag{1.48}$$

(b) The total kinetic energy

$$K = K_{\perp} + K_{\|}, \tag{1.49}$$

is a conserved quantity if only magnetic forces are present.
(c) Setting in the last identity $K_{\perp} = \mu\, B$ we can write

$$K = \mu\, B + K_{\|}. \tag{1.50}$$

(d) By taking the derivative of both sides with respect to time, we get

$$\left(\frac{d}{dt}\mu\right) B + \mu\,\frac{dB}{dt} + \frac{d}{dt}K_{\|} = 0. \tag{1.51}$$

(e) Taking into account that

$$\begin{aligned} \frac{dB}{dt} &\cong v_{\|}\frac{\partial B}{\partial z}, \\ \frac{d}{dt}K_{\|} &= -\mu\,\frac{dB}{dt}, \end{aligned} \tag{1.52}$$

we infer, from equation (1.51), the conservation of μ provided that the condition (1.48) is satisfied.

The conservation of energy corresponds to the conservation of the total velocity, namely

$$v_0^2 = v_{\perp}^2 + v_{\|}^2, \tag{1.53}$$

which, along with the adiabatic invariance of the magnetic moment, yields

$$v_{\|}^2 = v_0^2 - \frac{B(z)}{B_0} v_{\perp,\,0}^2, \tag{1.54}$$

and thus, according to the previous identity, the longitudinal velocity is transferred to the transverse component and eventually vanishes at some point along the z-axis.

Therefore, the velocity is transferred to $v_\perp$, the particle moves faster around the magnetic axis and, rotating in the same direction, is reflected back. Such back and forth bouncing motion repeats indefinitely.

This confining mechanism is, however, inefficient for real-life plasma to be a promising candidate for fusion reactors. This is a consequence of either Coulomb collisions or of other instabilities, which lead to a continuous transfer of trapped particles into the loss region, as discussed in more detail below.

It is now important to recognize that the mirror effect occurs in the Tokamak plasma too and leads to the so-called 'banana orbits'. Their onset is easily recognized by just putting together what we discussed so far.

The charged particles inside the plasma move with velocity components parallel to the field lines ($v_{||}$) and orthogonal to them $v_\perp = \sqrt{2B(R)\mu/m}$, where B is the toroidal magnetic field at the radial coordinate (major radius) R (see figure 1.17(a)), μ is the particle magnetic moment, and m is the particle mass.

The parallel velocity varies with R, as can be seen from the following identity obtained combining the energy and magnetic moment

$$v_{||}^2 = v_0^2 - \frac{2B(R)}{m}\mu. \tag{1.55}$$

The trajectory of the charged particles is therefore influenced by:

(a) the R dependence of the magnetic field;

(b) the drift due to the field gradient and to the centrifugal force (see equations (1.34) and (1.35)).

The net result is that shown in figure 1.17(a). While the particles rotate around the axis, they experience a difference in the field intensity and move within regions where the force lines are more dense. It is therefore expected that some particles, with the

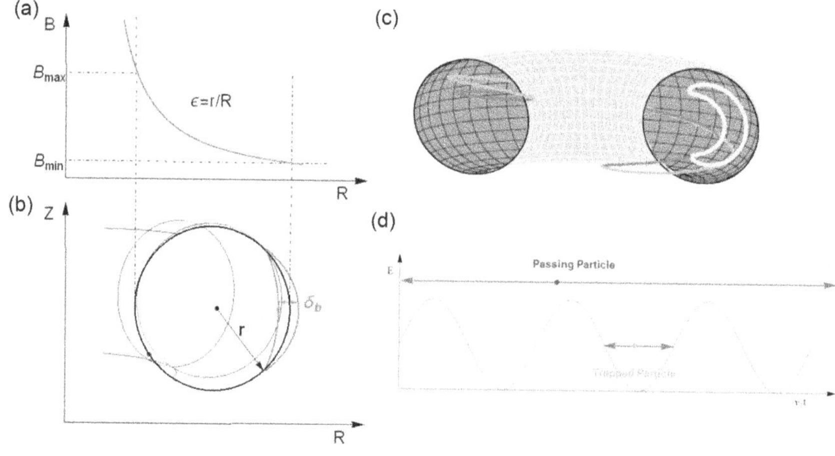

Figure 1.17. (a) Variation of the magnetic field versus the major radius R. (b) Projection of the particle orbit on the torus transverse section. (c) Full tridimensional picture of banana orbit. (d) Energy variation of trapped and untrapped particles.

appropriate kinematic conditions, transfer their longitudinal velocity in favor of the orthogonal counterpart and when it becomes zero they are reflected back. In the meantime the aforementioned drift becomes active and the trapped particles move along a shifted orbit (figure 1.17(b)). There are, however, particles with sufficiently large velocities which are not trapped and execute an almost circular orbit.

It must be understood that the 'banana' we have drawn in our figures, is just a 2-D projection on the poloidal plane of the motion occurring along the torus chamber (see figure 1.17(c)).

The number of trapped particles depends on their kinematic conditions (see figure 1.17(d)), and it is easily argued that only those particles satisfying the condition

$$\frac{v_{\|}^2}{v_{\perp}^2} < \frac{B_{\max} - B_{\min}}{B_{\min}} \tag{1.56}$$

are confined.

We will further comment on the issues associated with 'banana trapped' particles in the forthcoming chapter, when we will touch on the problems of plasma instabilities.

We close this section and the first part of this chapter, by introducing a first example of the dimensionless parameter, which, within the context of the fusion reactor design, plays a role of pivotal importance.

It has already been noted that the helical motion winding the field lines introduces a fundamental time unit associated with the cyclotron frequency and a fundamental unit length, namely the Larmor radius. We have also seen that the deviation in the motion occurs for inhomogeneous or time dependent fields with spatial length L and time scales ω. It is therefore natural to introduce the following dimensionless parameter

$$\delta = \frac{r_L}{L} \tag{1.57}$$

which is useful to quantify the deviation from the ideal orbit.

In the forthcoming parts of this chapter we will introduce other such parameters, yielding a set of reference quantities allowing the derivation of scaling formulae that are helpful to the initial design of magnetic confinement devices.

1.5 Plasma as a state of matter

Introducing disciplines like *plasma physics*, it is customary to start with answering the question 'What is a plasma?' which raises a complex ontological query rather than a simple pragmatic interrogation [20, 21].

In physics 'being' means 'behaving' in response to changes of the environmental conditions, which corresponds to parameters setting specific equilibrium conditions.

Matter is not defined by its constituents only, but also by the forces acting between them.

The states of matter are identified as solid, liquid, gas and plasma. Such an identification is associated with the correspondingly increasing freedom in motion of the constituent particles.

Regarding solids, atoms are disposed in periodic lattices and they can undergo a small jiggling, but do not move from place to place. Thus, the solids conserve such relevant attributes as shape and size.

The same does not occur for liquids, since their atoms and molecules have significant freedom of movement. Thus the liquids change their form, but, owing to the intensity of the inter-atomic forces, the volume remains unchanged.

In a gas, the atoms move freely, the relevant interactions are limited to occasional collisions between them. It, therefore, takes both the volume and shape of the container.

Plasma is a partially or fully ionized gas, which is characterized by electric neutrality and collective motion of both the electrons and ions under the action of the present (applied) electromagnetic fields. There is a fairly rich phenomenology associated with such a state of matter.

In table 1.1 we have summarized some plasma sources along with the relevant characteristic parameters, namely density, temperature, plasma and collision frequencies. In the following, we clarify their physical role, and underscore the huge variability (orders of magnitudes) of their parameters. Such variability is a very distinctive feature of the plasma with respect to the solid, liquid or gaseous states of matter.

The definition we have just given of plasma is that of a 'quasi-neutral gas' exhibiting a collective behavior, characterized by the density, temperature and plasma frequency; in the next sections we will provide a first non-qualitative description of how these quantities enter the game.

Before getting into the specifics of the topics to be treated here, let us underscore a few relevant points.

In table 1.1, we have reported a list of plasma sources and their characteristic parameters. The relevant role of any such parameter can be understood in terms of a 'wise' dimensional analysis.

According to the previous discussion and taking into account the presence of free charges and hence of Coulomb interactions, we can introduce three quantities ruling the physics of the system, namely the particle number density, the thermal velocity and the Coulomb interaction strength

Table 1.1. Plasma sources and relevant parameters.

Plasma type	Particle density ♯/m^3	Temp. T (K)	Debye length [m]	Plasma freq. ω_p [Hz]	Collision freq. ν [Hz]
Interstellar gas	10^6	10^4	10	6×10^4	10^{-4}
Solar corona	10^{12}	10^6	10^{-2}	6×10^7	3×10^9
Solar atm. gas discharge	10^{20}	10^4	10^{-6}	6×10^{11}	3×10^9
Diffuse lab plasma	10^{18}	10^6	10^{-4}	6×10^{10}	10^5
Dense lab plasma	10^{20}	10^6	10^{-6}	6×10^{12}	5×10^8
Thermonuclear plasma	10^{22}	10^8	10^{-5}	6×10^{12}	8×10^5

$$
\begin{array}{lll}
d_1 \equiv n & \text{number particle density} & [L^{-3}] \\
d_2 \equiv v_{\text{th}} & \text{thermal velocity} & [LT^{-1}] \\
d_3 = \dfrac{e^2}{m\,\varepsilon_0} & \text{interaction strength} & [L^3T^{-2}] \\
m \equiv m_e(m_i) & \begin{array}{l} m_e \text{ electron mass} \\ m_i \text{ ion mass} \end{array} & \\
e = 1.6 \cdot 10^{-19} C & & \\
\varepsilon_0 & \text{vacuum space dielectric permittivity} &
\end{array}
\tag{1.58}
$$

The quantity d_3 is a measure of the strength of the Coulomb interaction between the free charges moving in the plasma.

The use of the Buckingham's theorem allows to get a useful dimensionless constant. Such a quantity is determined by setting [22]

$$
\Pi = d_1^\alpha d_2^\beta d_3^\gamma = [L^{3(\gamma-\alpha)+\beta} T^{-(2\gamma+\beta)}] \tag{1.59}
$$

and requiring that

$$
\begin{aligned}
3(\gamma - \alpha) + \beta &= 0, \\
2\alpha + \beta &= 0.
\end{aligned}
\tag{1.60}
$$

Keeping, arbitrarily $\alpha = 1$, we obtain

$$
\Pi = n\left(\frac{d_3^{1/2}}{v_{\text{th}}}\right)^6, \tag{1.61}
$$

which needs to being further refined to capture the relevant physical meaning. The use of the explicit expression of d_3 yields

$$
\Pi = \frac{1}{n^2}\left(\frac{\left(\dfrac{n\,e^2}{m\,\varepsilon_0}\right)^{1/2}}{v_{\text{th}}}\right)^6. \tag{1.62}
$$

We are accordingly allowed to define the dimensionless plasma parameter

$$
\Lambda = \frac{1}{\sqrt[3]{\Pi}} = n\left(\frac{v_{\text{th}}}{\Omega}\right)^3, \tag{1.63}
$$

where

$$
\Omega = \left(\frac{ne^2}{m\varepsilon_0}\right)^{1/2}. \tag{1.64}
$$

Bearing in mind Ω has a dimension of a frequency, we can, therefore, rearrange equation (1.63) as

$$\Lambda = n\lambda_{\rm D}^3, \tag{1.65}$$

where

$$\lambda_{\rm D} = \left(\frac{v_{\rm th}}{\Omega}\right) = \left(\frac{mv_{\rm th}^2\varepsilon_0}{ne^2}\right). \tag{1.66}$$

The quantity $\lambda_{\rm D}$ is the so-called Debye length. It is a measure of the distance in a plasma over which the electric field of a charged particle is 'screened out' by the random thermal motions of the other charged plasma particles.

It is more instructive to frame the discussion within a more quantitative context. We have noted that a plasma is a 'quasi-neutral' ionized gas. A neutral gas hardly becomes ionized, unless the constituting atoms have kinetic energies comparable with the atom ionization energy, which, typically, amounts to tens of eV.

If we take into account that 1 eV $\cong 1.6 \cdot 10^{-19}$ J and use the relation (1.6) for the Boltzmann constant $k_{\rm B}$, a gas with atoms having such a kinetic energy E_K is characterized by a temperature $T = E_K/k_{\rm B}$ of the order of 10^5 Kelvin.

We have so far assumed a homogeneous plasma density, which conflicts with the presence of moving charges determining the relevant internal dynamics.

The charge distribution can be determined using a very simple argument. It is evident that such distribution inside the plasma determines an electric field which induces a force acting on the plasma particles. The plasma temperature and the number density per unit volume n can be used to define the energy density (equivalent to a kinetic pressure)

$$p = k_{\rm B}T\, n. \tag{1.67}$$

Moreover, the non-homogeneity of the charge density determines a force, associated with the spatial distribution of n. An equilibrium occurs when the electric forces and the pressure balance. Assuming, for simplicity, a one-dimensional configuration, the equilibrium condition is written as

$$-enE = k_{\rm B}T\,\frac{\partial n}{\partial z}. \tag{1.68}$$

This identity, albeit elementary, can be exploited to get an idea of the density spatial distribution. Denoting by ϕ the plasma potential, originating the intra-plasma electric field, we find

$$E = -\frac{\partial \phi}{\partial z}, \tag{1.69}$$

which, once inserted in equation (1.68), yields the differential equation

$$en\,\frac{\partial \phi}{\partial z} = k_{\rm B}T\,\frac{\partial n}{\partial z}. \tag{1.70}$$

Solving for n we find (for electrons and ions, respectively)

$$\begin{aligned} n_e(z) &= n_{0,e} \exp\left(\frac{e\phi}{k_B T}\right), \\ n_i(z) &= n_{0,i} \exp\left(-\frac{e\phi}{k_B T}\right), \end{aligned} \tag{1.71}$$

$n_{0,e}$ and $n_{0,i}$ being the initial number of particles per unit volume.

We can now derive an expression of the potential by the use of the Poisson equation, namely

$$\vec{\nabla} \cdot \vec{E} = -\nabla^2 \phi = \frac{\rho}{\varepsilon_0}. \tag{1.72}$$

By noting that for a small deviation from the equilibrium condition we have

$$\rho = e\,(n_e - n_i) \cong -\frac{2n_0\, e^2 \phi}{k_B T}, \tag{1.73}$$

which, once substituted in (1.72), yields, for the one-dimensional case, the equation

$$\frac{\partial^2 \phi}{\partial z^2} = \left[\frac{n_0}{\varepsilon_0} \frac{2e^2 \phi}{k_B T}\right]. \tag{1.74}$$

It should now be noted that as

$$m v_{\text{th}}^2 = 2 k_B T, \tag{1.75}$$

we obtain

$$\frac{n_0}{\varepsilon_0} \frac{2e^2}{k_B T} = \frac{1}{\lambda_D^2} \tag{1.76}$$

or

$$\lambda_D = \left(\frac{\varepsilon_0 T}{e^2 n_0}\right)^{1/2}. \tag{1.77}$$

The solution of equation (1.74) can be written as

$$\phi = \phi_0 \exp\left(-\frac{z}{\lambda_D}\right). \tag{1.78}$$

This result, albeit derived on the basis of very general considerations, yields an idea of the role of the Debye screening length. The obvious extension of equation (1.78) to the 3-D case is

$$\begin{aligned} \phi &= \phi_0 \exp\left(-\frac{r}{r_D}\right), \\ \phi_0 &= \frac{1}{4\pi\varepsilon_0} \frac{Q}{r}, \end{aligned} \tag{1.79}$$

where r_D is the Debye radius. It is evident that

$$\begin{aligned} \phi &\cong \phi_0, \qquad r \gg r_D, \\ \phi &\cong \exp\left(-\frac{r}{r_D}\right), \qquad r \leqslant r_D. \end{aligned} \tag{1.80}$$

Therefore, the plasma charges effectively screen out or shield the electric field of the test charge outside of the Debye sphere with a radius r_D.

The *Debye screening* or *shielding* has an analogy with the so called Yukawa potential, describing the strong interactions, which are mediated by heavy particles instead of photons (we are referring to the 'old' strong interaction theory and not to the current treatment based on quantum chromodynamics).

The equation describing the static potential of the strong interactions is simply given by

$$\begin{aligned} \nabla^2 \phi_Y &= \frac{1}{R_\pi^2} \phi_Y, \\ R_\pi &= \frac{\hbar}{m_\pi c}, \end{aligned} \tag{1.81}$$

with $\hbar$ being the reduced Plank constant, ϕ_Y the Yukava potential and R_π the Compton wavelength, associated with the strong force mediating particle (m_π denotes the mass of the pion).

We can now use the definition of Debye length in the context of the physical mechanism underlying the particle collisions in a plasma, which are characterized by an impact parameter $b \cong \lambda_D$. At larger values of b, collisions do not occur because the charge is shielded. Within this context the quantity

$$\Lambda = \frac{\lambda_D}{b_{\min}} \tag{1.82}$$

plays a crucial role to define the collision time τ_c. Namely, the average time necessary for the particle to be deflected by 90°, without entering the details of the derivation, it is given by

$$\tau_c = \frac{12\varepsilon_0^2 \sqrt{m_e(\pi k_B T)^3}}{\sqrt{2} n_i Z^2 e^4 \ln(\Lambda)}, \tag{1.83}$$

where $\ln(\Lambda)$ is usually defined as Coulomb logarithm (with values ranging between 10 and 20 for typical fusion plasma) and T is the energy equivalent of the plasma temperature. The use of the above equation with $T_e = 10$ keV, $n_e = 10^{20}$ m^{-3} yields a collision time $\tau_c \cong 10^{-4}$ s, which is quite a short time and explains why the mirror confining devices are inefficient.

In passing, we note that the presence of a strong magnetic field is associated with a magnetic pressure

$$p_M = \frac{B^2}{2\,\mu_0}, \tag{1.84}$$

which can be compared with the plasma kinetic pressure

$$p_K = n\,k_B T, \tag{1.85}$$

through the 'β ratio'

$$\beta = \frac{p_K}{p_M}. \tag{1.86}$$

This is one more example (see equation (1.57)) of dimensionless parameters useful to define semi-empirical scaling formulae for the design of Tokamak devices, as discussed in the forthcoming chapters.

1.6 Plasma kinetic theory

According to the discussion of the previous sections, the plasma can be viewed as an ensemble of charge particles (electrons and ions) and neutrals. The analysis of the dynamics of such a complex system is characterized by the change of the relative coordinates and velocities under the action of the external fields as well as of the internal processes associated with ionization, Coulomb scattering, charge exchange, and so on.

The number of particles composing the plasma is so large that a set of differential equations, tracking the particle orbits evolution, is beyond the capabilities of the present computers [23–25].

A most appropriate and effective analysis can be achieved through the study of the evolution of the relevant distributions (in terms of position and velocities) whose average values provide macroscopic plasma parameters like current density, pressure, temperature, etc.

It is customary to represent the particle distribution in the so-called phase space, which is widely used within the framework of the Hamiltonian mechanics.

If an ensemble of particles is ruled by a not explicitly time-dependent Hamiltonian $H(q, p)$, where q, p are the canonical variables (position and momentum, respectively) one can follow the particle evolution by assigning to each of them the relevant initial conditions and then by solving the Hamilton equations of motion.

A less computationally expensive procedure uses the equations defining the particle evolution in terms of the canonical variables.

We assume that the particle evolution in the phase space is defined by a density function $F(q, p, t)$, linked to the number of particles N by

$$\begin{aligned} F(q, p, t) &= \frac{1}{N} f(q, p, t), \\ f(q, p, t) &= \frac{dn}{dq\,dp}, \end{aligned} \tag{1.87}$$

where $dq\,dp$ denotes the phase space area element.

It is evident that as

$$\int\int f(q, p, t)dpdq = N \tag{1.88}$$

the function $F(q, p, t)$ is normalized to unity.

It is implicit in the assumption that the Hamiltonian, ruling the evolution of the system of particles, is time-independent since the forces acting on the system are external. We exclude, therefore, the description of processes involving intra-beam scattering radiation losses, particle absorption, and so on.

The number of particles under the action of conservative forces is accordingly a constant quantity (according to Liouville's theorem) and, therefore, by keeping the total derivative of the F distribution with respect to time (we have omitted the arguments for brevity) we find

$$\frac{d}{dt}F = \dot{q}\,\frac{\partial F}{\partial q} + \dot{p}\,\frac{\partial F}{\partial p} + \frac{\partial F}{\partial t} = 0. \tag{1.89}$$

The use of the Hamilton equations of motion

$$\begin{aligned}\dot{q} &= \frac{\partial H}{\partial p},\\ \dot{p} &= -\frac{\partial H}{\partial q},\end{aligned} \tag{1.90}$$

finally yields[12]

$$\begin{aligned}&\frac{\partial F}{\partial t} = \hat{L}F,\\ &\hat{L} = -\frac{\partial H}{\partial p}\frac{\partial}{\partial q} + \frac{\partial H}{\partial q}\frac{\partial}{\partial p}.\end{aligned} \tag{1.91}$$

The operator $\hat{L}$ is called Liouville's operator and its relevant properties will be exploited later in this section. The solution of Liouville's equation is accordingly equivalent to that of the Hamilton equations of motion.

In plasma, however, the conditions allowing the derivation of equation (1.91) do not hold. We cannot indeed exclude intra-beam forces or radiative effects.

We relax the assumption that our system is constrained within the framework of a Hamiltonian system, but retain the assumption of the particle density conservation. The phase space, we will deal with, is defined by the position-velocity coordinates which, in general, are not canonical. This last point has important consequences, as shown below.

[12] We recall that Liouville's equation can also be cast in the form $\frac{\partial F}{\partial t} = -\{H, F\}$ where the braces denote the Poisson brackets (PBs) defined as $\{Q(q, p), G(q, p)\} = \frac{\partial Q}{\partial q}\frac{\partial G}{\partial p} - \frac{\partial Q}{\partial p}\frac{\partial G}{\partial q}$; within this formalism the canonical variables q, p are those with $PB = 1$.

According to the conservation of the number of particles, we can write the equation ruling the distribution as

$$-\frac{\partial F}{\partial t} = \vec{\nabla}_r \cdot (\vec{v}\, F) + \vec{\nabla}_v \cdot \left(\vec{a}\, F\right), \tag{1.92}$$

where $\vec{\nabla}_{r,v}$ are the position and velocity gradients, $\vec{v}, \vec{a}$ are the velocity and acceleration vectors.

We can now make the assumption that the velocity is independent of the position but we cannot exclude the friction mechanisms and therefore the acceleration is not independent of the velocity, we therefore write equation (1.92) as

$$\frac{\partial F}{\partial t} + \vec{v} \cdot \vec{\nabla}_r F + \vec{a} \cdot \vec{\nabla}_v F = -\left(\vec{\nabla}_v \cdot \vec{a}\right) F, \tag{1.93}$$

where the last term represents the contribution due to the difference with respect to a 'simple' Liouvillian system. The identity (1.93) is called Vlasov equation.

In the case in which the plasma particles are ruled by the Lorentz force only, namely $\vec{a} = q/m[\vec{E} + \vec{v} \times \vec{B}]$ we find

$$\vec{\nabla}_v \cdot \vec{a} = \frac{q}{m}\left[\vec{\nabla} \cdot \vec{E} + \vec{\nabla} \cdot \left(\vec{v} \times \vec{B}\right)\right] = 0. \tag{1.94}$$

Thus the Vlasov equation reduces to the Liouville form

$$\frac{\partial F}{\partial t} + \vec{v} \cdot \vec{\nabla}_r F + \frac{q}{m}\left[\vec{E} + \vec{v} \times \vec{B}\right] \cdot \vec{\nabla}_v F = 0. \tag{1.95}$$

The conclusion we may draw is that, at the assumed conditions, the kinetic model is equivalent to the single-particle description.

The importance of the formalism based on Vlasov equations will be appreciated in a forthcoming chapter, where we will develop the so-called MHD (magneto-hydrodynamic) model.

1.7 Ohmic heating

Before concluding this chapter we go back to the Tokamak physics. We have already mentioned the onset of a plasma current as a consequence of the poloidal field, induced by the solenoid placed in the center of the Tokamak chamber (see figure 1.15) [26].

The current circulating inside the chamber is a primary source for heating the plasma itself. Such a mechanism, however, becomes inefficient above a certain temperature, around 3 keV and, therefore, is not suitable to sustain an effective fusion plasma, with the required temperature higher than 10 keV.

The toroidal current induced in the plasma is due to the changing magnetic field, provided by ramping the current inside the primary winding at a constant rate, for a finite time. The field (and hence plasma current) have a limited duration, therefore, the associated heating process does not occur in a steady-state.

It is a matter of the elementary physics and mathematics to describe the underlying process

(a) The magnetic field inside the solenoid is

$$B = \mu_0 \mu_r \frac{N\, I}{L}, \tag{1.96}$$

with N the number of turns, L the turn length and I the current flowing in the primary winding.

(b) In the integral form, the Faraday–Lenz law gives

$$\oint \vec{E} \cdot d\vec{l} = -\frac{\partial}{\partial t} \int_{\Sigma} \vec{B} \cdot d\vec{\sigma}, \tag{1.97}$$

which yields

$$E = -\frac{\mu_0 \mu_r}{2\, \pi\, R} \frac{N}{L} \frac{\partial}{\partial t} I. \tag{1.98}$$

(c) The field drives a current density $\vec{J}_p$ (the subscript p stands for plasma) which depends on the resistivity η, through the generalized Ohm law

$$\vec{E} + \vec{v} \times \vec{B} = \eta\, \vec{J}_p. \tag{1.99}$$

By applying the curl operator to both sides of the previous equation, we find

$$\vec{\nabla} \times \vec{E} = -\frac{\partial}{\partial t}\vec{B} = \eta \vec{\nabla} \times \vec{J}_p. \tag{1.100}$$

Applying once again the curl operator to the second part of the previous identity, we obtain

$$\begin{gathered} -\frac{\partial}{\partial t}\vec{\nabla} \times \vec{B} = \eta \vec{\nabla} \times \left(\vec{\nabla} \times \vec{J}_p\right), \\ \nabla \times \vec{B} = \mu_0 \vec{J}_p. \end{gathered} \tag{1.101}$$

After exploiting the properties of the double vector product and taking into account the toroidal nature of the current density ($\vec{\nabla} \cdot \vec{J}_p = 0$), we obtain the following 'heat' type equation

$$\begin{gathered} \frac{\partial}{\partial t}\vec{J}_p = D \nabla^2 \vec{J}_p, \\ D = \frac{\eta}{\mu_0}, \end{gathered} \tag{1.102}$$

characterized by the diffusion coefficient D. It is defined in terms of the resistivity and evidently plays a central role in the ohmic heating mechanism. It is therefore important to clarify how η depends on the plasma parameters.

The charged particles in the plasma are accelerated by the electric field and decelerated by the collisions. The current induced by the toroidal electric field is mainly due to electrons, which have less inertia than the ions. The electron velocity v_δ, emerging as the combined effect of the collisions and the acceleration, is ruled by the equations

$$\frac{d}{dt}v_\delta = -\frac{e\,E}{m_e} - f_{e,i}\,v_\delta, \tag{1.103}$$

where $f_{e,i}$ is the frequency of the collisions between the electrons and ions. The electrons are here considered as a whole, therefore, the only source of momentum loss is associated with the frequency of the electron–ion collisions.

Assuming $v_\delta(0) = 0$, the solution of equation (1.103) reads

$$v_\delta(t) = -\frac{e\,E}{m_e f_{e,i}}\left(1 - e^{-f_{e,i}\,t}\right). \tag{1.104}$$

For times $t \gg 1/f_{e,i}$, the drift velocity reaches the stationary value

$$v_\delta^* = -\frac{e\,E}{m_e f_{e,i}}, \tag{1.105}$$

and the associated current density is

$$J_p = -n_e e\, v_\delta^* = \frac{n_e e^2\, E}{m_e f_{e,i}}. \tag{1.106}$$

The use of the identity (1.99) (without the inclusion of the magnetic term) yields for the resistivity the following expression in terms of the 'microscopic' quantities

$$\eta = \frac{m_e\, f_{e,i}}{n_e e^2}. \tag{1.107}$$

The collision frequency plays, therefore, a crucial role in the definition of the plasma resistivity. Its dependence on the temperature is crucial to understand the limitations of the ohmic heating. To appreciate this point we recall that the Debye length scales with the temperature as $\lambda_D \propto T^{\frac{1}{2}}$ and the number of particles contained in a Debye sphere increases with the temperature as $\lambda_D^3 \propto T^{\frac{3}{2}}$. This means that the number of shielded particles increases, they do not contribute to the collisions and we can, therefore, conclude that the resistivity exhibits the following dependence on the temperature $\eta \propto T^{-\frac{3}{2}}$. These simple but productive considerations allow us to express the resistivity in the following convenient form

$$\eta[\Omega \cdot m] \cong 5.2 \cdot 10^{-5}\frac{\ln(\Lambda)}{(T_e[eV])^{3/2}}. \tag{1.108}$$

This quantity is called Spitzer resistivity (for a rigorous derivation the reader is referred to the literature at the end of the chapter). Such a dependence on the temperature, characterizing also the diffusion coefficient in equation (1.102), is one of the key relations necessary for understanding (at a qualitative level) the limits of the ohmic heating. In passing we note that

$$D[m^2s^{-1}] \cong \frac{10^3}{(T_e[eV])^{3/2}}, \tag{1.109}$$

and therefore, for plasma length of the order of 1 m, the diffusion time at a temperature of 10 eV is around 10 ms, while at 1 keV it becomes tens of seconds. In chapter 3 we will reconsider the results concerning resistivity and diffusion within the framework of a more general treatment.

Let us summarize what has been discussed so far

(a) According to Faraday–Lenz law an electromotive force, proportional to the derivative of the current flowing in the central solenoid coils, is induced in the chamber containing the plasma (see figure 1.18, where the chamber radius and the torus major radius are shown)
(b) The associated plasma current produces ohmic heating power, which drops with the raising of the temperature.
(c) We can, therefore, conclude that such a mechanism is inefficient and other (additional) heating sources should be foreseen.

Before considering this aspect of the problem, we introduce some further quantitative considerations on the plasma current and its dependence on the field generated by the solenoid and plasma current itself. We have already mentioned that the plasma toroidal current induces a poloidal field, which is given by (π stands for poloidal)

$$B_\pi = \frac{\mu_0 I_p}{2\pi a}. \tag{1.110}$$

It combines with the toroidal counterparts, as shown in figure 1.19(a), and the pitch angle along the magnetic field line is given by

$$\frac{dl_t}{dl_p} = \frac{B_t}{B_p}. \tag{1.111}$$

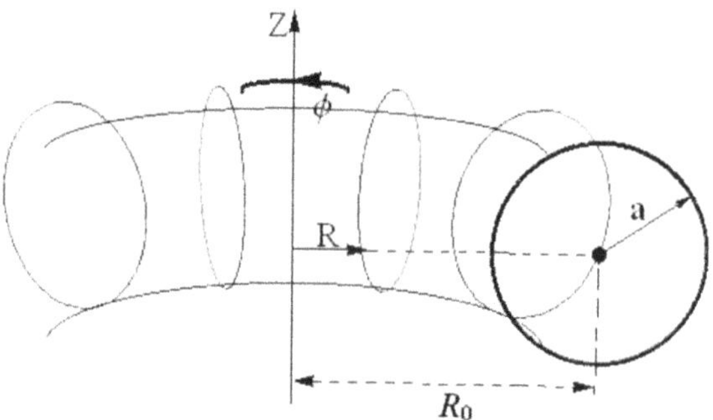

Figure 1.18. Major (R_0) and minor radius (a) of the toroidal chamber.

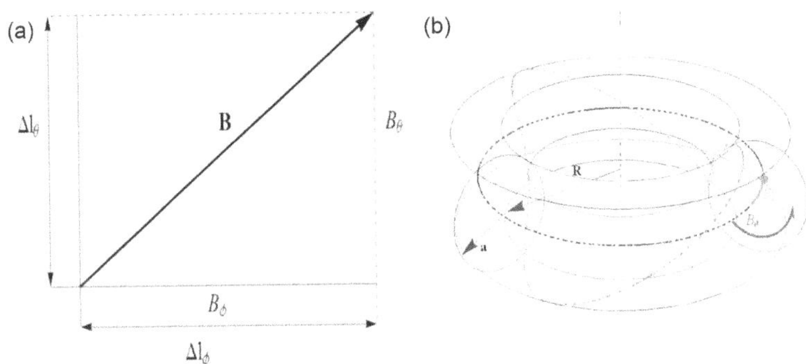

Figure 1.19. (a) Poloidal and toroidal field; (b) magnetic fields inside the toroidal chamber, the red line refers to the total field.

The length of the toroidal magnetic field line is linked to the pitch angle by the obvious identity

$$l_t = \int \frac{B_t}{B_p} dl_p. \tag{1.112}$$

In one poloidal turn (figure 1.19(b)), assuming the fields independent of the l_p we get

$$l_t = 2\pi a \frac{B_t}{B_p}. \tag{1.113}$$

The number of toroidal turns (figure 1.19(b)) in one poloidal turn is therefore given by

$$q = \frac{l_p}{2\pi R} = \frac{a}{R}\frac{B_t}{B_p}, \tag{1.114}$$

where the meaning of q is the winding number of helices provided by the ratio of the number of the toroidal magnetic field (at a given radius a) to the poloidal counterparts.

Further comments on this point are provided in the final section of the forthcoming chapters. Here we just mention that from the engineering point of view it plays a role of central importance, since in order to ensure the plasma stability, it should satisfy the condition $q > 2$ on the plasma edge, otherwise a plasma disruption[13] occurs.

After combining equations (1.110) and (1.114), the plasma current can be written in terms of the q parameter as

$$I_p = \frac{2\,\pi\,a^2}{R\,q\,\mu_0} B_t. \tag{1.115}$$

[13] The concept of plasma disruption should be understood in magnetohydrodynamics terms. Here we summarize the process as the instability mechanisms leading to a plasma inefficient for fusion processes.

This is a reference formula of paramount importance in the design of magnetically confined fusion devices and will be further commented on in this book.

It is now worth going back to the pulsed nature of the heating mechanism before ending up this section.

The limits of the ohmic heating are associated not only with the drop of the resistivity, but are determined by engineering issues too. It is evident that a constant electric field inside the plasma is sustained by a current rising linearly with the time. The current growth cannot last indefinitely and therefore the operation is limited to pulses of finite duration.

During the plasma startup, the current quickly diffuses into the plasma and the resistive heating raises the temperature. As the discharge goes on, the current tends to diffuse into the plasma core, I_p increases and the safety factor q falls, with consequent implications for the plasma stability.

In a steady state, $\eta\, J_p$ is constant and therefore $J_p = k/\eta$ and most of the current flows in the hot core of the plasma. The resistivity decreases with the temperature and the amount of energy, which can be fed to the plasma, quickly drops. As already stressed, for typical plasma parameters, the maximum temperature is 3 keV, which is insufficient to sustain self-burning plasma.

Additional heating is therefore necessary to achieve fusion and energy production in a Tokamak [27].

For these reasons external heating sources play a central role in the design of Tokamak devices. Some of them (like those providing heating via neutral beams injection) will be discussed in the first part of the book.

The second part is addressed to electromagnetic sources designed to provide additional heating and in order to reach the ignition and allow the self-sustained reactor operation. We will discuss in particular a number of high power, high efficiency free electron laser devices, including Undulator based free electron lasers, gyrotrons and cyclotron auto-resonance masers (CARMs). We will analyze the required performances to be candidates for plasma heating devices, discuss the relative pros and cons and finally the engineering issues to couple the radiation to plasma.

In this chapter we have touched on a number of issues associated with the nature of plasma itself.

We have dwelt on plasma single-particle and kinetic theories, we have commented on their equivalence in the collisionless regime. It has been underlined that kinetic models are, in principle, good candidates to describe plasma physics in detail. The associated methodology is tailored suited to describing the distribution evolution in space, momenta and time for any family of particles (electrons, ions, neutrals) contributing to the plasma dynamics. They can accordingly be exploited to infer multi-fluid models in which each species is treated as an independent fluid.

The applicability of these models is hampered by the 'limited' capabilities of the present computers. In the forthcoming chapter we will discuss a different approach, based on the MHD (magnetic hydrodynamic dynamics), which has the advantage of treating the plasma as a fluid (without any distinction among the species), characterized by a current density. The relevant interaction with the magnetic field is thereby treated without resorting to the Maxwell equations.

References

[1] Stathis Michaelides Efstathios E 2011 *Alternative Energy Sources* (Berlin: Springer)
[2] Dattoli G, Artioli M and Tuccillo A A 2017 Understanding the limits of the development *ENEA Technical Report* RT/2017/17/ENEA http://hdl.handle.net/20.500.12079/6787
[3] Brown J H *et al* 2011 Energetic limits to economic growth *Bioscience* **61** 19–26
[4] Atzeni S and Meyer-Ter-Vehn J 2004 *The Physics of Inertial Fusion* (Oxford: Oxford University Press)
[5] Miyamoto K 1980 *Plasma Physics For Nuclear Fusion* (Cambridge, MA: MIT Press)
[6] Maisonnier D *et al* 2005 A conceptual study of commercial fusion power plants *EFDA Technical Report* (05)-27/4.10 https://www.euro-fusion.org/fileadmin/user_upload/Archive/wp-content/uploads/2012/01/PPCS_overall_report_final.pdf
[7] Lawson J D 1957 Some criteria for a power producing thermonuclear reactor *Proc. Phys. Soc.* B **70** 6–10
[8] Harms A H, Schoepf K F, Miley G H and Kingdon D R 2014 *Principle of Fusion Energy* (Singapore: World Scientific)
[9] Einstein A 1905 Ist die trägheit eines körpers von seinem energieinhalt abhängig *Ann. Phys., Lpz.* **18** 891–921
[10] Bodanis D 2000 *$E = mc^2$: A Biography of the World's Most Famous equation* (New York: Walker Company)
[11] Borchardt G 2009 The physical meaning of $E = mc^2$ *Proc. of the Natural Philosophy Alliance (Storrs, CN)* **6** 1–5
[12] Balantekin A B and Takigawa N 1998 Quantum tunneling in nuclear fusion *Rev. Mod. Phys.* **70** 77
[13] Miley G H, Towner H and Ivich N 1974 Fusion Cross section and Reactivities *University of Illinois Nuclear Engineering Technical Report* C00-2218-17
[14] Hutchinson I 2003 Chapter 2: Motion of charged particles in fields *Introduction to Plasma Physics, Lecture Notes* https://ocw.mit.edu/courses/nuclear-engineering/22-611j-introduction-to-plasma-physics-i-fall-2003/lecture-notes/chap2.pdf
[15] Babusci D, Dattoli G and Sabia E 2011 Operational methods and Lorentz-type equations of motion *J. Phys. Math* **3** 1–17
[16] Dattoli G, Doria A, Sabia E and Artioli M 2017 *Charged Beam Dynamics, Particle Accelerators and Free Electron Lasers* (Bristol: IOP Publishing)
[17] Pucella G and Segre S E 2014 *Fisica dei Plasmi* (Bologna: Zanichelli)
[18] Peter T 2013 *Introduction to Plasma Physics Lecture. Lection 5: Plasma Mirroring* https://www.tcd.ie/Physics/people/Peter.Gallagher/lectures/PlasmaPhysics/Lecture5_single_particle.pdf
[19] Jackson J D 1998 *Classical Electrodynamics* 3rd edn (New York: John Wiley & Sons, Inc)
[20] Hutchinson I 2003 Chapter 1: Introduction *Introduction to Plasma Physics, Lecture Notes* https://ocw.mit.edu/courses/nuclear-engineering/22-611j-introduction-to-plasma-physics-i-fall-2003/lecture-notes/chap1.pdf
[21] Johnson P W 1979 *Lecture in Plasma Physics* http://mypages.iit.edu/~johnsonpo/plasmaweb.pdf
[22] Petty C C 2006 Sizing up plasmas using dimensionless parameters *48th Annual Meeting of the Division of Plasma Physics, Philadelphia, Pennsylvania, 30 October* https://fusion.gat.com/pubs-ext/APS06/Pettyvgs.pdf
[23] Bittencourt J A 2004 *Fundamentals of Plasma Physics* 3rd edn (Berlin: Springer)
[24] Thompson W B 2013 *An Introduction to Plasma Physics* 2nd edn (Oxford: Pergamon)

[25] Klimontovich Y L 1967 The statistical theory non-equilibrium processes in a plasma *International Series of Monographs in Natural Philosophy* vol 9 (Oxford: Pergamon)

[26] Suprunenko V A, Sukhomlin E A and Reva N I 1965 Ohmic heating and the electrical conductivity of a plasma in strong electric fields *J. Nucl. Energy* C **7** 3

[27] Kikuchi M 2010 A review of fusion and Tokamak research towards steady-state operation: a JAEA contribution *Energies* **3** 1741–89

IOP Publishing

High Frequency Sources of Coherent Radiation for Fusion Plasmas

G Dattoli, E Di Palma, S P Sabchevski and I P Spassovsky

Chapter 2

MHD models, plasma equilibrium and instabilities

2.1 Introduction

The previous chapter has been devoted to outlining what fusion is and we have argued that it relies on two pillars:

(a) Under appropriate conditions, light elements fuse and a significant amount of energy is released in terms of the kinetic energies of the particles born during the fusion reaction.

(b) The commercial use of a fusion reactor demands that the amount of output energy be controlled and be larger than that used to trigger the process itself.

The previous two points constitute the premise and the final goal, respectively. The problem appears therefore well posed, even though, the road map in the middle and the associated strategy are far from being straightforward [1].

This chapter is aimed at providing a broader view of plasma physics and the technical issues, put forward to progress from (a) to (b). Before addressing the specific details of the chapter, we would like to summarize the discussion developed so far by organizing the few notions, we have acquired, to infer some design criteria for a Tokamak device. It has already been argued that there are dimensionless global parameters, which play a key role in the Tokamak design, since they are useful tools to get a first idea of the relevant dimensions and performances. We have introduced, without explaining the reason for its importance, the parameter β, defined as the ratio between the plasma (outward) and the magnetic (inward) pressure force. It emerges in a fairly natural way from the conditions of the plasma equilibrium, summarized by the identity [2]

doi:10.1088/978-0-7503-2464-9ch2

$$nk_BT + \frac{B_i^2}{2\mu_0} = \frac{B_e^2}{2\mu_0}, \tag{2.1}$$

where B_i is the internal magnetic field (if any) due to charges of internal motion. It exerts a pressure force opposing the external field B_e, associated with poloidal and toroidal components. We can therefore write (see equation (1.86))

$$\beta = 1 - \left(\frac{B_i}{B_e}\right)^2. \tag{2.2}$$

If the plasma is perfectly diamagnetic[1], then $\beta = 1$ and this would be the most efficient use of the external magnetic field, entirely devoted to the control of the internal kinetic pressure (for a more complete discussion see section 2.5). The previous identities, albeit elementary, are sufficient to underscore the central role of β, which links together the magnetic field, plasma density and temperature. If, according to specific considerations, its value is given as an external (target) parameter, it can be used to determine the output power and the B value. We consider a fusion plasma with the following (typical) parameters

$$\begin{aligned} &n = n_D = n_T = 10^{20}\ \mathrm{m}^{-3}, \\ &T = 10\ \mathrm{keV},\ p \cong 3.2 \cdot 10^5\ \mathrm{Pa}. \end{aligned} \tag{2.3}$$

The external magnetic field B_e, counteracting the internal pressure, can be written in terms of β as

$$B_e = \sqrt{2\frac{\mu_0 nk_BT}{\beta}} = \sqrt{\frac{2\mu_0 p}{\beta}} \cong \frac{0.9}{\sqrt{\beta}}\mathrm{T}, \tag{2.4}$$

which, for $\beta = 1$, yields 0.9 T, that is, a pretty low intensity for the present technological capabilities. In the actual devices it is assumed that $\beta = 0.05$ (see below), with a corresponding field strength $B_e \cong 4$ T, still an easily achievable value.

The role of β in the definition of the output power, can be understood by using the first part of equation (2.4) to express the plasma density in terms of the external magnetic field

$$n = \beta\frac{B_e^2}{4\mu_0 k_BT}. \tag{2.5}$$

The output fusion power has been reported in chapter 1 (equation (1.10)), therefore, the last identity yields the following link between the fusion power and β

$$P_f \propto \beta^2 B_e^4. \tag{2.6}$$

[1] The plasma diamagnetism is an essential property. When a plasma is immersed in a magnetic field, the gyrating particles generate a magnetic field opposing the external one. In idealized conditions this effect would imply perfect diamagnetism. In practice, however, an additional opposite magnetic moment, due to gyrating particles reflected by the walls, causes (partial) loss of the diamagnetic properties.

The impact of β is, therefore, crucial for the design and the commercial realization of any Tokamak reactor. However, its choice is indeed an extremely delicate matter, to be carefully determined by taking into account various physics, engineering and economical constraints. We outline below a few simple, but effective arguments, helpful to fix a convenient design value of this important and critical quantity.

We first recall that the external magnetic field has two components (toroidal and poloidal) such that $B^2 = B_p^2 + B_t^2$ and for later convenience we introduce the quantity

$$\beta_p = \frac{nk_B T}{\frac{B_p^2}{2\mu_0}}, \tag{2.7}$$

and write

$$\beta = \frac{\beta_p}{1 + \left(\frac{B_t}{B_p}\right)^2}. \tag{2.8}$$

The β parameter depends on the ratio between the toroidal and poloidal field components, which, according to equation (1.114), is provided by the identity (see figure 1.19)

$$\frac{B_t}{B_p} = A \cdot q, \tag{2.9}$$

where $A = R/a$ is the plasma aspect ratio [3].

The combination of the last two equations yields the parameterization

$$\beta = \frac{\beta_p}{1 + (A\,q)^2}, \tag{2.10}$$

which is important since it includes the most essential parameters that need to be considered in order to choose a convenient value of β.

As will be discussed later in this chapter, considerations regarding the plasma stability suggest $q(R) > 1$ and $\beta_p \geqslant 0.5$, choosing $A = 4$, $q(a) = 3$, $\beta_p = 0.5$ we find a low β value ($\beta \cong 3.45 \cdot 10^{-3}$), which makes fusion commercially unattractive. The current design trends quote a β value around 5% as commented later in this and in the forthcoming chapter.

We can acquire further confidence in handling the few elements in our hands by considering the following example, which is based on the previously outlined essential parameters used to evaluate the performance and the dimensions of a D–T Tokamak with the plasma properties given in equation (2.3) and characterized by

$$\beta = 5\%,\ R = 6.6\ \text{m},\ Z_e = 1,\ f_p = 2, \tag{2.11}$$

being f_p the profile factor, whose role is clarified below.

The questions we raise are:***Are we able to define***

(a) ***The confinement time (*** τ_E ***)?***
(b) ***The Tokamak size to reach the break-even?***
(c) ***The required on-axis B-field intensity?***
(d) ***The Fusion power at the ignition?***
(e) ***The wall power load?***

The answers, albeit hampered by the limited notions at our disposal, are given below and will be further commented on in the forthcoming parts of the book, when we acquire further awareness on the topics under study.

Confinement time is an indispensable constituent of the Lawson balance criterion. We can accordingly refer to the discussion developed in section 1.2, with the further assumption that the power losses due to bremsstrahlung are not negligible, namely the power radiated by the electrons, when they are decelerated during the collision with ions, will be taken into account in the energy balance equations. It depends on the particle velocity and in practical units can be written as

$$P_b[\mathrm{W\ m^{-3}}] \cong 5.4 \cdot 10^{-37} n^2 Z \sqrt{T\ [\mathrm{keV}]}. \tag{2.12}$$

By taking into account this contribution, we modify the Lawson criterion as

$$n\tau_E = \frac{3\ k_B T}{\frac{1}{4}\langle \sigma\ v\rangle E_\alpha f_p - 5.4 \cdot 10^{-37}\sqrt{T\ [\mathrm{keV}]}}, \tag{2.13}$$

or in a more convenient form as

$$n\tau_E = \frac{L}{1 - R}, \tag{2.14}$$

where

$$L = \frac{12\ k_B T}{\langle \sigma v\rangle E_\alpha f_p},\ R = \frac{2.16 \cdot 10^{-36}\sqrt{T\ [\mathrm{keV}]}}{\langle \sigma v\rangle E_\alpha f_p}. \tag{2.15}$$

Before proceeding further we have to justify the presence of the extra-factor f_p which is simply a multiplicative term, accounting for the plasma density and temperature dependence on the radial coordinate. It is called *profile factor* and its typical values are between 1.5 and 2.

Now, we have all the information that is needed in order to evaluate the confining time.

Using these parameters and recalling that at 10 keV (see figure 1.7) $\langle \sigma v\rangle = 1.1 \cdot 10^{-22}\mathrm{m^3\,s^{-1}}$ we find $\tau_E \cong 1.64$ s. It is worth noting that at 10 keV the correction due to bremsstrahlung is about 5.5%, which is not totally negligible. The behavior of the reaction rate $\langle \sigma v\rangle$ versus T, shown in figure 2.1, confirms that for typical plasma fusion temperature the bremsstrahlung losses are not harmful (for further comments see the discussion in the next chapter).

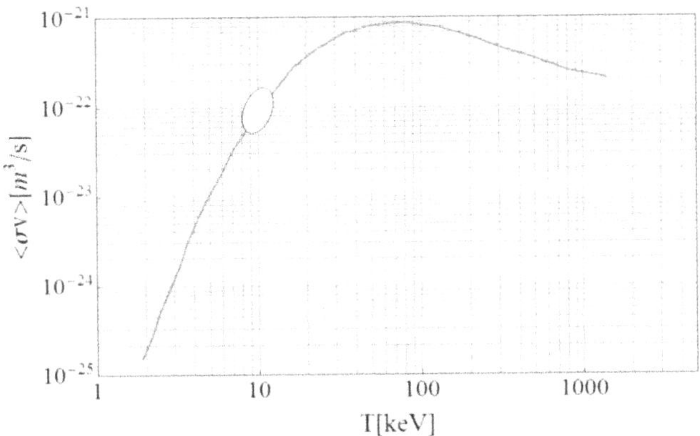

Figure 2.1. D–T reactivity versus temperature and Tokamak magnetic fusion working point (white ellipse).

Let us now come to the second point, namely the ***Tokamak size to reach the break-even***.

Although we have not yet developed any argument relating plasma density, confinement time and plasma radius, there is, an empirical relation (called Alcator scaling formula) which removes this deficiency. It links the confinement time, plasma radius and density through the identity

$$n\,\tau_E \cong 6 \cdot 10^{-21} n^2 a^2, \tag{2.16}$$

which eventually yields the answer as follows

$$a \cong 0.41 \cdot 10^{11} \sqrt{\frac{\tau_E}{n}} \cong 1.65 \text{ m}. \tag{2.17}$$

The empirical scaling laws play a role of paramount importance in the Tokamak design. We have quoted Alcator scaling without any physical justification, which is difficult to find since these scaling relations are the result of extensive numerical simulations and of experimental observations. In chapter 3 we will discuss the methodology underlying the plasma scaling 'laws' and provide an account of their importance in the design strategy.

Regarding the ***required on-axis field intensity***, the question is straightforwardly answered by noting that by combining equations (2.7) and (2.8) we find

$$\begin{aligned} B_p &= \sqrt{2\frac{\mu_0 n k_B T}{\beta_p}}, \\ B_p &= \frac{B_t}{\sqrt{\frac{\beta_p}{\beta} - 1}}. \end{aligned} \tag{2.18}$$

The first equation yields $B_p \cong 4T$ and the second one, for $\beta_p = 0.5$ allows us to fix the value of the poloidal field around 1/3 of the toroidal counterpart (assuming $\beta = 5\%$).

The derivation of ***the fusion power at the ignition*** is just a matter of straightforward computation, multiplying, indeed the fusion power density by the torus volume we get

$$P_{\text{fus}} = \frac{1}{4} n^2 \langle \sigma v \rangle E_{\text{fus}} V_T, \tag{2.19}$$

where

$$V_T = 2\pi R\,(\pi a^2), \tag{2.20}$$

which, along with equation (2.14), yields

$$P_{\text{fus}} = \frac{R\pi^2}{12} 10^{21} n\tau_E \langle \sigma v \rangle E_{\text{fus}}. \tag{2.21}$$

The use of (2.19) (or (2.21) as well) provides the numerical value $P_{\text{fus}} \cong 549$ MW.

We can finally come to the last question regarding ***the wall power load***, according to the discussion of the previous chapter (equation (1.5)). Almost 80% of the fusion power is carried by neutrons, namely 439 MW, distributed on the surface of the torus, the wall load L_W is therefore estimated to be

$$L_W = 0.8\,\frac{P_{\text{fus}}}{4\pi^2 Ra} \cong 1\ \text{MW m}^{-2}. \tag{2.22}$$

The message we wanted to convey is that one can use simple means to draw important relationships, helping to determine the Tokamak design and working points. It should be understood that when we say simple, we mean relationships, albeit straightforward, which are the result of years of experience and of deep understanding of the underlying physics.

To summarize what has been achieved in this section, we note that we have started from an equilibrium relation, reminiscent of hydrodynamic stability conditions, and, using common sense arguments, we have been able to draw conclusions of practical interest. Stability in magnetic fusion plasma is a topic of utmost importance. The discussion of the forthcoming sections is devoted to the magneto-hydrodynamic aspects of plasma physics and to its consequences.

2.2 Fusion reaction in the Sun and associated energy production

There is no doubt that our planet has benefited from fusion processes occurring in the Sun for billions of years. Albeit it is an active fusion reactor, the Sun cannot be viewed as a Tokamak and it might be misleading to make analogies between the fusion processes at the stars and in man-made reactors for controlled thermonuclear fusion[2].

The existence of fusion processes in the Sun can be argued on the basis of a simple consideration. A few reference numbers are, however, in order both to catch the

[2] The authors express their sincere appreciation to Dr Franco Alladio who warned us about this common but misleading analogy.

point and to properly frame the forthcoming discussion. We, therefore, recall that the temperatures at the surface and in the core of the Sun are $T_s = 5.8 \cdot 10^3$ K, $T_c = 1.55 \cdot 10^7$ K, respectively. By keeping an average temperature of 10^6 K and taking into account that the number of atoms composing the Sun is

$$\frac{M_s}{m_p} \cong 10^{57},$$

where $M_S = 2 \cdot 10^{30}$ kg, $m_p = 1.7 \cdot 10^{-27}$ kg are the Sun and proton mass, respectively, we can infer that the total energy contained in it is 10^{40} J. The radiated power (W_r) can be calculated by the use of the Stefan–Boltzmann law [4]

$$W_r = 4\pi\ R_S^2 \sigma T^4 \cong 3.9 \cdot 10^{26}\ \mathrm{W}, \tag{2.23}$$

where $\sigma = 5.67 \cdot 10^{-8}$ W ($\mathrm{m}^{-2}\,\mathrm{K}^{-4}$) and $R_S \cong 7 \cdot 10^8$ m. With such an energy loss rate, the Sun energy reservoir, if it were a fixed budget of 10^{40} J, would be lost in only 10^6 years, too short if compared to its age of 4.6 billion years.

The mechanism which replaces the lost energy and keeps the Sun at constant temperature is due to some energy production, now identified as fusion processes, in which the reactants are not confined by an external magnetic field.

As a matter of fact, in massive objects, the gravitational pressure is the key factor providing the conditions to confine the fusion elements, the energy produced in their cores balances the effect of the outward pressure force. This is not the only and most significant difference. The most abundant element in the Sun is the ionized hydrogen, that can fuse to produce deuterons (nuclei of deuterium atoms).

The point is 'how do protons come close and with sufficient energy to overcome the Coulomb repulsion?'

In an environment like a massive object, the most significant force exerted on protons is that due to gravity. The kinetic energy it acquires under its influence is inferred from the Newton law (in MKS unit) [5]

$$\frac{3}{2}\, k_{\mathrm{B}} T = \frac{G M_S m_p}{R_S}, \tag{2.24}$$

where gravitational constant

$$G = 6.67 \cdot 10^{-11}\ \mathrm{N\, m^2\, kg^{-2}}, \tag{2.25}$$

which yields $T \cong 1.67 \cdot 10^7 K$ and a corresponding energy of a few keV. It is worth stressing that the actual value is $1.55 \cdot 10^7$ K (not so bad for such naive derivation) and that it is consistent with that discussed for the Tokamak configuration.

The gravitational pressure can be determined again from Newton's law (see below), thus getting

$$P = \frac{3G\ M_S^2}{8\ \pi\ R_S^4} \cong 1.3 \cdot 10^{14}\ \mathrm{Pa}, \tag{2.26}$$

which underestimates by about two orders of magnitude the experimental value. What we have got, as inferred from the observational data, is anyway larger than the values of 6–7 atm quoted for the magnetic confinement[3].

The inward gravitational forces counteract the outward pressure and this competition gives rise to a self-regulating mechanism: if fusion activity increases, the outward force prevails, thus determining an expansion of the core volume, which, is in turn responsible for the decrease of the fusion process rates.

If no energy generation occurred inside stars, the gravity pressure would induce an increase of the core temperature, with a consequent energy flow out, because of radiation or convection losses. Without any feedback, the gravity pressure further increases, the core becomes hotter, more energy is further lost and eventually the core collapses.

The inside star energy production actually compensates the effect of flow out radiation and prevents the core collapse. The protons come close to each other and the following reaction takes place

$$p + p \rightarrow D + e^{+} + \upsilon_e, \tag{2.27}$$

which is the first step of the reaction chain reported in figure 2.2.

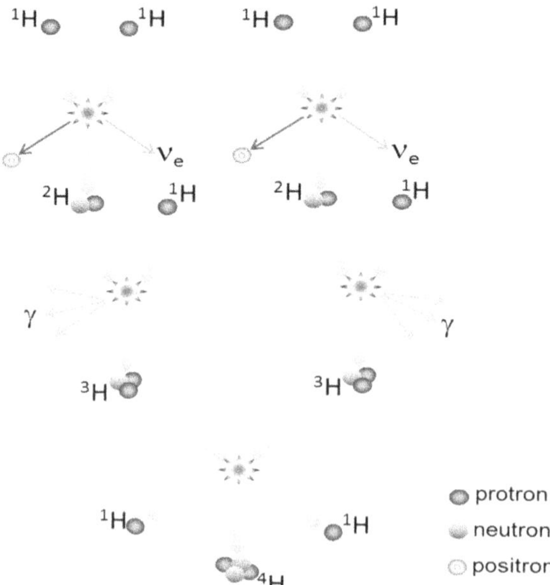

Figure 2.2. The proton–proton reaction chain in the Sun with the release of the gamma(γ) ray energy.

[3] If we trust the first equation (2.1) and use the expression of the magnetic pressure (see equation (1.84)) we obtain the following gravitational pressure equivalent magnetic field $B \cong \frac{M_S}{R_S^2}\sqrt{\frac{\mu_0 G}{2\pi}}$ which, using the Sun's parameters, corresponds to values of the magnetic intensity larger than 10^4 T!!!.

The actual reaction should be further complemented with an intermediate step in which a (p, p) state is formed and yields the final decay product, thanks to the weak reaction $p \to n + e^{+} + v_e$. The probability that the reaction in equation (2.27) occurs is very low, only one event per 10^{25} collisions, which, at the previously specified physical conditions, occurs with a rate of 10^{8} Hz, namely one every 10^{17} s, i.e., about 10 Gy, an extremely large time. However, considering the internal Sun density compatible with 10^{32} protons per cubic meter we end up with 10^{15} successful reactions of the type (2.27) in one cubic meter.

The rest of the reactions occur in a much faster time-scale and, cutting off any further discussion, we note that all the 'game' reduces to the formation of a nucleus of He starting from four protons. More important, however, is that $4\,m_p > m_{\mathrm{He}}$ which is in fact the source of the solar fusion energy.

We can naively comment on this discussion by noting that fusion occurs at star level by the concurrence of all the fundamental forces (strong, weak, electromagnetic and gravitational). In order to trigger fusion, the Sun has to synthesize deuterium, this is the reason why it (and the stars belonging to the main sequence as well) burns so slowly. Humans do not have this type of problem and find deuterium on the Earth in abundance, (tritium is available through self-breeding) this allows mankind to skip one step and avoid the contribution of gravity.

The digression on the gravity triggered fusion mechanisms has a two-fold motivation:

(a) It marks the differences with its magnetic counterpart.
(b) It underscores that a kind of hydrodynamic equilibrium is one of the key mechanisms regulating the relevant dynamics.

Regarding, the equilibrium, the situation is that reported in figure 2.3 displaying the internal gradient pressure counteracted by the outward forces, associated with the gradient pressure, which in the case of gravity reads

$$\vec{\nabla}_r p = \frac{dP}{dr} = -G\frac{M(r)}{r^2}\rho(r) \tag{2.28}$$

for a star with a constant density (ρ) we can integrate the previous equation and express the pressure as

$$P_{\text{central}} = G\frac{M}{2R}\rho. \tag{2.29}$$

Regarding the magnetic confinement we find that (see next sections for further comments)

$$\vec{\nabla} p = \vec{J} \times \vec{B}, \tag{2.30}$$

where $\vec{J}$ is the volumetric plasma current density.

The plasma description, we have developed till now, is based on kinetic models, which are in principle the most accurate way of depicting both plasma evolution and equilibrium. Its inevitable drawback, however, is determined by the significant

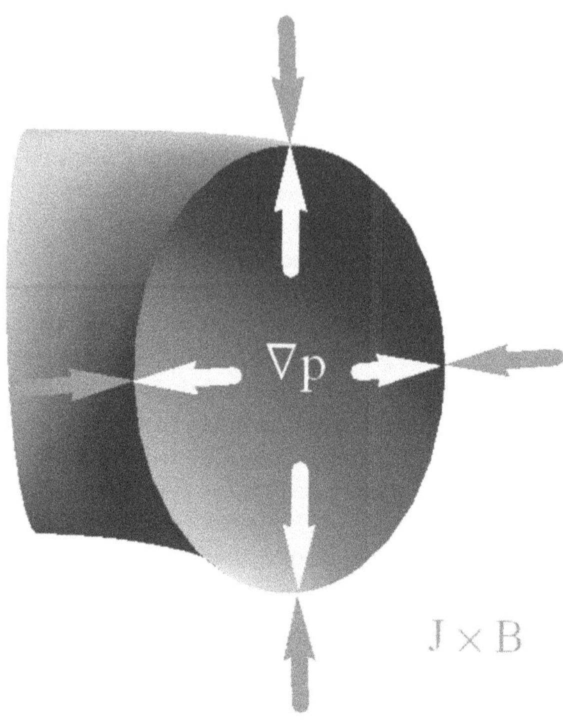

Figure 2.3. Internal pressure gradient and external magnetic force balance.

computational effort needed to track a large number of particles as well as derive the relevant distribution and the associated macroscopic quantities in terms of the moments of the distribution itself.

A less computer intensive treatment is the magnetohydrodynamics (MHD) model in which the plasma is treated as a fluid (of electrons and ions), coupled to Maxwell's equations and capable of providing an accurate definition of macroscopic equilibrium, stability, the onset of instabilities, heating and so on. The forthcoming sections of this chapter are devoted to the relevant discussion on this matter.

2.3 Elements of magnetohydrodynamics and plasma physics

There are many authoritative textbooks in which MHD, the topic we are going to summarize here, is treated with the necessary rigor and depth. The forthcoming sections of this chapter are aimed at giving a phenomenological treatment only, useful for the purposes of this book [6, 7].

Here, we treat the plasma as a fluid, namely a continuous medium characterized by mass density ρ, and pressure p in an element of volume with velocity v and temperature T. Regarding this last characteristic, we note that such a fluid model of plasma is highly collisional and the corresponding velocity distribution function is Maxwellian.

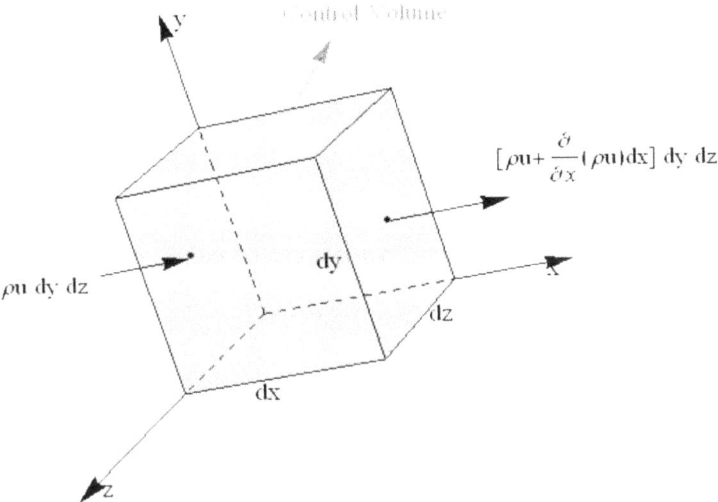

Figure 2.4. Flow in and out in a volume element.

One of the key equations of the fluid plasma model is the mass continuity equation, whose derivation is detailed below. In figure 2.4 we have reported a region inside the plasma with a volume

$$V = l \cdot A. \tag{2.31}$$

The number of particles inside the volume is given by

$$N = n \cdot l \cdot A, \tag{2.32}$$

and those leaving the box in a time T is just

$$\frac{N}{T} = \frac{n \cdot l \cdot A}{T} = n \cdot A \cdot v, \tag{2.33}$$

where $v = l/T$.

The number of the particles and mass density fluxes are given, respectively, by

$$\frac{N}{AT} = n \cdot v, \tag{2.34}$$

and

$$m_p \frac{N}{AT} = m_p \cdot n \cdot v = \rho v, \tag{2.35}$$

with m_p the mass of the individual particles constituting the whole ensemble. The rate of change of the mass flux, flowing in (sub-index i) and out (sub-index o) through the six faces of the cube, can be written as [8]

$$\begin{aligned} \frac{dm}{dt} = &\left((\rho_o v_o)_x - (\rho_i v_i)_x\right)dzdy + \left((\rho_o v_o)_y - (\rho_i v_i)_y\right)dxdz \\ &+ \left((\rho_o v_o)_z - (\rho_i v_i)_z\right)dxdy, \end{aligned} \tag{2.36}$$

where

$$m = \rho\, dx\, dy\, dz, \tag{2.37}$$

we can write (note the change of the total time derivative into partial)

$$\frac{dm}{dt} = \frac{\partial}{\partial t}\rho\, dx\, dy\, dz, \tag{2.38}$$

thus eventually getting, for the density variation, the equation

$$\frac{\partial \rho}{\partial t} = -\left[\frac{\partial}{\partial x}(\rho v)_x + \frac{\partial}{\partial y}(\rho v)_y + \frac{\partial}{\partial z}(\rho v)_z\right]. \tag{2.39}$$

The operator form (which is certainly less cumbersome and more transparent) is

$$\frac{\partial \rho}{\partial t} = -\vec{\nabla} \cdot (\rho \vec{v}), \tag{2.40}$$

which is popularly known as the *mass continuity equation*.

The use of material, or convective derivative

$$\frac{d}{dt} = \frac{\partial}{\partial t} + \vec{v} \cdot \vec{\nabla}, \tag{2.41}$$

providing a transition between the total and the partial derivatives, allows us to cast the continuity equation in the form[4]

$$\frac{d}{dt}\rho + \rho \vec{\nabla} \cdot \vec{v} = 0. \tag{2.42}$$

Furthermore, we specify the MHD, which is summarized by the equations reported below.

The previous general remarks can be exploited to write the Newton equation

$$\rho \frac{d\vec{v}}{dt} = \vec{F}, \tag{2.43}$$

which in terms of the convective derivative

$$\rho\left(\frac{\partial \vec{v}}{\partial t}\right) + \rho\,(\vec{v} \cdot \vec{\nabla})\,\vec{v} = \vec{F}, \tag{2.44}$$

where v is the velocity of the fluid element and F is the force per unit volume acting on the element. Without entering into subtle distinctions between the volume and surface forces, we specify F as

$$\vec{F} = n_q(\vec{E} + \vec{v} \times \vec{B}) - \vec{\nabla} p, \tag{2.45}$$

[4] Note the following operatorial identity $\vec{\nabla} \cdot \left(\alpha\, \vec{b}\right) = \alpha(\vec{\nabla} \cdot \vec{b}) + \left(\vec{b} \cdot \vec{\nabla}\right)\alpha$.

where

$$n_q \equiv n_e, \vec{J} = n_q\vec{v}. \tag{2.46}$$

We are now sufficiently well prepared to derive the so called *Ideal MHD equations*. To this aim we recall the already quoted Ohm's generalized law written as

$$\vec{E} + \vec{v} \times \vec{B} = \eta \vec{J}, \tag{2.47}$$

and with η being the plasma resistivity one can see that in an ideal collisionless plasma (i.e. with zero resistivity) the following equation holds

$$\vec{E} + \vec{v} \times \vec{B} = 0. \tag{2.48}$$

The pressure exerted by the magnetic field on the plasma can be written (using equation (2.45) in the absence of the electric field) as

$$\vec{F} = \vec{J} \times \vec{B} - \vec{\nabla}p. \tag{2.49}$$

At equilibrium we find, for the pressure gradient

$$\vec{J} \times \vec{B} = \vec{\nabla}p, \tag{2.50}$$

where

$$\vec{J} = \frac{1}{\mu_0}\vec{\nabla} \times \vec{B}. \tag{2.51}$$

And from equation (2.50) it follows that

$$\frac{1}{\mu_0}(\vec{\nabla} \times \vec{B}) \times \vec{B} = \vec{\nabla}p. \tag{2.52}$$

The use of standard vector calculus rules eventually yields

$$\frac{1}{\mu_0}\left[\left(\vec{B} \cdot \vec{\nabla}\right)\vec{B} - \frac{1}{2\mu_0}\vec{\nabla}\left(|\vec{B}|^2\right)\right] = \vec{\nabla}p, \tag{2.53}$$

which can be rearranged in the following more compact form

$$\frac{1}{\mu_0}(\vec{B} \cdot \vec{\nabla})\vec{B} = \vec{\nabla}\left(p + \frac{|\vec{B}|^2}{2\mu_0}\right). \tag{2.54}$$

The first term in equation (2.54) is the magnetic tension and is not vanishing if the magnetic field lines are curved. To better explain its role we consider the picture in figure 2.5 and note that if $\hat{B}$ denotes the unit vector along the $B-$ direction, one gets

$$\begin{aligned} \hat{B} \cdot \vec{\nabla}\hat{B} &\approx \frac{\Delta\hat{B}}{\Delta l}, \\ \hat{B}(l + \Delta l) - \hat{B}(l) &\approx -\hat{n}\Delta\theta, \\ \Delta l &= R\Delta\theta, \end{aligned} \tag{2.55}$$

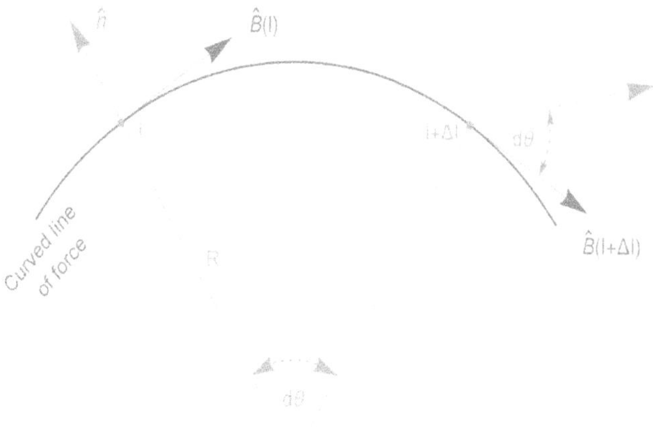

Figure 2.5. Magnetic field lines with a curved geometry.

from which, therefore, we get

$$(\vec{B} \cdot \vec{\nabla})\vec{B} \approx |\vec{B}|^2 \frac{\Delta \hat{B}}{\Delta l} \approx -|\vec{B}|^2 \frac{\hat{n}}{R}. \tag{2.56}$$

Now, if the lhs term in equation (2.54) is zero, namely if the field does not change in its own direction, we end up with

$$p + \frac{|\vec{B}|^2}{2\mu_0} = \text{const}, \tag{2.57}$$

which is a hydrodynamic equilibrium law, similar to the Bernoulli conservation theorem of the elementary fluidodynamics.

In the further analogy with what happens in a conducting fluid, where, in the absence of a narrowing of the radius of the conductor itself, the fluid velocity increases thus leading to a decrease of the pressure; if in some points of the plasma column, assumed to be cylindrical (see figure 2.6), there is a reduction of the radius r, the azimuthal field (inversely proportional to r) increases, with an induced decrement of the internal plasma pressure. This determines a kind of a dangerous positive feedback. The plasma tends to restore its shape and such a mechanism could take place in another part of the plasma [9–11].

All this dynamics is not beneficial to the plasma itself, it might lead to an instability known as *sausage instability*, which results in particle losses.

A further example of trouble is provided by the physical situation in which the column bends (see figure 2.7) and the azimuthal field lines become closer, the magnetic pressure is higher in the upper part, where the field is more intense and further bends towards the lower pressure region.

The plasma tends to restore the equilibrium conditions by propagating the bending effect in other regions of the plasma, this may induce another unwanted mechanism known as *kink instability*, potentially harmful for the plasma stability.

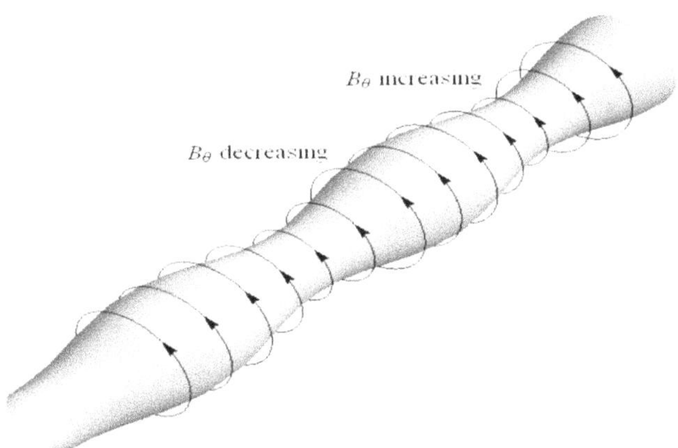

Figure 2.6. Sausage instability.

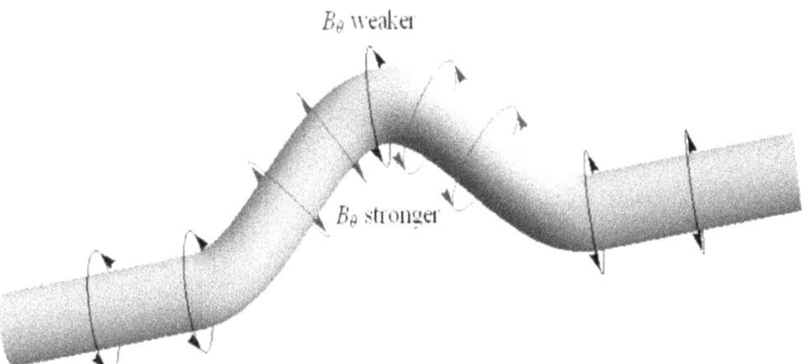

Figure 2.7. Kink instability.

In this section we have described from a very elementary point of view based on the MHD the mechanisms of instabilities. The forthcoming parts of this chapter provide a slightly more detailed look into this kind of problems.

2.4 Liouville, Vlasov and Boltzmann equations and ideal MHD

The discussion we have presented so far, in this and in the previous chapter, has not paid sufficient attention to the role of the interplay between the particles and the fields. In this section we follow a slightly more rigorous treatment aimed at filling such a gap. A more appropriate treatment can be found in [12].

A proper treatment should therefore deal with the Newton equations of the single particles constituting a plasma ensemble, by taking into account the *microscopic* field acting on the single particle. If we denote by

$$N_s = \sum_{i=1}^{N_0} \delta(r - R_i(t))\, \delta(v - V_i(t)), \tag{2.58}$$

the number of particles with velocity $\vec{v}$ at the location $\vec{r}$, with N_0 being the total number of the particles, we can write the relevant equations of motion as

$$\begin{aligned}\frac{\partial N_s}{\partial t} = &- \sum_i \vec{V}_i(t) \cdot \vec{\nabla}_r \delta(\vec{r} - \vec{R}_i(t))\delta(\vec{v} - \vec{V}_i(t)) + \\ &- \sum_i \frac{q_s}{m_s} \left\{ \vec{E}^{(m)}\left[\vec{R}_i(t), t\right] + \vec{V}_i(t) \times \vec{B}^{(m)}\left[\vec{R}_i(t), t\right] \right\} \\ &\cdot \vec{\nabla}_v \delta[\vec{r} - \vec{R}_i(t)]\delta[\vec{v} - \vec{V}_i(t)],\end{aligned} \tag{2.59}$$

where q_s, m_s are the charge and mass of the number of particles N_s, the superscript (m) denotes the *'microscopic fields'*, acting on the ith particle and generated by the superposition of the external fields and of those induced in a self-consistent way by the other particles. They are rapidly oscillating fields, which can in principle be derived from the Maxwell equations, coupled to the evolution of the charge and current microscopic distributions.

It is evident that $\vec{R}_i(t)$, $\vec{V}_i(t)$ are Lagrangian variables and $\vec{r}$, $\vec{v}$ represents their Eulerian counterparts. The problem can be simplified, considering the evolution in the $\vec{r}$, $\vec{v}$ space thus getting from equation (2.31) the alternative form

$$\frac{\partial N_s}{\partial t} = -\vec{v} \cdot \vec{\nabla}_r N_s - \frac{q_s}{m_s}\left[\vec{E}^{(m)} + \vec{v} \times \vec{B}^{(m)}\right] \cdot \vec{\nabla}_v N_s, \tag{2.60}$$

known as Klimontovich equation. Its solution is particularly awkward since it involves an enormous number of individual particles. To further simplify the problem we consider the transition (see figure 2.8) from the untractable ensemble of N_s particles to the reduced distribution function

$$f_s = \frac{\iint_{\Delta\Omega} N_s(\vec{r}\,', \vec{v}\,'; t)}{\Delta\Omega}, \tag{2.61}$$

where $\Delta\Omega = d^3r\, d^3v$, which can be interpreted as the number of particles in a given phase space interval.

It is now convenient to introduce the quantities

$$\begin{aligned} N_s &= f_s + \delta f_s, \\ \vec{E}^{(m)} &= \vec{E} + \delta\vec{E}^{(m)}, \\ \vec{B}^{(m)} &= \vec{B} + \delta\vec{B}^{(m)}, \end{aligned} \tag{2.62}$$

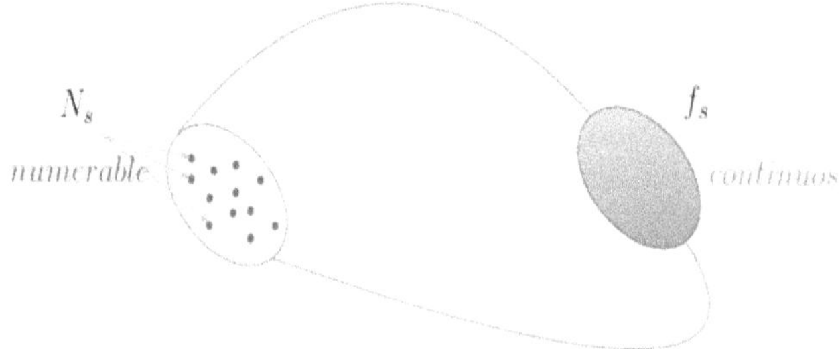

Figure 2.8. Transition from 'particle ensemble' to a continuous distribution.

where the δ-terms are the fluctuating parts, such that their time averages on the plasma characteristic time (see below) vanish, we are left with

$$\frac{\partial f_s}{\partial t} + \vec{v} \cdot \vec{\nabla}_r f_s + \frac{q_s}{m_s}[\vec{E} + \vec{v} \times \vec{B}] \cdot \vec{\nabla}_v f_s = \langle \ldots \rangle, \tag{2.63}$$

where the rhs contains terms including averages of the product of the field and of the distribution function fluctuations. If we assume that these averages are vanishing too, the previous equation is nothing but the L–V discussed in the previous chapter. If not, the equation can be written as

$$\frac{\partial f_s}{\partial t} = -\vec{v} \cdot \vec{\nabla} f_s - \frac{q_s}{m_s}[\vec{E} + \vec{v} \times \vec{B}] \cdot \vec{\nabla}_v f_s + \left(\frac{\delta f_s}{\delta t}\right)_{\text{coll}}, \tag{2.64}$$

where the subscript 'coll' stands for collisional. The relevant role of this term will be discussed below. For the moment, however, we assume that it can be neglected.

We have loosely stated that Liouville and Vlasov equations are equivalent. Such correspondence should, however, be more correctly formulated. In this specific case, the Vlasov equation written in a Liouville form applies to an ensemble of non-interacting particles in an external macroscopic field. The inclusion of the self-consistency is automatically included. The averages are indeed evaluated by the instantaneous values of the distribution f_s, modified by the fields themselves.

These equations are formulated in the six-dimensional $(\vec{r}, \vec{v})$ phase-space, which in the previous chapter has been indicated with the canonical variables (q, p). We will assume that the previous two pictures coincide, without entering any further subtleties regarding the canonicity of the system.

In deriving equation (2.64) we have assumed that the electric field is not affected by the particles' evolution itself. This is not the case, since both the electric and magnetic fields depend on the space charge distribution. To make the analysis self-consistent, it is therefore necessary to couple the charge particle dynamics (equations of motion) to the Maxwell equations for the fields. Put in these terms, solutions for such a coupled system of differential equations can be obtained for a very restricted number of problems, mostly limited to one-dimensional cases. For more realistic geometries, using suitable averaging on the Larmor orbits, leads to a set of equations which are tractable on supercomputers. Alternatively, instead of dealing with the full solution for the particles and fields, the use of the so-called velocity moments

$$\begin{aligned} \langle f_s \rangle_n &= \int v^n f_s \, dv, \\ \langle f_s \rangle_0 &= n(x, t) \propto \rho, \end{aligned} \tag{2.65}$$

may give important information on the most significant quantities characterizing the plasma dynamics. The 0th order is indeed linked to the density, the first to the average velocity, the second to the energy, the third to the heat flux and so on. Technical mathematical difficulties associated with the fact that each moment depends on the next makes the relevant use doubtful in view of the assumptions to be done to solve the associated equations.

The one possible assumption is that the chain can be truncated by assuming that those of order $n + 1$ vanish, which is true only for 'smooth' dynamical behaviors. Enormous computational resources necessary to deal with the coupled system of plasma field equations is not always justified. On the contrary, an analytical approach (although limited to the most simple situations and configurations) is often able to provide insightful physical analysis.

If we assume that the plasma is adiabatic, namely that the pressure and density are linked by[5]

$$\frac{d}{dt}\frac{p}{\rho^{\gamma}} = 0, \tag{2.66}$$

where γ is the ratio between specific heats, the previous equation can be written in partial derivatives as

$$\left(\frac{\partial}{\partial t} + \vec{v} \cdot \vec{\nabla}\right) p = -\gamma \, \rho \, \vec{\nabla} \cdot \vec{v}, \tag{2.67}$$

where the rhs term represents the heating or cooling corresponding to an adiabatic compression or expansion.

The whole system of equations, which has to be considered, is therefore,

$$\textbf{\textit{Zero resistivity equation}} \quad E = -\vec{v} \times \vec{B}, \tag{2.68}$$

$$\textbf{\textit{The continuity equation}} \quad \frac{\partial \rho}{\partial t} + \vec{\nabla} \cdot (\rho \, \vec{v}) = 0, \tag{2.69}$$

$$\textbf{\textit{The Faraday law}} \quad \frac{\partial \vec{B}}{\partial t} = -\vec{\nabla} \times \vec{E} = \vec{\nabla} \times (\vec{v} \times \vec{B}), \tag{2.70}$$

$$\textbf{\textit{The momentum equation}} \quad \rho \left(\frac{\partial \vec{v}}{\partial t} + \left(\vec{v} \cdot \vec{\nabla}\right)\vec{v}\right) = -\vec{\nabla} p + \vec{J} \times \vec{B}, \tag{2.71}$$

$$\textbf{\textit{The adiabatic energy equation}} \quad \left(\frac{\partial}{\partial t} + \vec{v} \cdot \vec{\nabla}\right) p = -\gamma \, \rho \, \vec{\nabla} \cdot \vec{v}. \tag{2.72}$$

They summarize the equations of the *ideal MHD*, sometimes used in our previous descriptions.

[5] The state equation for perfect gas under the adiabatic condition is pV^{γ} = constant, with $\gamma = C_p/C_V$ (where C_p, C_v are the thermal capacity, pressure and volume constant, respectively). Noting that $V \propto 1/\rho$ we get the analogous relation.

The physical regime they describe is limited to particular conditions, hereafter specified:

(a) Time scales longer than characteristic times like the inverse of plasma and cyclotron frequencies for the ions and electrons.
(b) Scale length larger than the Debye length and electron/ion gyro-radii.

In view of (b), the ideal MHD is applicable to conditions when quasi-neutrality holds.

We have so far introduced the mathematical tools and in the forthcoming section we will provide a few examples of their usage in order to get further information on the plasma physics.

2.5 Plasma MHD phenomenology: a qualitative picture

In this section we apply the *corpus* of the ideal MHD equations to equilibrium problems in plasma. Some of the results, we have already obtained, will be re-derived here within a more rigorous context and in particular we will deal with early concepts of plasma confinement like θ-pinches and Z-pinches.

The plasma is diamagnetic since its particles generate a current, producing a magnetic field which opposes the applied external magnetic field.

The diamagnetic current flowing inside the plasma is associated with the charged particle velocity, depending on their thermal energy, and, in turn, defines the field opposing the external counterpart (see below). The diamagnetic current can be derived straightforwardly but is worth underscoring as an important physical quantity [13].

With reference to figure 2.9(a) we write the momentum equation by including also the electric field, namely

$$\rho\left(\frac{\partial \vec{v}}{\partial t} + \left(\vec{v}\cdot\vec{\nabla}\right)\vec{v}\right) = -\vec{\nabla}p + qn\left(\vec{v}\times\vec{B} + \vec{E}\right). \tag{2.73}$$

The equilibrium condition implies that the convective derivative can be neglected and, therefore,

$$\vec{\nabla}p\times\vec{B} = qn\left(\vec{v}\times\vec{B} + \vec{E}\right)\times\vec{B} \tag{2.74}$$

where $(\vec{v}\times\vec{B})\times\vec{B} = -v_\perp\,|\vec{B}|^2$, we get

$$v_\perp = \frac{\vec{E}\times\vec{B}}{|\vec{B}|^2} - \frac{\vec{\nabla}p\times\vec{B}}{qn\,|\vec{B}|^2} = \vec{v}_d + \vec{v}_D. \tag{2.75}$$

The first term is just the drift velocity $\vec{v}_d$, the second $\vec{v}_D$ is the velocity component, giving rise to the diamagnetic current which depends on the charge of the particle. Accordingly, we find

$$\vec{J}_\perp = qn(\vec{v}_{D,\,i} - \vec{v}_{D,\,e}) = (k_B T_e + k_B T_i)\,\frac{\vec{B}\times\vec{\nabla}n}{|\vec{B}|^2} = 2k_B T\,\frac{\vec{B}\times\vec{\nabla}n}{|\vec{B}|^2}. \tag{2.76}$$

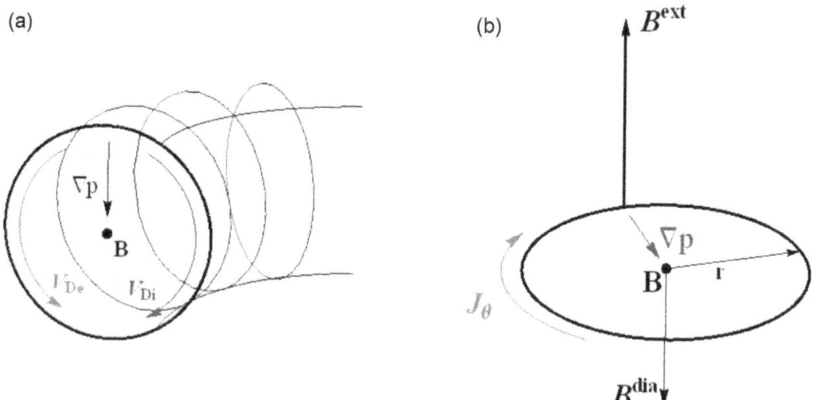

Figure 2.9. Pressure gradient induced by the electron–ion density gradient (a). The diamagnetic field B^{dia} induced by the external magnetic field B^{ext} (b).

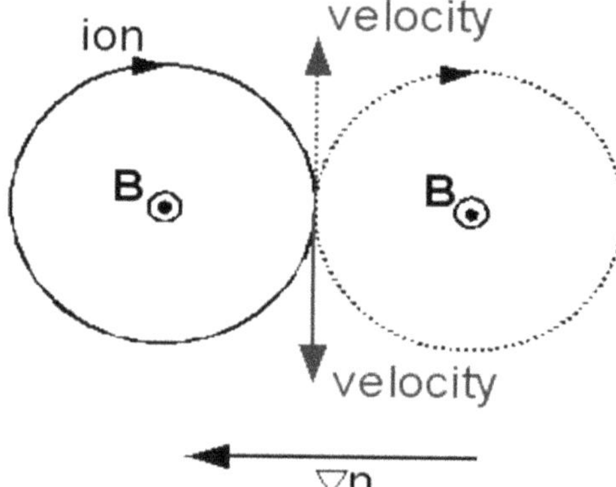

Figure 2.10. Ion and electron trajectories around the applied magnetic field and associated particle density gradient.

According to figure 2.9(b) the pressure force is perpendicular to $\vec{J}$ and $\vec{B}$, which lie on a plane of constant pressure.

Along with the considerations leading to equation (2.76), we propose the following alternative derivation of $J_\perp$, useful to better understanding its role. With reference to figure 2.10 we infer that it can be expressed as

$$\vec{J}_\perp = qv_\perp[n(x+R) - n(x-R)] \cong 2qv_\perp R\frac{\partial n}{\partial x}, \tag{2.77}$$

where $R = mv_\perp/(qB)$ and where we have implicitly assumed that a density gradient does exist.

Putting everything together and recalling that $\langle mv_\perp^2\rangle = k_B T$, we find the same expression given in equation (2.76).

We have already seen that the plasma and magnetic pressure can be combined to get equation (2.57), which holds if the field lines are straight and parallel, i.e. if $\vec{B} \cdot \vec{\nabla} = 0$.

Furthermore, relaxing the last condition and considering an axisymmetric magnetic field having the following expression in cylindrical coordinates

$$\vec{B} = [0, B_\theta(r), B_z(r)], \tag{2.78}$$

from equation (2.54) we get[6]

$$\frac{d}{dr}\left(p(r) + \frac{B_z^2(r) + B_\theta^2(r)}{2\mu_0}\right) = -\frac{1}{\mu_0}\frac{B_\theta^2(r)}{r}. \tag{2.79}$$

We have a differential equation for the equilibrium with three unknown $p(r)$, $B_\theta(r) B_z(r)$; fixing two of them we get from equation (2.79) the third unknown for a given boundary condition. There are infinite equilibrium configurations, from which significant for the confinement are the so called *pinches* achieved assuming $B_\theta(r) = 0$ or $B_z(r) = 0$ [14].

Regarding the case we have just examined, known as θ-pinch, from equation (2.54) with the condition $B_\theta = 0$ and a fixed external on-axis magnetic field B_e we get

$$p(r) + \frac{B_z^2(r)}{2\mu_0} = \frac{B_e^2}{2\mu_0}. \tag{2.80}$$

In order to better understand its physical content we provide an ad hoc derivation.

As already underscored in the θ-pinches, the external magnetic field is ensured by a current flowing in the azimuthal direction, which is in turn responsible for the onset of a magnetic field pointing in the z-direction, as shown in figure 2.11(a).

The magnetic field is due to the effect of an externally generated field and of the contribution induced by the diamagnetic current. We make the assumption that all these quantities are depending on the radial coordinate. It is easily checked that the

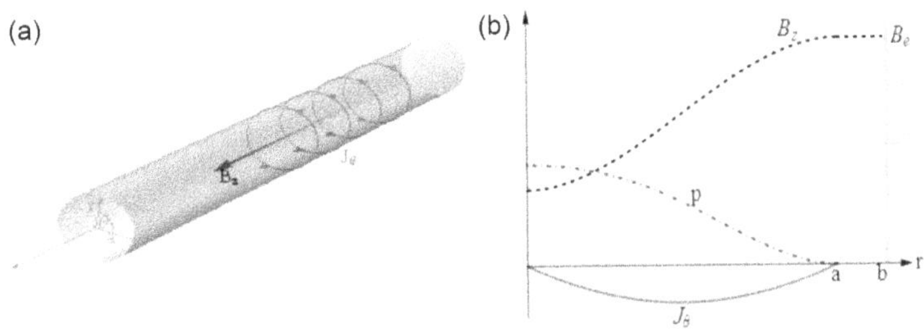

Figure 2.11. A sketch of the θ-pinch configuration for a plasma column (a). Magnetic (dash-dot) and kinetic pressure (dash) profile for a θ-pinch configuration with B_e the fixed external magnetic field $B_z(a)$ and a parabolic plasma current profile (continuous) picked at the peripheral of the plasma column (b).

[6] For the $(\vec{B} \cdot \vec{\nabla})\vec{B}$ operator in cylindrical coordinates see the introductory table.

Lorentz pressure force is directed toward the interior of the plasma, and we get indeed

$$\vec{J} \times \vec{B} = -J_\theta B_z \hat{e}_r. \tag{2.81}$$

Ampère law

$$\vec{\nabla} \times \vec{B} = \mu_0 \vec{J} \rightarrow \frac{dB_z(r)}{dr} = -\mu_0 J_\theta. \tag{2.82}$$

Equilibrium equation

$$\vec{J} \times \vec{B} = \vec{\nabla} p \rightarrow \frac{d}{dr} p = J_\theta B_z, \tag{2.83}$$

combining equations (2.82) and (2.83) we end up with

$$\frac{d}{dr}\left(p(r) + \frac{B_z^2(r)}{2\,\mu_0}\right) = 0, \tag{2.84}$$

which once integrated yields equation (2.80), whose handling, even though trivial, requires a few words of caution. It contains indeed two unknowns $p(r)$ and $B_z(r)$, the only constraint is that $B_z(a) = B_e$ when $p = 0$, namely externally to the fluid.

Finally, to determine the behavior of the pressure we can fix, e.g., the magnetic field, by specifying the form of the current, as reported in figure 2.11(b), in which the current $J_\theta(r)$ has been assumed to exhibit a quadratic dependence on the radial coordinate.

A further discussion regarding this property may be useful to understand the interplay between the plasma pressure and magnetic field.

The behavior of the magnetic field in the radial direction r is shown in figure 2.12 along with the corresponding azimuthal current. When the magnetic field is raised by increasing the current in the coils the magnetic pressure consequently increases without being balanced by the kinetic counterpart. The plasma is therefore compressed and the consequent work against the compression forces produces its heating.

It is worth noting that this confining mechanism is stable, as illustrated below (see figure 2.13). If a bend occurs, the magnetic tension tends to restore the original configuration, in the case of a squeezing the increase of the field intensity is counteracted by the increase of the internal pressure which tends to bring the system to the original configuration.

We can summarize the θ-pinch phenomenology as follows: a diamagnetic current, arising in a plasma with finite pressure immersed in a magnetic field, is the manifestation of the stable and efficient plasma confinement.

Its use in a reactor configuration, however, is hindered by the large end losses. As already underscored, the Tokamak configuration prevents such a drawback, since it is, in fact, a closed θ-pinch configuration. Along the torus curved path, the surfaces of constant pressure are realized by $\vec{J}$ and $\vec{B}$ vectors, as indicated in figure 2.14

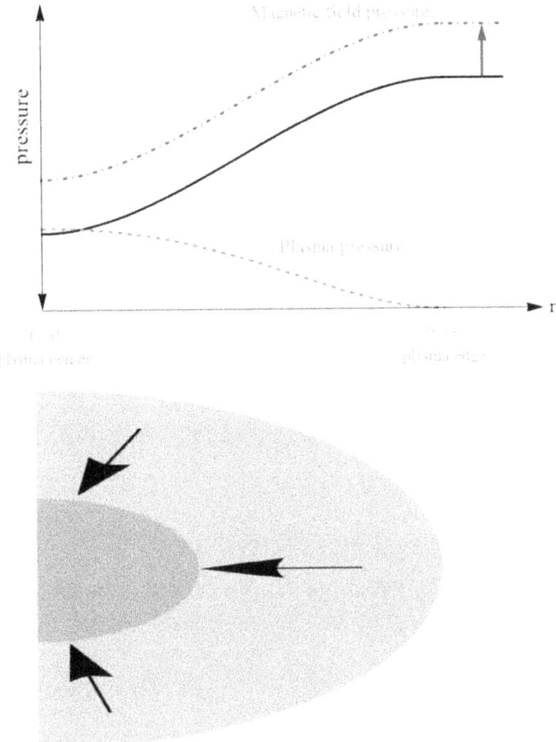

Figure 2.12. The kinetic (red-dashed) pressure profile versus plasma column radius increasing the external magnetic field (continuous black and dot-dashed blue) (top). The magnetic pressure picked at the plasma core (bottom).

The Z-pinch is the complementary configuration of the θ-pinch, as shown in figure 2.15(a), assuming $B_z(r) = 0$ in equation (2.78). In this case the magnetic and current component exchange their role: the current flows in the z-direction, inside the plasma column and the induced magnetic field encircles the plasma along the poloidal direction. The conditions in equation (2.84) follow from the geometry characterizing the problem under study.

$$\vec{B} \equiv B_\theta(r)\hat{e}_\theta, \qquad \vec{J} \equiv (0,\, 0,\, J_z(r)). \tag{2.85}$$

The radial dependence of the field, current and pressure stems from the fact that the magnetic field along the closed circular lines is constant. The Ampère law yields

$$\mu_0 J_z(r) = -\frac{1}{r}\frac{d}{dr}(rB_\theta(r)). \tag{2.86}$$

Putting everything in the stationary momentum equation, we find

$$-\vec{\nabla}p + \vec{J} \times \vec{B} = -\frac{d}{dr}p - \frac{B_\theta(r)}{r}\frac{d}{dr}(rB_\theta(r)) = 0, \tag{2.87}$$

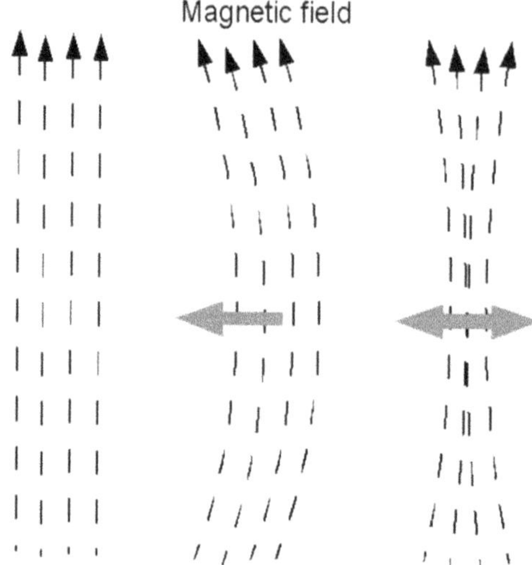

Figure 2.13. θ-pinch stability. Credit: Arthur Peeters.

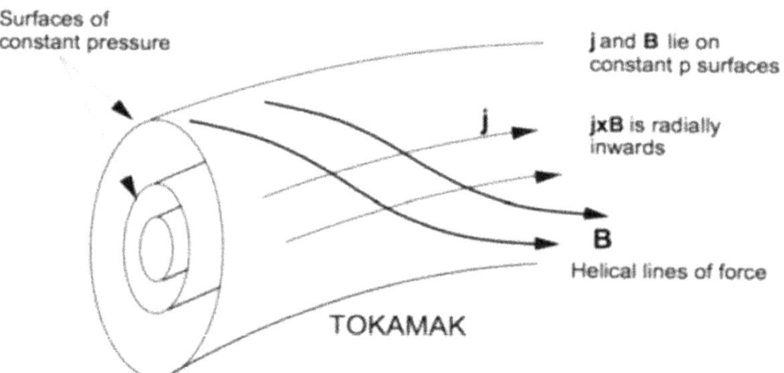

Figure 2.14. Magnetic induction and electrical density vector on the nested magnetic surface for a Tokamak configuration. Reprinted from [1] by permission of Springer Nature.

which eventually yields

$$\frac{d}{dr}\left(p(r) + \frac{B_\theta^2(r)}{2\mu_0}\right) + \frac{B_\theta^2(r)}{r\mu_0} = 0. \tag{2.88}$$

The previous equation contains an extra-term which is absent in the case of the θ-pinch. The last term is called the magnetic tension (see below for further comments).

The presence of this additional contribution prevents the integration in simple terms of the differential equation (2.88), unless we have reasonable arguments to provide the radial dependence of the flowing current.

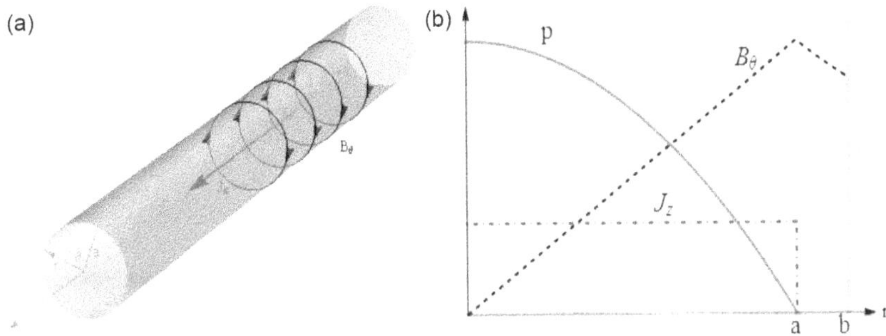

Figure 2.15. A sketch of the Z-pinch configuration for a plasma column (a). Magnetic (dash black) and kinetic pressure (continuous red) profile for a Z-pinch configuration with a constant current density profile (dot-dashed blue) (b).

We can, however, understand how this thing goes by the use of a naive calculation.

As already noted, the circulating current is the source of the magnetic field, which in terms of the current density, assumed to be independent of the coordinates, reads

$$B_\theta(r) = \mu_0 \frac{Jr}{2} \text{ for } r \leqslant a,$$
$$B_\theta(r) = \mu_0 \frac{a^2 J}{2r} \text{ for } r > a, \tag{2.89}$$

which, once substituted in equation (2.88) yields the differential equation

$$\frac{d}{dr}\left(p + \frac{\mu_0 r^2 J^2}{8}\right) + \frac{\mu_0 r J^2}{4} = 0, \tag{2.90}$$

that is straightforwardly integrated and gives a parabolic profile for the pressure

$$p(r) = \frac{\mu_0 J^2}{4}(a^2 - r^2). \tag{2.91}$$

The physical content of this discussion is summarized in figure 2.15(b) and can be commented as follows.

The uniform current density induces a magnetic field, increasing with the radius.

The pressure generated by the cross product of the magnetic field and the current exhibits a parabolic profile.

The pressure is maximum at the center of the plasma and scales with the square of the current itself. In this case the magnetic pressure and tension contribute in the same way to the plasma confinement. The situation is different if the current density has a parabolic profile, as reported in figure 2.16(a). In this latter case, the tension force is seen to confine the plasma in the outer region (see figure 2.16(b)).

Making reference to the Z-pinch configuration, we derive a relation between current and plasma temperature.

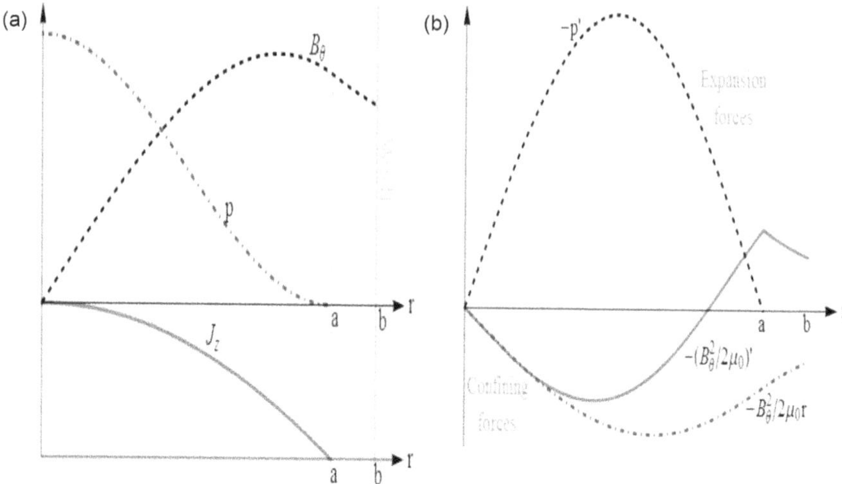

Figure 2.16. Magnetic (dashed black) and kinetic pressure (dot-dashed blue) profile with a parabolic current density profile (continuous red) for Z-pinch configuration (a). The radial forces for a Z-pinch configuration (b).

Therefore, equation (2.88) can be rearranged as

$$\frac{d}{dr}p(r) = -\frac{1}{\mu_0 r}B_\theta(r)\frac{d}{dr}(rB_\theta(r)), \tag{2.92}$$

multiplying both sides with r^2 and integrating from 0 to a yields

$$2\int_0^a rp(r)dr = \frac{1}{2\mu_0}[aB_\theta(a)]^2. \tag{2.93}$$

Considering the plasma as an ideal gas for which the relation $p(r) = n(r)kT$ holds, the lhs term becomes

$$\frac{1}{\pi}\int_0^a 2\pi r(n(r)kT)dr = \frac{kT}{\pi}\int_0^a 2\pi rn(r)dr = \frac{N_l kT}{\pi}, \tag{2.94}$$

with N_l being the number of particles per unit length of the plasma column.

Furthermore, the current intensity flowing into the plasma column can be written in the following

$$I = \int_0^a 2\pi rJ_z(r)dr, \tag{2.95}$$

and using the Ampère law (see equation (2.86)) we get

$$I = \frac{2\pi}{\mu_0}aB_\theta(a). \tag{2.96}$$

If we couple equations (2.93), (2.95), and (2.96) we obtain the so-called Bennett relation

$$I^2 = \frac{8\pi}{\mu_0}N_l kT. \tag{2.97}$$

The relation (2.97) gives an idea of the current amount we need to reach the fusion temperature. The main problem with this configuration is that when increasing the current above a certain value the equilibrium became unstable.

In fact, the Z-pinch unlike θ-pinch is not stable and the main source of instability is the already discussed kink effect. The column plasma can be discretized as an infinite straight wire where the current flows in the same direction; the increase of the current determines an increment of the wire attraction till the kink instability occurs, as summarized in figure 2.17, showing that any motion in the outward direction creates a magnetic field gradient enhancing the perturbation.

The combination of Z-pinches and θ-pinches is known as screw-pinch. In this case in the equation (2.79) longitudinal and azimuthal magnetic fields are simultaneously present as well as the associated currents. The characterization of the relevant MHD equilibrium follows the previously outlined procedures.

The concept of poloidal beta (β_p), marginally touched on in the previous discussion, can be now expanded making reference from equation (2.79).

If we introduce the volumetric average

$$\langle f \rangle = \frac{2}{a^2} \int_r^a f(r) dr, \tag{2.98}$$

from equation (2.79), after taking the average on both sides and doing some algebra, we get

$$\beta_p = 1 + \frac{B_z^2(a) - \langle B_z^2 \rangle}{B_\theta^2(a)}, \tag{2.99}$$

where

$$\beta_p = \frac{2\mu_0 \langle p \rangle}{B_\theta^2(a)}, \tag{2.100}$$

The above identity, although apparently innocuous, is extremely important. The parameter β_p is a measure of the magnetic nature of the fluid; if $\beta_p > 1$, then $\beta_z^2(a) > \langle \beta_z^2 \rangle$, the plasma tends to expel the field and therefore it is diamagnetic. When $\beta_p < 1$ the plasma is said to be paramagnetic and $\beta_z^2(a) < \langle \beta_z^2 \rangle$. The role and

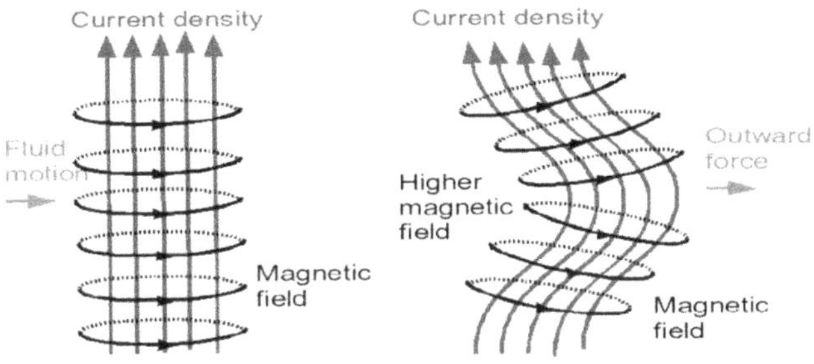

Figure 2.17. Instability mechanism in the case of Z-pinch. Credit: Arthur Peeters.

importance of the β_p in the design of magnetic fusion devices will be discussed in the forthcoming chapter.

The analysis we have developed so far will be exploited in the following to study more complicated configurations, like the Tokamak equilibrium.

We conclude this section by using the MHD model to treat the so-called *resistive diffusion*. It has already been underlined, that in the context of ideal MHD, the plasma is a perfect conductor. Knowing, however, that the ohmic heating is incompatible with this assumption, we allow therefore a finite plasma resistivity and state that the interplay between plasma and magnetic pressure leads to a diffusion of the magnetic field inside the plasma fluid.

The analysis we develop is elementary, but, owing to its importance, is reported here in detail.

The resistivity contribution becomes relevant if collisional terms are important. We accordingly write the electric field as

$$\vec{E} = \eta\vec{J} - \vec{v} \times \vec{B}. \tag{2.101}$$

The Faraday law (without the contribution of the displacement current) can therefore be written as

$$\frac{\partial\vec{B}}{\partial t} = -\vec{\nabla} \times (-\vec{v} \times \vec{B} + \eta\vec{J}) = -\frac{\eta}{\mu_0}\vec{\nabla} \times (\vec{\nabla} \times \vec{B}) + \vec{\nabla} \times (\vec{v} \times \vec{B}). \tag{2.102}$$

After expanding the first cross product, we end up with

$$\frac{\partial\vec{B}}{\partial t} = \frac{\eta}{\mu_0}\nabla^2\vec{B} + \vec{\nabla} \times (\vec{v} \times \vec{B}), \tag{2.103}$$

which is a diffusive equation with a source term for the magnetic field and is the finite resistivity version of the ideal MHD equation (2.70).

The diffusivity term $D = \eta/\mu_0$ has the dimensions $[L^2T^{-1}]$ and we can therefore introduce a characteristic time

$$\tau_d = \frac{\mu_0}{\eta}l^2. \tag{2.104}$$

The Ampère law on the same scale length yields

$$J \cong \frac{B}{\mu_0 l}. \tag{2.105}$$

Putting together equations (2.104)–(2.105) we can determine the energy density flowing inside the plasma as

$$\eta\, J^2\tau_d = \eta\left(\frac{B}{\mu_0 l}\right)^2\frac{\mu_0}{\eta}l^2 = \frac{B^2}{\mu_0}. \tag{2.106}$$

This last relation states that the energy density flowing inside the plasma is just the magnetic field (induced by the diamagnetic current) energy density.

Another hint can be obtained by writing equation (2.103) in terms of the dimensionless variables, namely

$$\frac{\partial \vec{B}}{\partial \xi} = \tau_d \frac{\eta}{\mu_0 l^2} \nabla_\sigma^2 \vec{B} + \vec{\nabla}_\sigma \times (\vec{v} \times \vec{B}),$$
$$\vec{\sigma} = \frac{\vec{r}}{l}, \ \xi = \frac{t}{\tau_d}, \ \vec{v} = \frac{d\vec{\sigma}}{d\xi}. \tag{2.107}$$

Within the present formulation, the significance of one term with respect to the other depends on the coefficients appended to them. We define accordingly the dimensionless number $R = \mu_0 l^2/(\eta \tau_d)$, and assuming $l/\tau_d \cong v$ write

$$R \cong \frac{\mu_0 v\, l}{\eta}, \tag{2.108}$$

which is reminiscent of the Reynolds number, provided that η/μ_0 is recognized as a kind of kinematic viscosity. For large values of η the diffusive part is dominating, while for perfectly conducting plasma it reduces to

$$\frac{\partial \vec{B}}{\partial t} = \vec{\nabla} \times (\vec{v} \times \vec{B}), \tag{2.109}$$

which is the Faraday law for a perfectly conductive plasma. The flux of the magnetic field line enters the surface of the plasma moving with velocity v (see figure 2.18).

The figure could be commented as it follows:

(a) the contour curve that encloses the surface S moves with the plasma velocity;

(b) during this surface translation, the arc element $d\vec{s}$ describes a surface of an area $\vec{v}\delta t$;

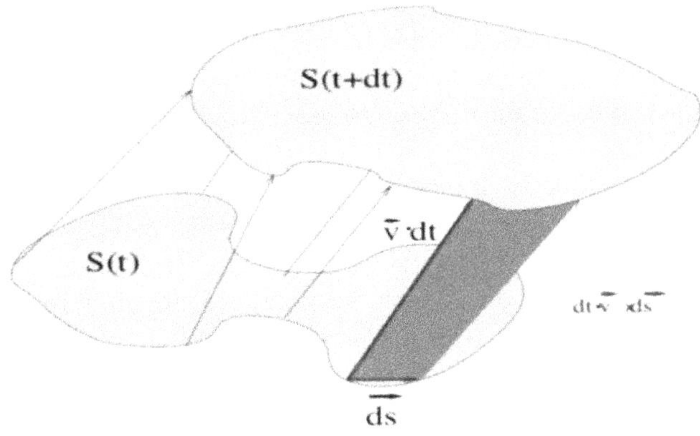

Figure 2.18. Plasma surfaces and integration contours.

(c) the magnetic flux crossing this area portion is $\vec{B} \cdot (d\vec{s} \times \vec{v}\,\delta t) = -\delta t\vec{v} \times \vec{B} \cdot d\vec{s}$;
(d) the flux through the surface S is $\iint_S \vec{B} \cdot d\vec{S}$.

We accordingly find that the total flux variation is written as

$$\frac{\partial \Phi}{\partial t} = \iint_S \frac{\partial}{\partial t}\vec{B} \cdot d\vec{S} - \oint_s \left(\vec{v} \times \vec{B}\right) \cdot d\vec{s}. \tag{2.110}$$

The use of the Stoke's theorem yields for the second integral on the rhs of the previous equation

$$\frac{\partial \Phi}{\partial t} = \iint_S \left[\frac{\partial}{\partial t}\vec{B} - \vec{\nabla} \times (\vec{v} \times \vec{B})\right] \cdot d\vec{S}. \tag{2.111}$$

The condition (2.109) ensures that the magnetic flux is a conserved quantity. In other words, the magnetic field lines are *frozen* inside a perfectly conducting plasma. This statement is known as Alfvèn frozen flux theorem.

When a finite conductivity is present the theorem does not hold anymore and the so-called magnetic reconnection arises. This mechanism is addressed briefly in the forthcoming section.

2.6 Magnetic field Hamiltonian and rotational transform

In the previous section we have just considered three examples of MHD equilibria for configurations like θ-pinches, Z-pinches and the screw-pinches. We have mentioned that these examples assume an infinitely straight cylinder. In this geometry the MHD equilibrium is made possible because the plasma cannot flow across the magnetic field. The finiteness of the cylinder length makes the confinement inefficient because the plasma flows along the axis and eventually is lost outside the open ends. We have underscored that the idea of closing the end losses is an ingenious solution to solve the problem of losses. The price to be paid is that of transforming the cylinder into a torus.

The field is accordingly bent and this has major consequences in terms of MHD stability. It is evident that the magnetic field topology plays within this context a major role. In order to frame the discussion in general terms, it is worth adding a few remarks.

The bending of a cylinder into a torus determines the breaking of the conditions determining the MHD equilibrium. A first macroscopic consequence is due to the fact that the external surface of the torus is larger than its internal part, the plasma tends to move towards the outer part because the pressure forces are equally distributed on the external part with larger radius (see figure 2.19)

A second reason is of strict physical nature; the anti-parallel currents tend to repel each other, as shown in figure 2.20. The Lorentz force associated with the magnetic field generated in point 2 by current 1 tends to repel the plasma in the outward direction, as does the force associated with the configuration 1–2. The plasma is

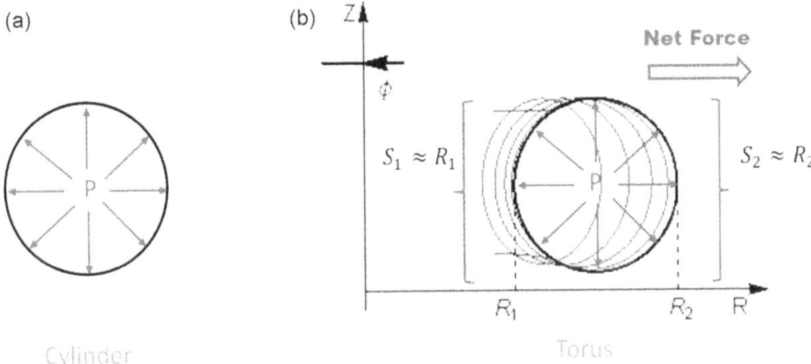

Figure 2.19. Equal lateral surfaces in a cylindrical configuration (a). Inhomogeneous lateral surfaces and emergence of a net outer force in toroidal geometry (b).

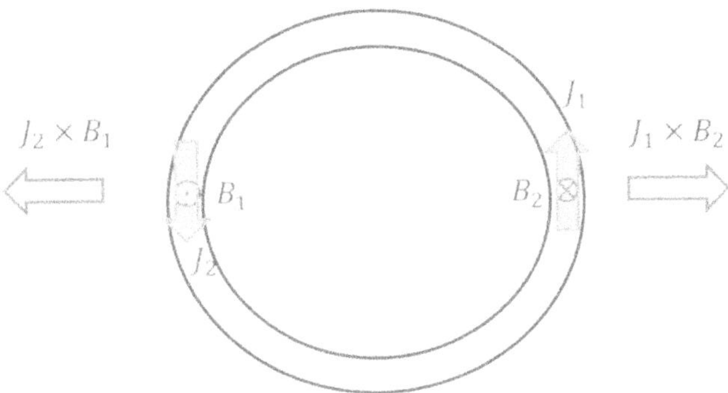

Figure 2.20. Current induced Lorentz force and plasma stretching.

therefore stretched out (figure 2.20). Thus the use of external coils can be effective to counteract this effect.

Let us summarize the features characterizing the cylindrical pinches:

(a) in an infinitely long cylinder the regions of constant pressure are individuated by concentric cylindrical surfaces, therefore $p = p(r)$;

(b) the equilibrium condition $\vec{\nabla} p = \vec{J} \times \vec{B}$ implies that $\vec{B} \cdot \vec{\nabla} p = 0$, which in turn ensures that the pressure is constant along the direction of the field lines;

(c) the same holds for the current density, with $\vec{J} \cdot \vec{\nabla} p = 0$, the current flows on these surfaces;

(d) currents and fields do not overlap, unless the pressure gradient vanishes.

Regarding the toroidal geometry, we assume axial symmetry (independence of the physical quantities of the toroidal angle ϕ). The constant pressure surfaces are nested toroidal surfaces, wrapping the torus axis as shown in figure 2.21, where we have also reported the distribution of field lines and currents on them.

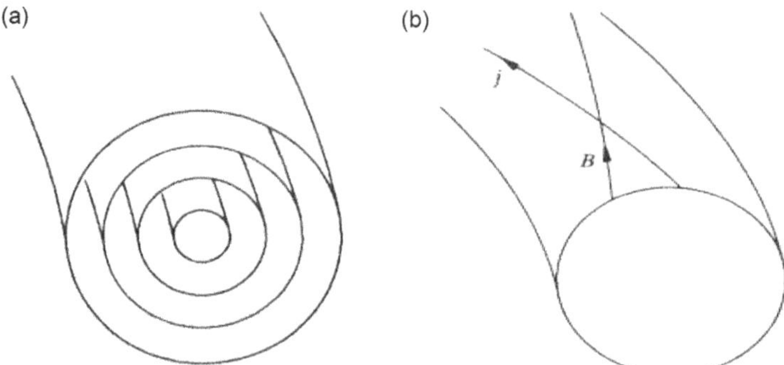

Figure 2.21. Nested constant pressure surfaces (a). Field lines and currents along the pressure surface (b).

In the forthcoming section we treat the problem in general terms and successively it will be specialized to the cylindrical coordinates.

Before entering into the specifics of the discussion and to better appreciate the (non-trivial) problems one may be faced with, we like to provide a more appropriate description of the topological structure of the field lines.

Even though already discussed, it is worth summarizing the following points:

- toroidal devices need a rotational transform so that the drift induced by the magnetic field gradient will not be responsible for a charge separation;
- the toroidal current flowing inside the chamber ensures this possibility;
- different solutions can however be conceived.

Let us now fix some mathematical notions to be exploited in the following.

- The magnetic field lines can be 'traced' as the solution of the differential equation

$$\frac{d\vec{x}}{ds} = \alpha\, \vec{B}(\vec{x}), \tag{2.112}$$

where ds denotes an element of a curvilinear coordinate implicitly labeling the position along the field line trajectory. From equation (2.112) it also follows that if $\vec{x} \equiv (l_x, l_y, l_z)$ then

$$\frac{dl_x}{B_x} = \frac{dl_y}{B_y} = \frac{dl_z}{B_z} = \alpha\, ds, \tag{2.113}$$

where $B_k = \vec{B} \cdot \hat{k}$ with $k = x, y, z$ and $\hat{k}$ the unit vector in the k direction.

- Any vector defined on a torus can be expressed as

$$\vec{A} = \vec{\nabla} g + \psi_t \vec{\nabla}\left(\frac{\theta}{2\pi}\right) - \psi_p \vec{\nabla}\left(\frac{\phi}{2\pi}\right), \tag{2.114}$$

where all the quantities appearing on the rhs of the previous equation are functions of the spatial coordinates. Furthermore ϕ, θ are toroidal and poloidal coordinates, respectively. The interpretation of $\psi_{t,p}$ needs just a few more details to be understood.

If $\vec{A}$ is interpreted as a vector potential, the associated magnetic field reads

$$\begin{aligned} 2\pi\vec{B} &= \vec{\nabla} \times \vec{A} = \vec{\nabla}\psi_t \times \vec{\nabla}\theta - \vec{\nabla}\psi_p \times \vec{\nabla}\phi, \\ 2\pi B_\theta &= \vec{\nabla}\psi_t \times \vec{\nabla}\theta, \\ 2\pi B_\phi &= -\vec{\nabla}\psi_p \times \vec{\nabla}\phi. \end{aligned} \tag{2.115}$$

When keeping the curl of $\vec{\nabla}g$ its contribution disappears from the definition of the magnetic field, therefore, within this context, it plays the role of a Gauge vector.

Going back to equation (2.112) we note that

$$\alpha\, ds = \frac{d\,\psi_t}{\vec{B}\cdot\vec{\nabla}\psi_t} = \frac{d\,\theta}{\vec{B}\cdot\vec{\nabla}\theta} = \frac{d\,\phi}{\vec{B}\cdot\vec{\nabla}\phi}, \tag{2.116}$$

from which we get

$$\alpha\vec{B}\cdot\vec{\nabla}\psi_t\, ds = d\psi_t, \tag{2.117}$$

and

$$\begin{aligned} \frac{d\psi_t}{d\phi} &= \frac{\vec{B}\cdot\vec{\nabla}\psi_t}{\vec{B}\cdot\vec{\nabla}\phi} = -\frac{\left(\vec{\nabla}\psi_p \times \vec{\nabla}\phi\right)\cdot\vec{\nabla}\psi_t}{\left(\vec{\nabla}\psi_t \times \vec{\nabla}\theta\right)\cdot\vec{\nabla}\phi}, \\ \frac{d\theta}{d\phi} &= \frac{\vec{B}\cdot\vec{\nabla}\theta}{\vec{B}\cdot\vec{\nabla}\phi} = \frac{\left(-\vec{\nabla}\psi_p \times \vec{\nabla}\phi\right)\cdot\vec{\nabla}\theta}{\left(\vec{\nabla}\psi_t \times \vec{\nabla}\theta\right)\cdot\vec{\nabla}\phi}. \end{aligned} \tag{2.118}$$

The use of the cyclic properties of the product $\left(\vec{a}\times\vec{b}\right)\cdot\vec{c}$ allows us to cast the equation (2.118) in the suggestive form

$$\begin{aligned} \frac{d\psi_t}{d\phi} &= -\frac{\partial\psi_p}{\partial\theta}, \\ \frac{d\theta}{d\phi} &= \frac{\partial\psi_p}{\partial\psi_t}, \end{aligned} \tag{2.119}$$

which allows the interpretation of ψ_p as a kind of Hamiltonian and of ψ_t, θ as the associated conjugate coordinates, with ψ_t the canonical momentum.

Accordingly, the equations (2.119) are Hamiltonian equations with respect to the time-like variable ϕ [15].

A more 'physical' interpretation of ψ_t follows from the evaluation of the flux of the vector $\vec{B}$ across the surface with an element

$$d\vec{S}_\phi = J\vec{\nabla}\phi\, d\psi_t\, d\theta, \tag{2.120}$$

where J is the Jacobian of the transformation [16]

$$J = \det\left(\frac{\partial(\psi_t, \phi, \theta)}{\partial(x, y, z)}\right) = \left(\frac{\partial \vec{x}}{\partial \psi_t} \times \frac{\partial \vec{x}}{\partial \theta}\right) \cdot \frac{\partial \vec{x}}{\partial \phi},$$
$$J^{-1} = \det\left(\frac{\partial(x, y, z)}{\partial(\psi_t, \phi, \theta)}\right) = (\vec{\nabla}\psi_t \times \vec{\nabla}\theta) \cdot \vec{\nabla}\phi. \tag{2.121}$$

We obtain therefore

$$\int_S \vec{B} \cdot d\vec{S}_\phi = \int_S J\vec{B} \cdot \vec{\nabla}\phi \, d\psi_t \, d\theta. \tag{2.122}$$

From equation (2.115) we find

$$2\pi\vec{B} \cdot \vec{\nabla}\phi = 2\pi B_\phi = \left(\vec{\nabla}\psi_t \times \vec{\nabla}\theta\right) \cdot \vec{\nabla}\phi, \tag{2.123}$$

thus reducing the integral equation (2.122) to

$$\int_S J\vec{B} \cdot \vec{\nabla}\phi \, d\psi_t \, d\theta = \frac{1}{2\pi}\int_S J\left[\left(\vec{\nabla}\psi_t \times \vec{\nabla}\theta\right) \cdot \vec{\nabla}\phi\right] d\psi_t \, d\theta. \tag{2.124}$$

On account of the second of equations (2.121) we end up with

$$\frac{1}{2\pi}\int_S J\left[\left(\vec{\nabla}\psi_t \times \vec{\nabla}\theta\right) \cdot \vec{\nabla}\phi\right] d\psi_t \, d\theta = \frac{1}{2\pi}\int_S JJ^{-1} \, d\psi_t \, d\theta, \tag{2.125}$$

and eventually we get

$$\int_S \vec{B} \cdot d\,\vec{S}_\phi = \psi_t. \tag{2.126}$$

The variable ψ_t is therefore understood as the magnetic field flux flowing along the surface S in the direction ϕ. An analogous interpretation holds for ψ_p, which is interpreted as the flux along the θ direction (see figure 2.22).

The general framework we have so far developed will be specialized to the cylindrical coordinates in the forthcoming section. Within this context we discuss the MHD Tokamak equilibrium and touch on the Grad–Shafranov equations [17, 18].

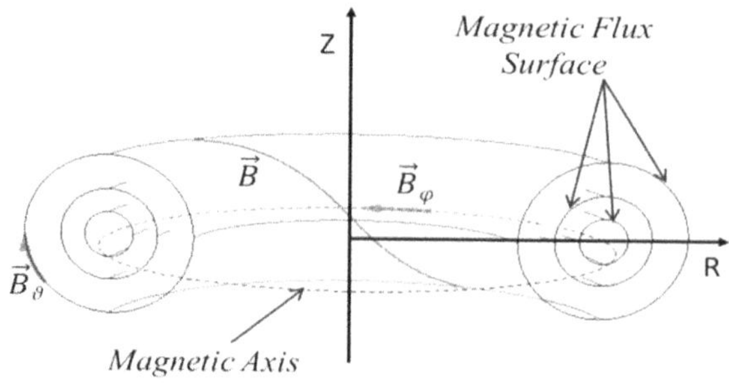

Figure 2.22. Magnetic surfaces and relevant sections inside the torus.

2.7 Toroidal MHD equilibrium

The transition from the previous general discussion to the cylindrical coordinates is not particularly complicated, but requires care. We first recall the coordinate frame sketched in figure 1.18.

The field line equations (2.113) read

$$\frac{dr}{B_r} = \frac{r\, d\phi}{B_\phi} = \frac{dz}{B_z}. \tag{2.127}$$

We follow the same steps adopted to study the cylindrical pinch configurations

$$\vec{\nabla} \cdot \vec{B} = 0 \rightarrow \frac{1}{r}(r\, B_r) + \frac{\partial B_z}{\partial z} = 0. \tag{2.128}$$

According to the definition of the magnetic field in terms of vector potential yields

$$\vec{B} = \vec{\nabla} \times \vec{A} \rightarrow \begin{cases} B_r &= -\dfrac{\partial A_\phi}{\partial z}, \\ B_z &= \dfrac{1}{r}\dfrac{\partial}{\partial r}\left(r\, A_\phi\right). \end{cases} \tag{2.129}$$

It is evident that equation (2.128) is naturally satisfied after setting

$$\psi_t = 2\pi\, r\, A_\phi. \tag{2.130}$$

It is also evident that

$$\begin{aligned} \vec{B}_p &= \frac{1}{r}\vec{\nabla}\psi_t \times \hat{e}_\phi, \\ \vec{B}_t &= \vec{B}_\phi = \frac{\mu_0 I}{2\pi\, r}\hat{e}_\phi. \end{aligned} \tag{2.131}$$

From the previous considerations it follows that rB_t is a function of ψ_t only, therefore

$$\frac{\mu_0 I}{2\pi} = rB_t = f(\psi_t). \tag{2.132}$$

This is a point of pivotal importance, we get indeed

$$\vec{\nabla} \times \vec{B} = \mu_0 \vec{J} \rightarrow \begin{cases} J_z = \dfrac{1}{\mu_0}\dfrac{\partial B_\phi}{\partial r} = \dfrac{1}{2\pi r}\dfrac{\partial I(\psi_t)}{\partial r}, \\ J_\phi = \dfrac{1}{\mu_0}\left(\dfrac{\partial B_r}{\partial z} - \dfrac{\partial B_z}{\partial r}\right). \end{cases} \tag{2.133}$$

This last quantity can be written in terms of the variable ψ_t, the use of the previous identities yields

$$J_\phi = -\frac{\Delta^* \psi_t}{2\,\pi\mu_0 r},$$
$$\Delta^* = \frac{1}{r}\frac{\partial^2}{\partial z^2} + \frac{\partial}{\partial r}\frac{1}{r}\frac{\partial}{\partial r}. \tag{2.134}$$

The MHD equilibrium condition $\vec{\nabla} p = \vec{J} \times \vec{B}$ eventually leads to

$$\frac{\Delta^* \psi_t}{2\pi\mu_0 r}B_z + \frac{1}{2\pi r}\frac{\partial I(\psi_t)}{\partial r} + \frac{\partial p(\psi_t)}{\partial r} = 0. \tag{2.135}$$

After a little algebra we end up with

$$\Delta^* \psi_t + \mu_0^2 II' + (2\pi R)^2 p' = 0, \tag{2.136}$$

known as Grad–Shafranov equation (the apex denotes the derivative $d/d\psi_t$) [19]. It is a generalization of the analogous conditions derived for the cylindrical pinches. The functions $p(\psi_t)$ and $I(\psi_t)$ are, in some sense, arbitrary and can be guessed from considerations associated with the experimental results or with transport calculations.

2.8 The Stellarator

In this section we introduce a further device, called Stellarator, which has been proposed in the past as an alternative to the Tokamak configuration. These machines, whose name is inspired by the confining dynamics occurring in the stars, realize a different concept with respect to the Tokamaks, based on more empirical conceptions [20]. In spite of their potentiality, they were not immediately recognized as important alternatives.

They were quite popular in the fifties and sixties of the last century, but the more easily manageable Tokamaks and the better results they achieved, made the Stellarator concept obsolete for a while. Renewed interest arose in the nineties because of the difficulties of achieving break-even with Tokamaks.

The Stellarator was conceived by Lyman Spitzer, at Princeton Plasma Physics Laboratory, at the beginning of the 1950s. He observed that there are three different ways to twist the magnetic field lines around a torus to confine the plasma, namely:

(a) making the magnetic axis non-planar;
(b) rotating the poloidal cross-section of stretched flux surfaces around the torus;
(c) creating a poloidal field by a toroidal electric current.

Whereas the Tokamak relies only on the third method, the Stellarator makes use of the first two concepts allowing for a large variety of non-axisymmetric helical confinement devices in which the confining magnetic field is solely produced by external coils without plasma current.

The geometry of the early experiment in 1961 proposed by Spitzer was achieved deforming the torus into a 'figure-of-eight' (figure 2.23) in order to induce a

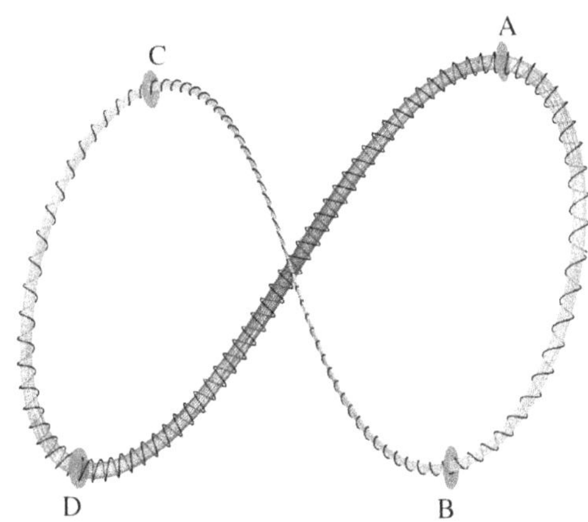

Figure 2.23. Figure-of-eight Stellarator.

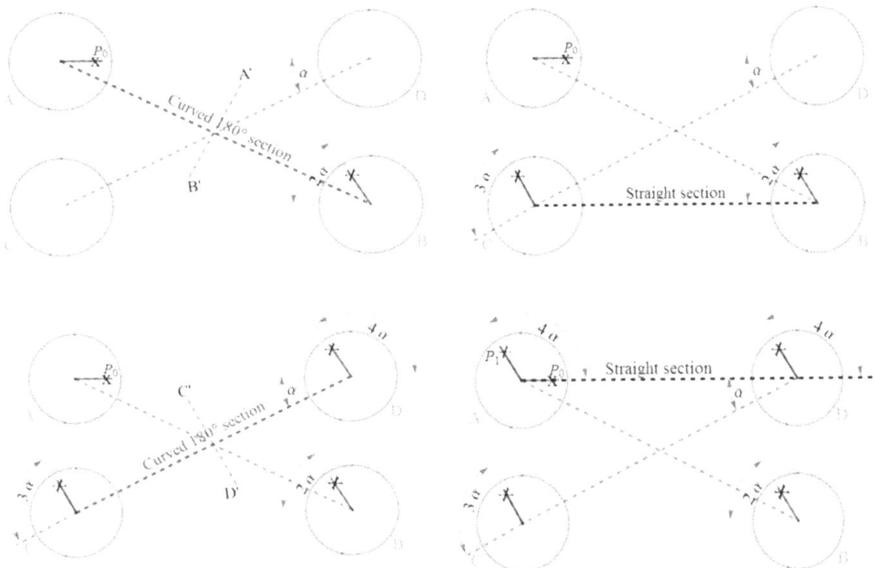

Figure 2.24. Angle of tilt and rotational transform.

rotational transform. In figure 2.23 the two curving end sections $\widehat{AB}$ and $\widehat{CD}$, which are both bent through an angle of 180°, are in separate planes, each tilted at an angle to the parallel planes in which are located the straight section $\overline{AD}$ and $\overline{BC}$. The existence of a rotational transform may be seen from an inspection of figure 2.24, which shows the cross-sectional planes at A,B,C and D (the red shadowed circles in figure 2.23). The angle between the planes containing *AB* and *AD*, and that between the corresponding planes containing *CD* and *BC*, is represented by α (the angle of tilt of the Stellarator). The lines connecting the four centers mark the path of the magnetic axis. The crosses indicate an arbitrary point where a single line of force intersects the plane at A, B, C and D. The complete circuit of the Stellarator tube brings the line of force back to P_1 in plane A, so that P_0OP_1 is the rotational transform angle.

The relationship between the rotational transform and the angle of tilt α can be obtained analyzing figure 2.24. The transformation of the cross section A into B, with the line of force traversing a 180° bend, is equivalent to reflection about $A'B'$, which is perpendicular to the magnetic axes in the middle of the section AB (figure 2.24 top-left). Transformation from B to C is an identity, since the line of force is in a straight section of the Stellarator tube (figure 2.24 top-right). The transformation from C to D is again a reflection about $C'D'$ (figure 2.24 bottom-left), whereas that from D back to A is an identity (figure 2.24 bottom-right). Starting at P_0, the completion of the circuit brings the line of force to P_1 and the rotational transform angle P_0OP_1 is seen to be four times the tilt angle of the Stellarator ($\iota = 4\alpha$) [21].

Some years later during the second half of the 1960s, at Princeton, the first experiment was upgraded by a Model-C Stellarator having a racetrack-type geometry (figure 2.25) with consecutive straight and curved sections.

The major drawback of these experiments was the small confinement time limited to few Bohm times $\tau_{\text{Bohm}} \approx a^2/D_{\text{B}}$, with a being the minor radius and $D_{\text{B}} = k_{\text{B}}Te/(16eB)$ the Bohm diffusion coefficient[7] and with $Te \approx 100$ eV. Such a deficiency became even more remarkable, because Russian Scientists had obtained in the same period much better performances with Tokamaks. Namely, confinement time an order of magnitude larger and at higher plasma temperatures ($Te \approx 1$ keV).

As was understood later, the poor confinement was most likely caused by the large error fields induced by the transitions from straight to curved sections leading

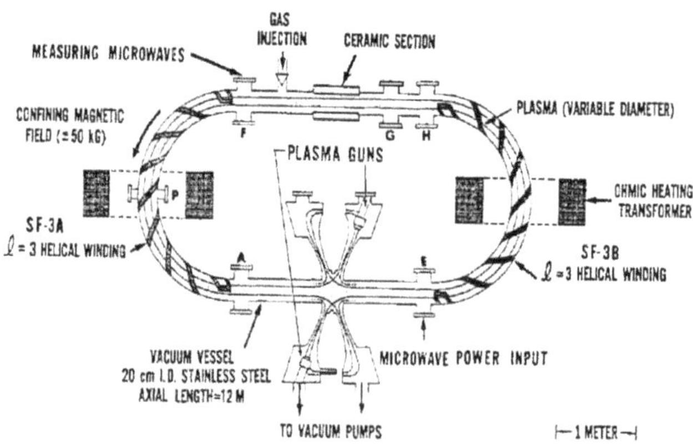

Figure 2.25. A simplified scheme of Model-C Stellarator [22].

[7] The diffusion coefficient of a plasma through a magnetic field has been experimentally found to be proportional to its temperature and inversely proportional to the magnetic field strength. It has been proven to follow the Bohm scaling $D_{\text{B}} = k_{\text{B}}T/(16eB)$ and the corresponding Bohm time is l^2/D_{B}, where l^2 is a characteristic surface area through which the plasma diffuses.

to a large fraction of unconfined particle orbits. Furthermore, in the absence of axial symmetry of a three-dimensional Stellarator, it was found that new classes of trapped particles (super-banana orbits) do exist. As in the Tokamak, the particles can be passing or trapped due to the presence of helical trajectory from inside (with high magnetic field) to outside (with low magnetic field) of the torus. During their bounce motion between the mirror points the particles suffer from radial drift with a velocity v_D, which leads to the banana orbits in the poloidal projection of their bounce motion. Due to the toroidal symmetry of the Tokamak magnetic field the inward and outward drift balance each other. In contrast, in the Stellarator these particles trapped in local minima (figure 2.26) are confined to regions on the upper or lower half of the flux surface, therefore, their ∇B drift does not cancel out (see figure 1.17), and they drift straight out of the machine.

Furthermore, the Tokamak axisymmetry along the ϕ angle makes the Hamiltonian, describing the magnetic field line (see equation (2.118) for more details), time independent and therefore integrable without singularity, which guarantees proper trajectories, without chaos, and good particles confinement. Even if the Tokamak till now provided a better plasma confinement the magnetic configuration is almost fixed to a 2D space while with the Stellarator we can have the possibility to optimize the magnetic configuration in a 3D space.

Following the previous considerations, different types of Stellarator have been built to induce a rotational magnetic field line.

In the helical axis configuration, the twist is induced in the magnetic axis by displacing the toroidal field coils (see figure 2.27(a)).

In the classical Stellarator, the rotational transform is generated with a toroidal magnetic field produced by toroidal field coils (as in a Tokamak) and a poloidal field with l-fold symmetry is produced by a set of 2l toroidally continuous helical windings with currents flowing in opposite directions in adjacent winding (figure 2.27(b)). Furthermore, adding poloidal field coils to produce a vertical field it was possible to push the magnetic axis towards a minimum-B configuration, with the force lines concave toward the plasma reducing the MHD instabilities.

This solution has the advantage of a continuous current flow controlled from outside and flexibility to vary the toroidal and poloidal fields independently. At the same time it has the disadvantage due to the fact that the helical and toroidal coils are inter-linked, making both the assembly and maintenance difficult.

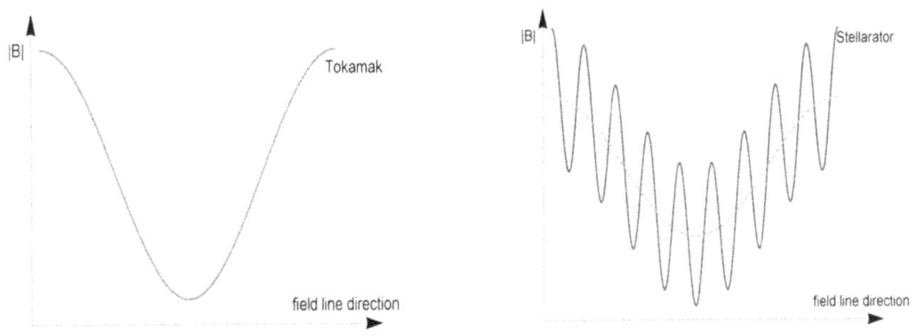

Figure 2.26. Magnetic field B along a magnetic field line in a Tokamak and Stellarator configuration.

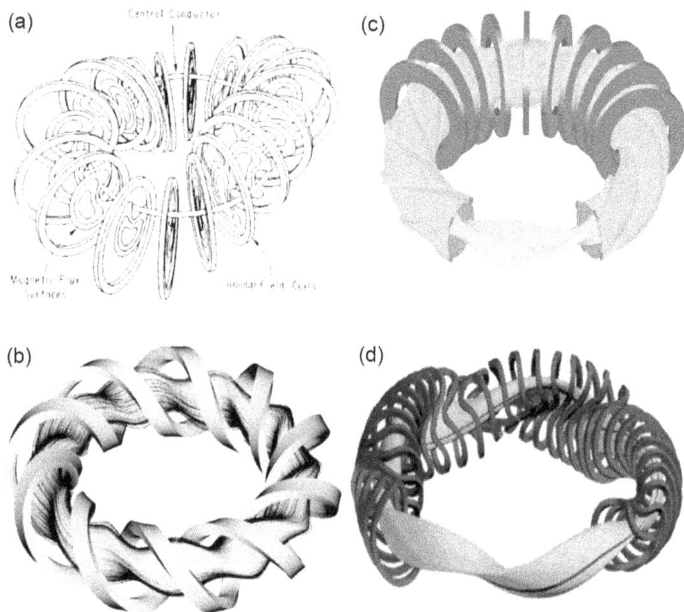

Figure 2.27. Magnetic field configuration for different Stellarator concepts: (a) heliac (reprinted from [23] with permission of AIP Publishing), (b) classical (reprinted with permission from [16], copyright (2005) by the American Physical Society), (c) torsatron (reproduced from [27], copyright The Author, CC BY 2.0), and (d) modular (reprinted with permission from [16], copyright (2005) by the American Physical Society).

Later, in Europe a new Stellarator was designed: the 'torsatron'; proposed by Gourdon from Fontenay-aux-Roses (independently by V F Aleksin from Kharkov-Ukraine in 1961), the design provides the helical current flowing in a unidirectional set of l-helical coils generating the rotational transform (figure 2.27(c)) without the toroidal-field coils. In this design, the poloidal coils are replaced specifying in the project the winding law: the way in which the pitch angle varies in the poloidal cross section. In this way the torsatron loses some of the flexibility of the classical Stellarator but gains considerably in engineering simplicity.

The engineering problems are related to the helical coils interlinked with each other and the plasma column which complicates the Stellarator geometry and increases the field strength.

In Garching, after studying the conventional Stellarator $W7-A$, Schluter proposed a totally different scheme based on a modular concept built with non-planar coils designed using sophisticated computational techniques, based on a genetic algorithm, that enabled specifying a field shape optimized for the plasma properties and then to design the coils that would produce the desired field (figure 2.27(d)). In this way, their design also has been able to reduce the Pfirsch–Schluter current (the longitudinal current equalizing the transverse charge separation) and the bootstrap current that affected the earlier Stellarator configuration. The first Wendelstein experiment $W7-A$ has been optimized (for low-shear field to avoid major resonances) in the next design of an 'Advanced Stellarator' $W7-AS$ at

Garching. More recently at Greifswald the Wendelstein-7X reactor, the largest Stellarator device, has been constructed to achieve operations of up to approximately 30 minutes of continuous plasma discharge in 2021 demonstrating an essential feature of a future fusion power plant.

Even if the early Stellarators did not reach the expected results they continued to make steady progress in terms of the underlying fusion plasma physics. Indeed, there was an upsurge of interest because of their potential advantages as steady-state reactors and because of the potential contributions that they could make to the understanding of toroidal confinement.

2.9 MHD plasma instabilities

Instabilities are a key phenomenon in plasma dynamics, which, according to the experimental evidence, often take place. MHD, albeit not always a good approximation, is, surprisingly, an efficient tool to explore the instability phenomenology [13, 14].

We use a very simplistic picture to introduce the instabilities (figure 2.28). A body at the bottom of the hole (or in more general terms in a potential bucket) is stable against any perturbation shifting it from the equilibrium position, around which the body oscillates. If the body is sitting on the top of a hill (or a repulsive potential) it will definitively move away from the equilibrium.

This is the simplest way to figure out the equilibrium of a physical system and offers a glimpse to its treatment. For a deeper investigation see [24–26].

In general, the instabilities grow in correspondence to the energy available to the system. In a confined plasma, an instability is driven by the free energy contained in the equilibrium configuration. In a Tokamak, there are two main sources of free energy: the kinetic energy of the plasma and the energy of the magnetic field generated by the plasma. The instabilities can therefore be driven by the radial gradient of either the pressure or the current profile.

The plasma instabilities realize a kind of *bestiary* which occupies a long list, they can be compiled into different types, according to the energetic conditions which

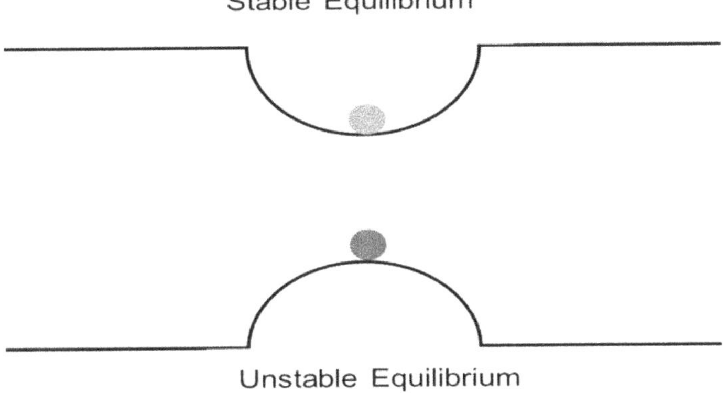

Figure 2.28. Equilibrium conditions.

characterize their onset. The MHD instabilities can cause the plasma performance to detriorate due to various effects: large-scale instabilities can lead to a loss of plasma control (e.g. during a disruption, the plasma current collapses in an uncontrollable way), whereas small scale perturbations and MHD turbulence can significantly enhance the radial transport of particles and energy.

In general, the plasma instabilities may be divided in two: macro branch macroscopic and microscopic. The first class involves the physical (spatial) displacement of plasma and can be described within the framework of the MHD equations. The microscopic instabilities, instead, are analyzed on the basis of the kinetic theory since they arise from changes in the velocity distribution functions that are not taken into account in the MHD description. Although the microscopic instabilities can be very important, usually they are less catastrophic than the MHD instabilities.

As has been mentioned before, the Grad–Shafranov (equation (2.136)) describes a force equilibrium, but it does not tell us if the equilibrium is stable, i.e. if a small variation of the plasma parameters or the external control currents will lead to another equilibrium or to an unstable situation. Therefore, we have to analyze the stability properties of a configuration by a separate treatment.

From the MHD theory, there are different ways to analyze the stability of a given equilibrium. Often, the analysis is done by introducing a perturbation of the equilibrium configuration as a displacement ξ of a fluid element.

Two methods are especially useful to check the stability against such displacements:

(a) Eigenmode analysis: the time-dependent MHD equations are solved with an eigenmode series expansion for the displacement ξ. For the linearized MHD equations, this leads to an eigenvalue problem for each single Fourier mode

$$\xi = \xi_{mn}(r)e^{i(m\theta - n\phi)}e^{\gamma t}, \tag{2.137}$$

being r in the minor radius direction of the Tokamak, (m, n) the poloidal and toroidal mode number, respectively. The mode will be stable for real $\gamma < 0$, the mode oscillates about the equilibrium position for a pure imaginary γ and for real and positive γ the corresponding normal mode will grow exponentially and the equilibrium configuration is unstable. The saturated amplitude of an instability can only be found by introducing nonlinear effects such as changes in the equilibrium introduced by the perturbation.

(b) The energy principle: based on the energy W_{MHD} calculation of the configuration as a functional of the displacement vector. Stability is obtained if the change in the energy $\delta W_{\mathrm{MHD}}(\xi)$ is positive for an arbitrary ξ[8]. This method is useful to prove that a configuration is not stable, because it is sufficient to find one unstable ξ, but the stability is hard to prove, as in this case the motion equations will not be solved. Furthermore, this method is valid for a

[8] In other words if a displacement from an equilibrium condition induces a decrease of the potential energy of the system the equilibrium is unstable as the equilibrium starting point is a maximum for the potential energy (see figure 2.28).

close system without any dissipative effect like for the ideal MHD, for which the energy conservation allows us to identify the stable equilibrium configuration with a minimum in the potential energy W_{MHD}.

We now give a couple of instability examples using the eigenmode analysis. In order to simplify the discussion in terms of mathematical treatment, we consider a particle subject to a one-dimensional (identified by the radial direction of the Tokamak) potential $V(x)$ which will be ruled by the force equation

$$m\frac{d^2x}{dt^2} = F(x). \tag{2.138}$$

We make the assumption that a small perturbation shifts the body from its equilibrium position x_0, to an arbitrary point x, in such a way that $\left|\frac{x-x_0}{x_0}\right| \ll 1$. By expanding the function $F(x)$ around the equilibrium position $x = x_0$ and keeping $F(x_0) = 0$, we find

$$m\frac{d^2x}{dt^2} = F(x_0) + F'(x_0)(x - x_0) + \frac{1}{2}F''(x_0)(x - x_0)^2 \ldots \tag{2.139}$$

retaining the first order term only and setting $\xi = x - x_0$, we are left with

$$\begin{aligned} \frac{d^2\xi}{dt^2} &= -\omega^2\xi, \\ \frac{F'(x_0)}{m} &= -\omega^2, \end{aligned} \tag{2.140}$$

and assuming the initial condition $(\xi'|_{t=0} = 0)$ the solution reads

$$\xi = \xi_0 e^{i\omega t}. \tag{2.141}$$

The motion is accordingly stable if $F'(x_0) < 0$, in this case ω is real and the solution is oscillatory. In contrast, if $F'(x_0) > 0$, ω is imaginary and any perturbation induces an exponential growth.

We apply this procedure to analyze the so called '*stream instability*'. The physical environment favoring its growth is a beam of energetic particles moving inside a plasma or a current driven through it, in such a way that the various species acquire different drift velocities with respect to each other. The different energies excite plasma waves and oscillation energy is gained at the expense of the drift energy of the unperturbed configuration.

We describe a simple example, known as two-stream (or Buneman) instability, regarding the case of a uniform cold plasma (both electrons and ions temperature is zero) in which the ions are stationary and the electrons move with respect to them with a constant velocity.

The equations describing this system are those of ideal MHD. We expand, at first order, variables like velocities and electric field, by setting

$$\vec{E} \cong \vec{E}_0 + \vec{E}_1,$$
$$\vec{v} \cong \vec{v}_0 + \vec{v}_1, \tag{2.142}$$

and after plugging them inside equations (2.43)–(2.45) and preserving first order terms only (namely, neglecting second order contributions due to 1-1 products) we end up with

$$\rho_e\left(\frac{\partial \vec{v}_{1,e}}{\partial t}\right) + \rho_e\,(\vec{v}_0 \cdot \vec{\nabla})\,\vec{v}_{1,e} = -\,en_{0,e}\vec{E}_1,$$
$$\rho_i\left(\frac{\partial \vec{v}_{1,i}}{\partial t}\right) = en_0\,\vec{E}_1, \tag{2.143}$$

with $\rho_i = n_{0,i}m_i$, $\rho_e = n_{0,e}m_e$, where the subscripts i and e stand for ion and electrons, respectively (note that $\vec{v}_{0,i} = 0$).

Regarding the densities, we find from the continuity equation

$$\frac{\partial}{\partial t}n_{1,i} + n_{0,i}\left(\vec{\nabla} \cdot \vec{v}_{1,i}\right) = 0,$$
$$\frac{\partial}{\partial t}n_{1,e} + \left(\vec{v}_0 \cdot \vec{\nabla}\,n_{1,e}\right) + n_{0,e}\left(\vec{\nabla} \cdot \vec{v}_{1,e}\right) = 0. \tag{2.144}$$

It is assumed that the result of the interaction is the excitation of a plasma wave with an associated electric field which is written as (we limit our analysis to a one-dimensional treatment which captures the essential features of the underlying physics)

$$\vec{E}_1 = E_0 e^{i(kx-\omega t)}\hat{x}. \tag{2.145}$$

If we replace the time and space derivatives, respectively, with $-i\omega$ and ik we can cast equations (2.143–2.144) as

$$-\,m_e(i\omega v_{1,e}) + im_e(v_0 k)v_{1,e} = -\,eE_0,$$
$$-\,m_i(i\omega v_{1,i}) = eE_0,$$
$$-\,i\omega n_{1,i} + in_{0,i}\,(kv_{1,i}) = 0,$$
$$-\,i\omega n_{1,e} + in_{1,e}\,(kv_0) + n_{0,e}\,(ik \cdot v_{1,e}) = 0, \tag{2.146}$$

where both velocities are in the x-direction.

This yields the following expression for the velocities induced by the excited field

$$v_{1,e} = -\,i\frac{e\,E_0}{m_e}\frac{1}{(\omega - v_0 k)},$$
$$v_{1,i} = i\frac{e\,E_0}{m_i\,\omega}, \tag{2.147}$$

and for the density perturbations

$$n_{i,1} = \frac{n_0}{\omega}(k\, v_{i,1}) = i n_0\left(k \frac{e\, E_0}{m_i\, \omega^2}\right),$$
$$n_{e,1} = \frac{n_0(k v_{e,1})}{(\omega - k v_0)} = -\frac{n_0 k e\, E_0}{m_e}\frac{1}{(\omega - v_0 k)^2}. \tag{2.148}$$

The only unknown is the excited field part $\vec{E}_1$, that can be derived from the Poisson equation

$$\varepsilon_0 \vec{\nabla} \cdot \vec{E}_1 = e\,(n_{1,i} - n_{1,e}), \tag{2.149}$$

which after a few algebraic steps yields the dispersion relation

$$\omega_p^2\left[\frac{m_i}{m_e}\frac{1}{\omega^2} + \frac{1}{(\omega - k\, v_0)^2}\right] = 1, \tag{2.150}$$

which is a fourth degree algebraic equation for the frequency ω and can be handled by the use of standard means to infer the complex roots giving rise to the growth of the instability.

The *Taylor–Rayleigh instability* occurs when a fluid with a density ρ_1 is above another with density $\rho_2 < \rho_1$. In the case of plasma, the two fluids are composed by ions and electrons. According to figure 2.29 the gravity force $\vec{g}$ is perpendicular to the magnetic field and charged particles acquire in this geometry a drift velocity given by

$$\vec{v}_d = \frac{m_i \vec{g} \times \vec{B}}{e\, |\vec{B}|^2}, \tag{2.151}$$

and the ions move as illustrated in figure 2.29 (an electron current is also generated but is much less intense because $m_e \ll m_i$).

If a perturbation, in the form of a wave, modulates the plasma interface, the combination with the drift current causes a migration of the positively charged ions towards the left hand side with a corresponding negatively charge accumulated on

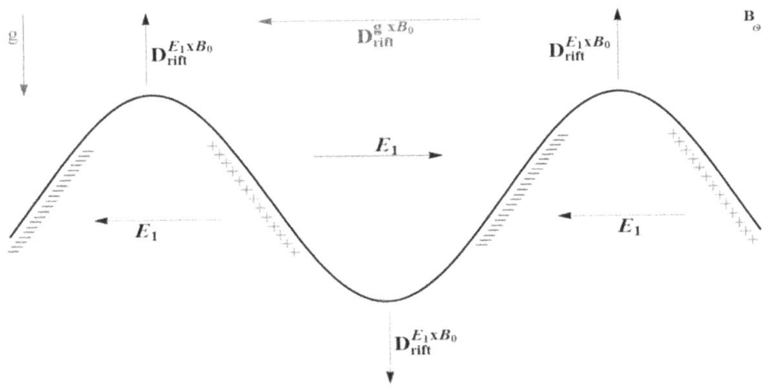

Figure 2.29. Drift induced by the electric field and by the gravitational force (for the ions).

the other side. This effect is replicated along the modulating separatrix. Alternating electric fields, associated with the charge separations, induce further velocity drifts, responsible for the enhancement of the modulation, hence for the instability growth.

The argument can be developed in analytical terms, by just following the paradigmatic procedure we have envisaged, namely first order perturbative expansion of the equilibrium equations and search for a dispersion relation.

The discussion we have just developed has pointed out that within this context the energy plays a central role. The use of a variational principle can therefore be a more rigorous starting point to frame the previous analysis. We start from the kinetic energy K, given by $1/2\rho_0\dot{\xi}^2$, integrated over the whole plasma volume

$$K(\vec{\xi}, \cdot\vec{\xi}) = \frac{1}{2}\int \rho\dot{\vec{\xi}} \cdot \dot{\vec{\xi}} d\vec{r} = -\frac{\omega^2}{2}\int \rho\vec{\xi} \cdot \vec{\xi} d\vec{r} = \frac{1}{2}\int \xi \cdot \vec{F}(\vec{\xi}), \tag{2.152}$$

obtained after replacing the time derivative with $-i\omega$ and using equation (2.140). In equation (2.152) $\vec{F}(\vec{\xi})$ is the force operator described below. By energy conservation we have

$$\delta W(\vec{\xi}, \vec{\xi}) = -\frac{1}{2}\int \xi \cdot \vec{F}(\vec{\xi}) d\vec{r}, \tag{2.153}$$

and get from equations (2.152) and (2.153) the variational formulation of the stability problem

$$\omega^2 = \frac{\delta W(\vec{\xi}, \vec{\xi})}{K(\vec{\xi}, \vec{\xi})}. \tag{2.154}$$

The expression of the operator $\vec{F}$ is necessary in order to minimize δW and to characterize the equilibrium of the system.

The operator $\vec{F}$, after tedious calculus, can be derived by linearizing the ideal MHD equation around the equilibrium point and using ***the momentum equation*** equation (2.71) we achieve the expression for $\vec{F}$. The integral in equation (2.153) can be split into three terms representing the changes in potential energy within the plasma (δW_P), at the surface (δW_S) and in the vacuum (δW_V). The last two terms will be null in the absence of the surface current and the perturbation of the vacuum field. Finally, the use of vector identities allow us to express the integrand in equation (2.153) as reported below:

$$\begin{aligned}
\delta W_p = \frac{1}{2}\int \Bigg[& \frac{|\vec{B}_{1\perp}|^2}{\mu_0} + && \text{Field} - \text{line bending} \geqslant 0 \\
& + \frac{|\vec{B}_0|^2}{\mu_0}|\vec{\nabla} \cdot \vec{\xi}_\perp + 2\vec{\xi}_\perp \cdot \vec{\kappa}|^2 + && \text{Magnetic compression} \geqslant 0 \\
& + \gamma P_0 |\vec{\nabla} \cdot \vec{\xi}|^2 + && \text{Plasma compression} \geqslant 0 \\
& - 2\left(\vec{\xi}_\perp \cdot \vec{\nabla} P_0\right)\left(\vec{\kappa} \cdot \vec{\xi}_\perp\right) - && \text{Pressure/curvature drive, + or} - \\
& - j_{\|}\left(\vec{\xi} \times \vec{b}\right) \cdot \vec{B}_{1\perp} \Bigg] d\vec{r} && \text{Parallel current drive, + or} -
\end{aligned} \tag{2.155}$$

The terms with subscripts 1 and 0 refer to perturbed and equilibrium quantities, respectively.

The first three terms in equation (2.155) are positively defined and stabilize the equilibrium position, and the last two contributions with negative sign act in the opposite direction. The instabilities they induce are said to be pressure and current-driven, respectively, but since $\vec{\nabla} P_0 = \vec{J}_{0\perp} \times \vec{B}_0$ both are driven by the energy associated with different components of the current.

Furthermore, at low β, the magnetic energy is much higher than the kinetic energy and the instabilities will mainly be current-driven; at high β, we expect the pressure driven instabilities to become significant.

The few elements of the discussion given so far will be amplified in the forthcoming chapter, where, among other things, we will discuss the phenomenology associated with the additional heating and how it can be helpful to counteract the instability onset.

References

[1] Chen F F 2018 *Introduction to Plasma Physics and Controlled Fusion* 3rd edn (Berlin: Springer)

[2] Miyamoto K 2005 *Plasma Physics and Controlled Nuclear Fusion* (Berlin: Springer)

[3] Bribiesca Argomedo F, Witrant E and Prieur C 2013 *Safety Factor Profile Control in a Tokamak* (Berlin: Springer)

[4] Schroeder D V 2000 *An Introduction to Thermal Physics* (San Francisco, CA: Addison-Wesley Longman)

[5] Lamers H J G L and Levesque E M 2017 *Understanding Stellar Evolution* (Bristol: IOP Publishing)

[6] Nakariakov V 2002 *Introduction to MHD, Lecture Notes* https://warwick.ac.uk/fac/sci/physics/research/cfsa/people/valery/teaching/khu_mhd/KHU_mhd_handout.pdf

[7] Kallen J D 2003 Chapter 6: Plasma description II *Fundamentals of Plasma Physics, On-line Book*

[8] Poedts S 2009 *Introduction to MHD Theory* (This reference is particularly recommended for generality, rigor and clarity)

[9] Cap F F 1976 *Handbook of Plasma Instabilities* (New York: Academic)

[10] Mikhailovskii A B 1974 *Theory of Plasma Instabilities, Volume 2: Instabilities of an Inhomogeneous Plasma* (Berlin: Springer)

[11] Mikhailovskii A B 1998 *Instabilities in a Confined PlasmaPlasma Physics Series* (Boca Raton, FL: CRC Press)

[12] Lee H J 2019 *Fundamentals of Theoretical Plasma Physics* (Singapore: World Scientific)

[13] Body T J M and Sanderson J J 2003 *The Physics of Plasmas* (Cambridge: Cambridge University Press)

[14] Freidberg J P 2007 *Plasma Physics and Fusion Energy* (Cambridge: Cambridge University Press)

[15] Boozer A H 1985 Magnetic Field Line Hamiltonian *Technical Report* PPPL-2094R Princeton University, Plasma Physics Laboratory https://www.osti.gov/servlets/purl/5793830

[16] Boozer A H 2005 Physics of magnetically confined plasmas *Rev. Mod. Phys.* **76** 1071–141

[17] Shafranov V D 1957 On magnetohydrodynamical equilibrium configuration *J. Exptl. Theoret. Phys. (U.S.S.R.)* **33** 710–22

[18] Grad H and Rubin H 1957 Hydromagnetic equilibria and force-free fields *Proc. of the Second United Nations Int. Conf. on the Peaceful Uses of Atomic Energy (Geneva, Switzerland)* pp 190–7 https://inis.iaea.org/search/search.aspx?orig_q=RN:39082408

[19] Goedbloed J P, Keppens R and Poedts S 2010 *Advanced Magneto Hydrodynamics* (Cambridge: Cambridge University Press)

[20] Xu Y 2016 A general comparison between tokamak and stellarator plasmas *Matter Radiat. Extremes* **1** 192–200

[21] Miyamoto K 2007 *Controlled Fusion and Plasma Physics* (London: Taylor & Francis)

[22] Young K M 1973 The c-stellarator—a review of containment *Plasma Phys.* **16** 119–52

[23] Boozer A H 1998 What is a stellarator? *Phys. Plasmas* **5** 1647–55

[24] Abraham R and Marsden J E 1978 *Foundations of Mechanics* (New York: AMS Chelsea Publishing)

[25] Arnold V I 1978 *Mathematical Methods of Classical Mechanics* (Berlin: Springer)

[26] Arnold V I and Avez A 1968 *Ergodic Problems of Classical Mechanics* (Reading, MA: Benjamin)

[27] Wagner F 2013 Physics of magnetic confinement fusion *EPJ Web of Conferences* **54** 01007

IOP Publishing

G Dattoli, E Di Palma, S P Sabchevski and I P Spassovsky

Chapter 3

Plasma additional heating and Tokamak engineering issues

3.1 Introduction

In the previous chapters we have outlined the main elements of plasma physics and magnetic fusion. We have underscored that the ohmic heating is not sufficient to reach the self-sustained operation, because the resistivity decreases when the plasma temperature is close to 3 keV. We have furthermore noted that an additional injection of external power is necessary to accomplish the goal.

In this chapter we touch on the ancillary tools adopted to transfer power to the plasma to enhance the temperature and explore the physical mechanisms governing these processes. Before entering more specific details, it is worth providing an idea of what is the amount of the additional power needed in a realistic configuration.

The estimate follows almost straightforwardly from the observation that the necessary energy density to be transferred to raise the temperature of a plasma with density n by an amount ΔT is

$$E_{\Delta} = nk_{\mathrm{B}}\Delta T, \tag{3.1}$$

and the associated volumetric power density reads

$$P_{\Delta} = \frac{nk_{\mathrm{B}}\Delta T}{\tau_E}. \tag{3.2}$$

If we consider the set of parameters of the example discussed in the introductory section of the previous chapter (namely a plasma with final temperature of 10 keV and volumetric density 10^{20}) we find that the previous equation yields $E_{\Delta} \cong 10^5\ \mathrm{J\,m^{-3}}$. Assuming furthermore the confining time $\tau_E = 1.68$ s, calculated

doi:10.1088/978-0-7503-2464-9ch3

from equations (2.13) and (2.14), we infer that the additional power density is of the order of $P_\Delta \cong 5.9 \cdot 10^4$ W m^{-3}. The total power is obtained using equation (3.2)

$$P_\Delta = 0.34\,\pi\, 10^{22} k_B \Delta T\, R, \tag{3.3}$$

which for this example yields a total power inside the plasma of 94 MW. If the plasma absorption efficiency is ε, the power to be launched in the plasma is $P_S = P_\Delta/\varepsilon$, where S stands for source power.

Assuming an average efficiency of 50% for the additional power, a reference number for the present example is about 200 MW.

These values cannot be reached via the natural heating, associated with the toroidal current flowing in the Tokamak and the plasma resistance. The ohmic heating is due to the collisions of the conduction electrons with the ions, characterized by much less mobility.

In chapter 1 we have derived the plasma resistance using a model based on the evaluation of the electron mobility specified by (see equation (1.100)

$$\mu = \frac{e}{m_e f_{e,i}}, \tag{3.4}$$

where $f_{e,i}$ is the frequency of the electron–ion collisions. The electron current density associated with the electron mobility is (see equation (1.101))

$$J_p = ne\mu\, E = 4\,\pi n \frac{r_0 l_{e,i}}{Z_0} E, \tag{3.5}$$

with

$$r_0 = \frac{e^2}{4\pi\, \varepsilon_0 m_e c^2},\; l_{e,i} = \frac{c}{f_{e,i}},\; Z_0 = \varepsilon_0 c = 376.730\ \Omega, \tag{3.6}$$

Z_0 being the free space impedance, r_0 the electron classical radius and $l_{e,i}$ denoting some length associated with the ion–electron collision rate. The resistivity follows from equation (3.5) and is just given by

$$\eta = (ne\mu)^{-1} = \frac{Z_0}{4\pi\, n r_0 l_{e,i}}. \tag{3.7}$$

In chapter 1 we have given a heuristic argument which has allowed us to conclude that the resistivity goes like $T^{-\frac{3}{2}}$, this conclusion is, however, qualitative and is helpful to understand that the Joule heating efficiency decreases with the plasma temperature, but does not allow any quantitative statement.

The key quantity, allowing more substantive predictions, is $l_{e,i}$. Without entering the details of the derivation, which can be found elsewhere[1] we can state that the limits of the ohmic heating are summarized by the equation

[1] http://silas.psfc.mit.edu/introplasma/chap3.html#tth_chAp3.

$$\eta \cong 5.2 \cdot 10^{-5} \frac{\ln(\Lambda)}{(T[eV])^{3/2}}, \tag{3.8}$$

in which the Coulomb logarithm $\ln(\Lambda)$ plays a central role. It controls indeed the momentum loss of electrons to ions and the conversion of the electrons' kinetic energy into heat. The electron–ion scattering depends on two quantities, namely the deflection angle ϑ and the impact parameter b. The Coulomb logarithm defined in terms of these quantities reads

$$\ln(\Lambda) = \int \frac{db}{b} = \int \frac{d\,\sin(\vartheta)}{\sin(\vartheta)}. \tag{3.9}$$

A more accurate calculation of the external heating power amount, necessary to reach the ignition, can be derived starting from the fusion power balance and assuming different scaling regime for the confinement time (τ_E).

We have seen in the introductory chapter that the Lawson criterion offers a practical guide to fix the conditions for the onset of ***ignition***, namely when the power due to fusion processes overcome that feed-in to bring the system to this configuration.

The heating power necessary to compensate the losses is

$$P_H = \left[\frac{3nT}{\tau_E} - \frac{1}{4} n^2 \langle \sigma v \rangle \mathcal{E}_\alpha \right] V, \tag{3.10}$$

where

$$P_\alpha = \frac{1}{4} n^2 \langle \sigma v \rangle E_\alpha V, \tag{3.11}$$

and

$$P_{\text{loss}} = \frac{3nT}{\tau_E} V, \tag{3.12}$$

are the α-particles and loss power, respectively.

Taking into account the formula in equation (3.10), the dependence of the D–T reaction cross section ($\langle \sigma v \rangle$) versus the temperature, the energy of the α particles and the temperature and density profile, we end up with the following triple product inequality

$$nT\tau_E > 5 \cdot 10^{21} \frac{\text{keV s}}{\text{m}^3}, \tag{3.13}$$

which is a more stringent condition than the Lawson criterion (see equation (1.18)). It includes temperature T into the global parameter defining the critical regime for a fusion reactor.

Let us now look at the problem using a dynamical point of view, equation (3.13) is the result of a process in which power is delivered into the system till the equilibrium condition occurs. If we oversimplify the problem, we can describe the heating dynamics using the differential equation (the subscript p stands for plasma) [1]

$$\frac{dP_p}{dt} = \frac{P_H}{V} + \frac{1}{4}n^2\langle\sigma v\rangle\mathcal{E}_\alpha - \frac{P_p}{\tau_E(n, T)}, \tag{3.14}$$

with

$$P_p = 3\,nT. \tag{3.15}$$

We cannot solve analytically equation (3.14) because we do not know the dependence of the confining time vs temperature and density. We can however expect that, using an adiabatic external heating, the relevant solution is just a sequence of stationary states and the various contributions versus T can be drawn as indicated in figure 3.1(a). The figure shows the losses, for different assumptions about the dependence of the confinement time versus the temperature, namely:

(a) independence, in this case the corresponding value $\tau_E = \tau_E^*$ can be inferred from equation (3.10) at the ignition

$$\tau_E^* = \frac{12T_i}{n\langle\sigma v\rangle_i\mathcal{E}_\alpha}, \tag{3.16}$$

where T_i is the temperature at the ignition and $\langle\sigma v\rangle_i$ is evaluated for $T = T_i$;

(b) inverse proportionality to the temperature,

$$\tau_E = T_i\tau_E^*/T. \tag{3.17}$$

Figure 3.1(b) yields an idea of how the dependence of the confinement time on the temperature influences the amount of required additional power to reach the ignition, in particular: the maximum external power, delivered around 5 keV, is 40% of the α power at the ignition for $\tau_E = \tau_E^*$ and will be reduced to 15% for $\tau_E = T_i\tau_E^*/T$.

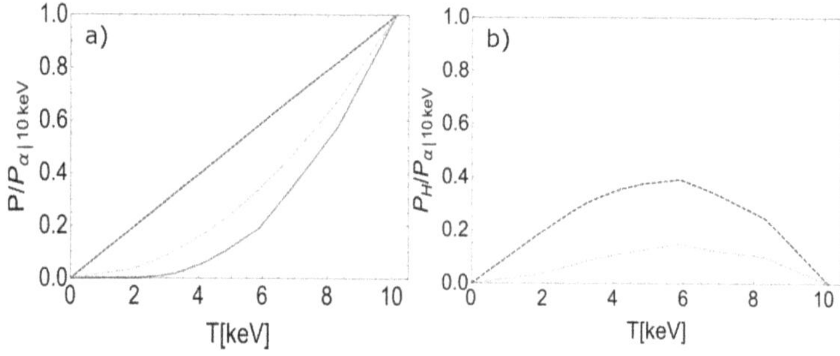

Figure 3.1. (a) Power loss (normalized to the α-particle power at the ignition) versus T for $\tau_E = \tau_E^*$ (blue-dashed) and $\tau_E \propto (1/T)$ (green dashed). The red curve yields the ratio of the α-particle power to the α-particle power at the ignition and has been introduced for comparison reasons. (b) Required additional power (normalized to the α-particle at the ignition) for different regime of the confinement time τ_E (same color convention of (a)).

The problem underlying this type of dynamics is that the equilibrium is unstable and an increase of the temperature can be determined by the imbalance of the heating against the losses.

To understand the role of this instability in the dynamics to the ignition we consider equation (3.14) for T close the break-even point and write [1]

$$3n\frac{dT}{dt} = \frac{1}{4}n^2\langle\sigma v\rangle\mathcal{E}_\alpha - 3n\frac{T}{\tau_E(T)}, \tag{3.18}$$

with the equilibrium condition, which is written as

$$\frac{T}{\tau_E} = \frac{1}{12}n\langle\sigma v\rangle\mathcal{E}_\alpha. \tag{3.19}$$

If we consider equation (3.18) around the equilibrium point after expending

$$\tau_E(T) \cong \tau_E + T\,\frac{\partial\tau_E}{\partial T}, \tag{3.20}$$

we find

$$\frac{d\Delta T}{dt} = \left[\frac{1}{12}n\langle\sigma v\rangle'\mathcal{E}_\alpha - \frac{1}{\tau_E}\left(1 + \frac{T}{\tau_E}\frac{\partial\tau_E}{\partial T}\right)\right]\Delta T, \tag{3.21}$$

where

$$\langle\sigma v\rangle' = \frac{d\langle\sigma v\rangle}{dT}. \tag{3.22}$$

If we use the stationary solution (3.19) we can eventually write

$$\frac{d\Delta T}{dt} = \frac{1}{12}n\langle\sigma v\rangle\frac{\mathcal{E}_\alpha}{T}\left[-1 + \frac{T}{\tau_E}\left(\frac{1}{\langle\sigma v\rangle}\frac{\partial\langle\sigma v\rangle}{\partial T} + \frac{1}{\tau_E}\frac{\partial\tau_E}{\partial T}\right)\right]\Delta T. \tag{3.23}$$

In order to prevent temperature growing exponentially, the term in square brackets should be kept negative, this brings the condition

$$\frac{T}{\tau_E}\frac{\partial\tau_E}{\partial T} < 1 - \frac{T}{\langle\sigma v\rangle}\frac{d\langle\sigma v\rangle}{dT} = \phi(T). \tag{3.24}$$

Taking into account that the experiments suggest that the confinement time temperature dependence is that in equation (3.17), we end up with the following more explicit stability condition

$$\phi(T) > -1. \tag{3.25}$$

It can be made quantitative by considering the dependence of the reactivity on the temperature as reported in chapter 1 (see figure 1.7). In figure 3.2 we have sketched graphically the stability condition, which occurs for $T \gtrsim 13.5$ keV.

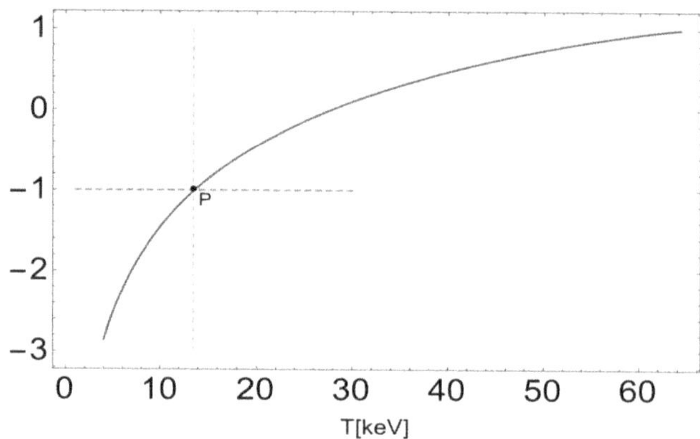

Figure 3.2. Graphical representation of the stability condition (3.24). The blue curve is the $\phi(T)$ function and the point $P \equiv (\approx 13.5, -1)$ marks the condition of equation (3.25).

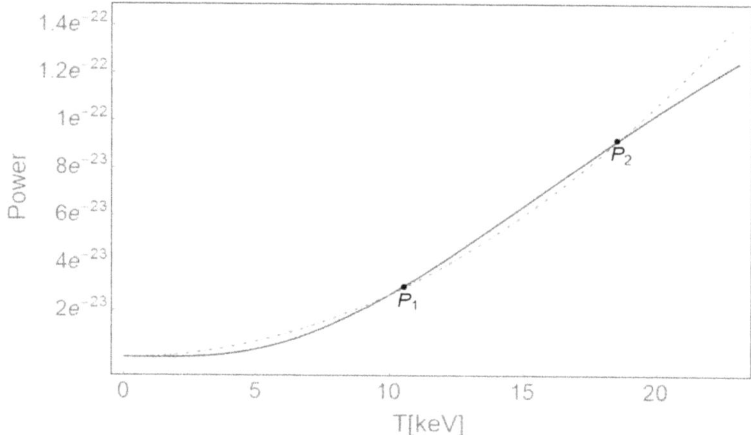

Figure 3.3. Power loss (red dot-dashed) and α-particles power (blue continuous) versus temperature (T). The intersections points below and above 15 keV mark the equilibrium conditions. It should be noted that the power values are normalized to $n^2\mathcal{E}_\alpha V$.

The power loss (equation (3.12)) specified for the confinement time ($\tau_E = T_i\tau_E^*/T$) leads to

$$P_{\text{loss}} = n^2\mathcal{E}_\alpha V\frac{T^2}{4T_i^2}\langle\sigma v\rangle_i. \tag{3.26}$$

In figure 3.3 we have plotted both P_{loss} and P_α versus the temperature. The curves intersect at two points P_1, P_2 corresponding to the ignition condition. The stability condition in (3.24) leads to considering only the point (P_2) for which the temperature is greater than 15 keV.

The forthcoming section will be dedicated to the plasma scaling formulae as we have seen before they are very crucial for example to estimate the amount of power necessary to achieve the ignition which is strictly related to the dependence of the confinement time on the temperature.

3.2 Plasma scaling formulae and ohmic heating

We have already anticipated the existence and the usefulness of the so-called scaling laws, which can be exploited for a quick evaluation of the engineering parameters of a fusion device.

The strategy underlying the methodology leading to the derivation of a scaling formula is straightforward. We have perhaps conveyed the idea that plasma physics is a very complicated subject and easily manageable formulae are hardly attainable via analytical means. It is furthermore not unlikely that quantities, e.g., the confinement time, are affected by almost all the plasma parameters (pressure, magnetic fields, plasma density$\cdots$) and also by the geometrical characteristics of the confining device itself.

What we have just mentioned can be mathematically stated as follows: given a plasma confining device depending on n parameters $x_i(i = 1, 2, \ldots, n)$, we make the assumption that by keeping one of them x_k its dependence on the others is expressed by the functional relationship

$$y = K \cdot \prod_i x_i^{\alpha_i}, \tag{3.27}$$

sometimes the constant K is written as $K = e^{\alpha_0}$. By keeping the logarithm of both sides we transform the products into a sum, namely

$$\begin{aligned} \eta &= \ln(K) + \sum_i \alpha_i \xi_i, \\ \xi &= \ln(x),\ \eta = \ln(y). \end{aligned} \tag{3.28}$$

The coefficients of the linear combination can be determined from a set of data obtained experimentally or from numerical computations and adjusted by wise physical assumptions. The reliability of the procedure is also linked to a proper normalization of the variables, statistical independence of the different quantities appearing on the rhs of equations (3.27)–(3.28) and definition of suitable errors and of confidence intervals.

The strategy is conceptually simple, but awkward in practice and takes a significant amount of work, not described here.

The following yields an idea of how the various parameters specifying either plasma and Tokamak machine are embedded to give a quantity of paramount importance like the plasma resistance (SI units)

$$R_p = \frac{10^{-3} R_0 Z_{\text{eff}}}{a^2 \kappa\, g(Z_{\text{eff}})} \left[1 + \left(\frac{a}{R_0}\right)^{1/2}\right] T^{-3/2}, \tag{3.29}$$

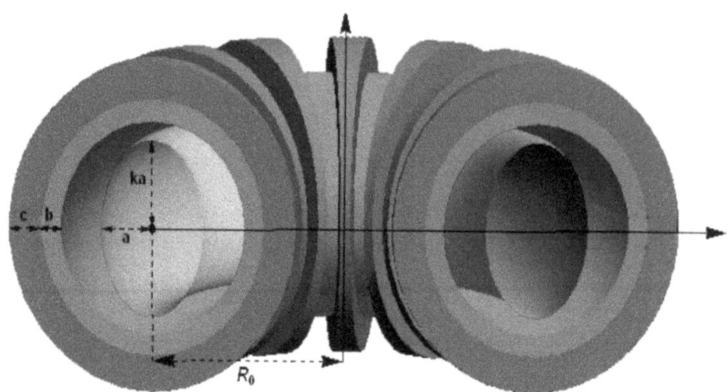

Figure 3.4. A sketch of the transverse Tokamak geometry with the toroidal coils (c), the breeding blanket (b) and a k factor elongated plasma.

where R_0, a, κ are the major radius, plasma radius and elongation, respectively (see figure 3.4), $Z_{\rm eff}$ is the ion effective charge[2], T is the average electron temperature and $g(Z_{\rm eff})$ is a function of the effective charge, which for a clean plasma can be approximated with 1/2. The terms in square bracket account for the corrections due to the trapped particles. The average electron temperature is given in eV. The ohmic power is

$$P_\Omega = R_p I_p^2, \tag{3.30}$$

according to equation (3.29) the resistance sharply decreases with increasing temperature, but from (3.30) it is not clear what happens if the plasma current I_p is strongly increased, in order to get a larger dissipated power.

The point is not of secondary importance, to settle the question we need to fix a dependence of the current on the temperature.

What follows is a genuine example of application of the previously quoted scaling formulae.

The amount of power transferred to the plasma by ohmic heating can be parameterized according to the following 'law'

$$P_{\Omega,SC} = 64 \cdot 10^3 M^{0.2} I_p^{0.8} R_0^{1.6} a^{0.6} \kappa^{0.5} N^{0.6}\ B_t^{0.35}, \tag{3.31}$$

where M is the relative isotopic mass, N is the (line averaged) plasma density expressed in units of 10^{20} m^{-3} and the current is expressed in MA.

By taking into account that the plasma power is

$$\begin{aligned} P_\Omega &= (3)2\pi^2\ \kappa a^2 R_0 n\ k_{\rm B} T, \\ n &= 10^{20} N. \end{aligned} \tag{3.32}$$

We can guess the dependence of the current on the temperature by comparing equations (3.13–3.14), thus inferring the following scaling formula for the temperature

[2] By $Z_{\rm eff}$ in contrast to Z is meant the effective nuclear ion charge including the effect of the electron charge shielding $Z_{\rm eff} = Z - S$, $S \equiv$ Shielding constant.

$$T = 68\ M^{0.2} I_p^{0.8} R_0^{0.6} a^{-1.4} \kappa^{-0.5} N^{-0.4} B_t^{0.35}. \tag{3.33}$$

The equation (3.33) suggests the following scaling between the plasma current and temperature

$$I_p \propto T^{1.25}, \tag{3.34}$$

and therefore

$$P_\Omega \propto T. \tag{3.35}$$

It could be further inferred that by suitably increasing the current the problems associated with the drop of the resistivity with the temperature should be overcome. Such an increase, however, imposes other effects associated with the instability which disrupt the plasma and nullify the possible benefits.

The upper limits imposed on the maximum plasma current will be discussed later in this chapter, here we just note that the following formula for the maximum current to be used as a reference design value

$$I_P[MW] = \frac{5a\ \kappa\ B_t}{2R_0}, \tag{3.36}$$

which for ITER-like parameters ($R_0 = 7.75$ m, $a = 2.8$ m, $\kappa = 1.6$, $B_t = 6$ T) yields 8 MA, which are values which cannot guarantee the fusion ignition with only ohmic heating [2].

The solution is therefore the use of external heating which will be discussed in the forthcoming sections.

3.3 Magnetic fusion heating devices: the neutral beam injection

In figure 3.5 we have reported the amount of power reached by magnetic fusion devices through the use of different heating systems

1. Ohmic
2. Radio-frequency (RF)
3. Neutral-beam injection—deuterium (NBI—D)
4. Neutral-beam injection—deuterium–tritium (NBI—D–T)

The associated technologies have characterized a certain decade and we can guess an interpolating logistic behavior between the fusion power they are allowed to reach. The power growth curve can be parameterized as

$$\begin{aligned} &P = P_0 \frac{e^{\alpha\, Y}}{1 + \frac{P_0}{P_F}(e^{\alpha Y} - 1)}, \\ &P_0 = P(1970) = 10^{-11}\ \mathrm{MW}, \\ &P_F = P(2000) = 10\ \mathrm{MW}, \\ &Y = year - 1970, \\ &\alpha \equiv \text{technological–growth rate}. \end{aligned} \tag{3.37}$$

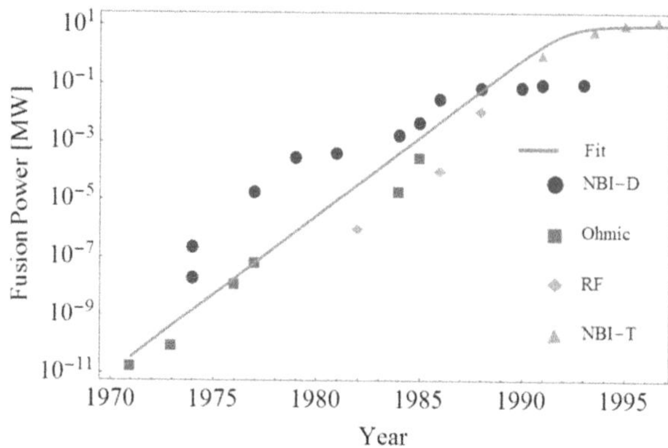

Figure 3.5. Magnetic fusion power during the last three decades of the XXth century for different additional heating systems and fitted (continuous-red line) with the logistic function (equation (3.37)) with the growth factor $\alpha = 1.25$.

The maximum power P_F can be considered the maximum power achievable with the actual technologies.

It is worth noting that the chart in figure 3.5 includes different technologies and each one has reached a kind of optimum performance, like, e.g., the case of NBI—D.

After this 'philosophical' remark useful to address the idea of how technological innovation in the field has allowed an improvement of about 12 orders of magnitude, we discuss more pragmatic aspects.

The primary heating mechanism in magnetic fusion is associated with the energy loss, inside the plasma, of the α-particles produced as fusion reaction products. The same mechanism occurs via the ohmic heating. Any other heating tool should therefore be based an energy transfer from some device to the plasma, with the only difference that it is said to be 'external' or 'additional' because it is not intrinsically built in, but externally added [3]. As happens in the case of NBI, brought inside the plasma with the tool sketched in figure 3.6. The physical reasons underlying such a design are understood, if it is kept in mind that [4, 5]:

1. External particles, to be injected inside the plasma, cannot be charged since they are deflected by the strong magnetic field confining the plasma
2. The injected particles should be characterized by large enough kinetic energy to be transferred to plasma
3. Energies of neutral beams around hundreds of keV can be reached by pre-accelerating, with electrostatic devices, ions which are successively neutralized.

According to figure 3.6, the first step of the injection system is the production of low energy ions. The choice of the species and of the ions to be accelerated deserves a few words of comment.

The use of means like plasma discharge does not imply only the production of D^+, but also not negligible fraction of molecular ions (D_2^+, D_3^+), they eventually dissociate into atoms with lower energies and thus with lower penetrating power. This is unlike since more heating power is deposited at the plasma edge, thus creating non-equilibrium distributions, eventually triggering unstable behavior.

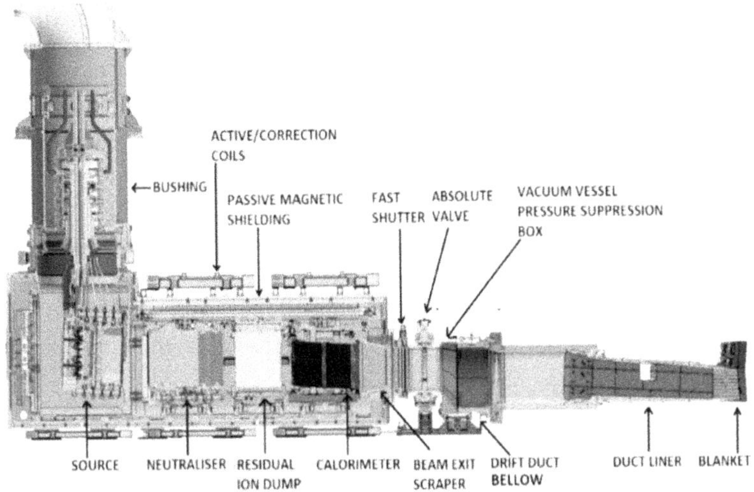

Figure 3.6. Neutral-beam injection device for ITER.

Such an effect (namely generation of molecular ions and less energetic secondary beams) can, e.g., be prevented by accelerating negative ions [6].

The above points clarify the different items in the external injection tool. The ions are partially neutralized after the acceleration step in a charge exchange neutralization chamber. The remaining charged fraction is magnetically intercepted and sent to a beam dump.

It is worth saying a few words about the neutralizer, which is just a gas cell containing molecular gas, usually of the same species of the accelerated ions.

The neutralization process is ruled at high energies by the so-called ***stripping reaction***, $D^{-}+D_2 \rightarrow D_0 + D_2 + e^{-}$ which converts an ion into an energetic neutral. The conversion efficiency η_{neutr} is provided in figure 3.6, which reports that, regarding positive ions, it significantly drops at higher energies, while it remains nearly 58% for energies within the interval 0.1–1 MeV for negative ions.

The power of neutrals injected inside the plasma can be estimated from the accelerated current, accelerating voltage and conversion efficiency as

$$P_{\text{neutr}}[\text{MW}] = \eta_{\text{neutr}} I_{\text{ions}}[\text{A}]\, E_{\text{ions}}[\text{MeV}], \tag{3.38}$$

the rest is lost and should eventually be recovered.

Postponing the considerations on efficiency and recovery of the power exploited to manage the NBI injection tool, we introduce a few elements of discussion to fix the power/energy of the NB to be injected inside the plasma.

It is not difficult to understand that the optimum kinetic energy of the neutral particles is fixed by the machine dimensions [7].

When the particles are penetrating the plasma their flux decreases exponentially according to the Lambert–Beer type law

$$\Phi(\Delta l) = \Phi_0 e^{-\frac{\Delta l}{L_{\text{NBI}}}}, \tag{3.39}$$

where L_{NBI} is the characteristic distance at which the flux is attenuated by a factor $1/e$. The processes responsible for the flux decay, thus for the energy deposition inside the plasma, are essentially due to ionization processes induced by different mechanisms due to collisions with plasma electrons, ions and impurities.

If n denotes the plasma density and σ the cross section of the ionization process it goes by itself that

$$L_{\mathrm{NBI}} = \frac{1}{n\sigma}. \tag{3.40}$$

Without entering the details of the calculation of the cross section and neglecting effects due to electron ionization impact, we note that, for energies such that

$$\begin{aligned} \bar{E} &= E/A_{\mathrm{NBI}} > 40 \text{ keV amu}^{-1}, \\ A_{\mathrm{NBI}} &\equiv \text{Atomic–Number}, \\ \text{amu} &\equiv \text{Atomic–Mass–Unit}, \end{aligned} \tag{3.41}$$

the cross section in practical units reads

$$\sigma[\mathrm{m}^2] \cong \frac{1.8 \cdot 10^{-18}}{\bar{E}[\mathrm{keV}]}, \tag{3.42}$$

which yields

$$L_{\mathrm{NBI}} \cong \frac{\bar{E}[\mathrm{keV}]}{180 \text{ N}}, \tag{3.43}$$

where N is defined in equation (3.32), for further details see figure 3.7.

We can now address a few remarks of practical usefulness.

The energy release to be effective should occur on an L_{NBI} comparable to the machine minor radius.

The required beam energy is therefore of the order

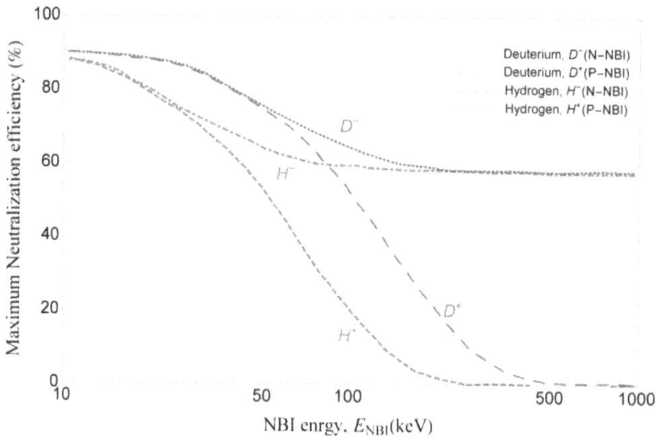

Figure 3.7. Ion neutralization efficiency versus energy for NBI.

$$E_{\mathrm{NBI}}[\mathrm{keV}] \cong 180\, A_{\mathrm{NBI}} Na. \tag{3.44}$$

In the case of a Tokamak with $n \cong 5 \cdot 10^{19}\mathrm{m}^{-3}$ and minor radius $a = 0.9$ m the required energy of a beam of neutral D ($A_{\mathrm{NBI}} = 2$) amounts to 162 keV.

For the same neutral beam and for $n \cong 10^{20}\mathrm{m}^{-3}$ and $a = 2$ m the required energy is above 600 keV. Accordingly, higher plasma densities and larger Tokamak devices require neutral beams with larger energies.

The choice of the optimum energy depends, however, on the chosen injection geometry.

The tangential injection (see figure 3.8) is chosen because the large propagation path minimizes the shine-through effect (namely the fraction of neutrals hitting the wall chamber; in section 3.5 further details of a practical nature are discussed).

In this case the choice of the optimum energy occurs by noting it should be avoided that a significant fraction (f) of the neutral energy be deposited on the length Δl_1, namely that

$$\frac{\Delta\Phi}{\Phi_0} = 1 - e^{-\frac{\Delta l_1}{L_{\mathrm{NBI}}}} < f, \tag{3.45}$$

which yields the following condition on the penetration length

$$L_{\mathrm{NBI}} > \frac{\Delta l_1}{|\ln(1 - f)|}. \tag{3.46}$$

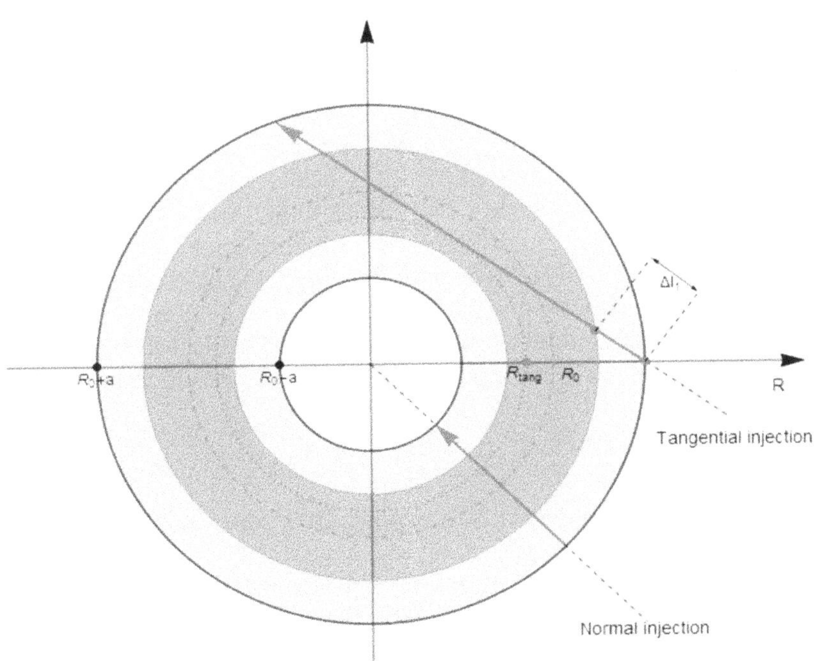

Figure 3.8. Neutral beam injection geometry.

by taking into account that (see figure 3.8)

$$\begin{aligned}
\Delta l_1 &= \frac{1}{2}\Delta l - \Delta l_2, \\
\Delta l_2 &= 2\sqrt{(R_0 + a)^2 - R_T^2}, \\
\Delta l &= \sqrt{\left(R_0 + \frac{a}{2}\right)^2 - R_T^2}, \\
R_T &= R_0 - \frac{a}{2}.
\end{aligned} \tag{3.47}$$

According to the previous discussion the energy of neutrals should satisfy the condition

$$E_{\mathrm{NBI}}[\mathrm{keV}] \geqslant 180 A_{\mathrm{NBI}} N \frac{\Delta l_1}{|\ln(1 - f)|}. \tag{3.48}$$

In the case of ITER-like parameters the above relation simplifies to [8]

$$E_{\mathrm{NBI}}[\mathrm{keV}] \geqslant 90\, A_{\mathrm{NBI}} N \sqrt{R_0 a}. \tag{3.49}$$

There are further technical issues which are to be taken into account in the design of an actual NB heating tool.

For the ITER case, equation (3.49) yields a value of the beam energy ranging around 600 keV, the actual design foresees, however, a larger value of 1 MeV of neutral D-beam. The stopping power of the plasma may be indeed enhanced by other effects like the multi-ionization mechanism, such an effect is heuristically accounted for by introducing a reducing factor $1 + \delta$ in equations (3.43), (3.48) and (3.49). This contribution may provide an increment of the cross section in the order of tens of percent. In the past, increments of a factor two have been observed for 350 keV hydrogen beam in JT-60U Tokamak ($\delta \approx 0.8$–1.05), in accordance with the theoretical predictions [8].

The amount of power associated with NBI heating devices is of the order of tens of MW. In the case of ITER, two different NBs are foreseen to be injected inside the plasma (deuterium 1 MeV/40 A, hydrogen 870 keV/46 A) and more than 16 MW per beam is planned to be deposited inside the plasma.

It is evident that the amount of power which has been lost in the NB injection process is not negligible, it is the complementary part of equation (3.38)

$$P_{\mathrm{lost}} = \frac{(1 - \eta_{\mathrm{neutr}})}{\eta_{\mathrm{neutr}}} P_{\mathrm{neutr}}, \tag{3.50}$$

it is sent to the beam dump and it is to be recovered, for example by powering the electrostatic accelerator.

Before closing this short description on the neutral beam plasma heating devices, we would like to complete the discussion by stressing the concept of critical energy.

It is evident that the slowing down process occurs because inside the plasma is a consequence of the Coulomb interactions induced after the neutral ionization.

A full understanding of the process requires the analysis of the neutral beam dynamics inside the plasma. The most appropriate tool is, within this context, the use of codes based on Fokker–Planck (FP) equations. Their detailed discussion goes beyond the scope of this section and therefore we limit our discussion to the physics emerging from these studies.

The FP dynamics is dominated by two mechanisms associated with:

1. diffusion in velocity space;
2. dynamical friction determining a deceleration of the distribution function.

Within this framework the concept of critical energy E_c plays a discriminating role, such a value is specified by

$$E_c \cong 14.8 A_{\mathrm{NBI}} T_e R, \tag{3.51}$$

with

$$R = \left[\sum_i \frac{n_i}{n_e} \frac{Z_i^2}{A_i} \right]^{2/3},$$

being A_i the mass number of the ion plasma.

Below this value the process is dominated by diffusion and above the dynamical friction (or drag) plays the central role.

The differential equation ruling the energy loss per unit length of a particle with initial energy E_0 is specified below

$$\frac{dE}{dx} = -\frac{\alpha}{E} - \beta \sqrt{E}, \tag{3.52}$$

where α, β are defined in such a way that

$$E_c = \left(\frac{\alpha}{\beta} \right)^{2/3}. \tag{3.53}$$

Equation (3.52) has been derived under the assumption that the injected particle energy is larger than the average energy of the plasma ions and less than that of plasma electrons. The explicit form of the α, β parameters can be found in [7, 9].

The first term in equation (3.52) accounts for the energy lost to ions while the second is for that lost to electrons. The first dominates the second if $E < E_c$ and vice-versa.

A convenient non-dimensional form of equation (3.52) is

$$\frac{y\, dy}{1 + y^{\frac{3}{2}}} = -d\xi, \tag{3.54}$$

where

$$\xi = \frac{x}{l_c}, \; l_c = \frac{E_c^2}{\alpha}. \tag{3.55}$$

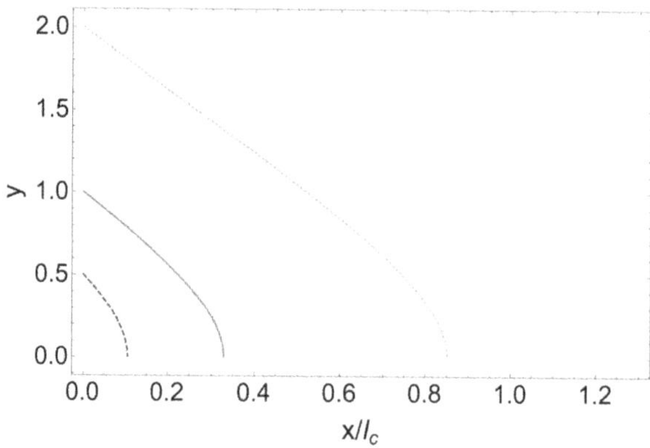

Figure 3.9. Energy loss versus x/l_c for $E_0/E_c = 2$ blue-dashed, 1 red-continuous, 0.5 green-dotted.

The behavior of the NBI energy loss is shown in figure 3.9, where we have reported $y = E/E_c$ for different values of y_0 versus ξ. The plots show that for $y_0 > 1$, the energy loss behavior is mostly linear.

The dependence of the energy loss time is straightforwardly derived by noting that, with the energy being a function of the propagation coordinate x, the relevant time derivative yields

$$\frac{dE(x)}{dt} = \frac{dx}{dt}\frac{dE}{dx} = -\sqrt{\frac{2E}{m}}\left[\frac{\alpha}{E} + \beta\sqrt{E}\right], \tag{3.56}$$

where

$$\frac{dx}{dt} = \sqrt{\frac{2E}{m}}, \tag{3.57}$$

which in terms of non-dimensional variables is written as

$$\frac{\sqrt{y}\,dy}{1 + y^{3/2}} = -d\tau, \tag{3.58}$$

where

$$\tau = \frac{t}{t_s},\ t_s = \frac{l_c}{v_c},\ v_c = \sqrt{\frac{2E_c}{m}}. \tag{3.59}$$

From equation (3.58) we find that the energy of the fast injected particle is transferred to plasma in a time interval

$$\Delta t = \frac{l_c}{v_c}\int_0^{y_0} \frac{\sqrt{y}\,dy}{1 + y^{\frac{3}{2}}} = \frac{2}{3}t_s \ln\left(1 + y_0^{3/2}\right). \tag{3.60}$$

This is a relation of practical interest, it provides the time of energy deposition inside the plasma in terms of the Spitzer time t_s, specified by

$$t_s[s] \cong 6.28 \cdot 10^{14} \frac{A_{\mathrm{NBI}} T_e[eV]^{3/2}}{Z^2 n_e[m^{-3}] \ln(\Lambda)}, \tag{3.61}$$
$$\ln(\Lambda) \equiv \text{Coulomb Logarithm} \cong 17.$$

Later, in this section and in section 3.5, we address further considerations of a practical nature. Let us, however, preliminarily note that, if we limit ourselves to plasma densities not larger than $5 \cdot 10^{19}\mathrm{m}^{-3}$ and $E_0 \cong 100$ keV, Δt is not larger than 100 ms which is small if compared with the particle confinement time; in section 3.5 we discuss more carefully these technical issues.

We have already mentioned that according to whether $E_{\mathrm{NBI}} > E_c$ or $E_{\mathrm{NBI}} < E_c$ the energy is lost towards ions or electrons, respectively. This is just a qualitative observation and a quantitative statement is in order.

This can be easily achieved by noting that from equation (3.56) we find (we use dimensional quantities, the Spitzer time is defined in a different way but easily reconciled with that reported in equation (3.61))

$$\frac{dE}{dt} = -\frac{E}{t_s}\left[\left(\frac{E_c}{E}\right)^{3/2} + 1\right], \tag{3.62}$$

with

$$t_s = \frac{1}{\beta}\sqrt{\frac{m}{2}}.$$

The first term is the power lost to the ions

$$P_i = -\frac{E}{t_s}\left[\left(\frac{E_c}{E}\right)^{3/2}\right]. \tag{3.63}$$

The use of equation (3.62) eventually yields

$$\left[\left(\frac{E_c}{E}\right)^{3/2} + 1\right]^{-1} \frac{dE}{dt} = -\frac{E}{t_s}, \tag{3.64}$$

and therefore

$$P_i = -\left(\frac{E_c}{E}\right)^{3/2}\left[\left(\frac{E_c}{E}\right)^{3/2} + 1\right]^{-1} \frac{dE}{dt}. \tag{3.65}$$

We can accordingly evaluate the fraction of energy flowing from fast particles to ions as

$$\frac{1}{E_0}\int_0^t P_i dt' = \frac{1}{E_0}\int_0^{E_0} \frac{E_c^{3/2}}{E_c^{3/2} + E^{3/2}} dE, \tag{3.66}$$

which, in dimensionless unit reads

$$G_i(y_0) = \frac{1}{y_0}\int_0^{y_0} \frac{dy'}{1 + y'^{3/2}}. \tag{3.67}$$

It goes by itself that the energy lost to electrons is just complementary to $G_i(y_0)$, namely

$$G_i(y_0) + G_e(y_0) = 1, \tag{3.68}$$

when $y_0 = 1$, almost 75% of the fast ions are transferred to the plasma ions. The fraction of energy going to electrons is just $1 - G_i(\alpha)$. It is accordingly evident that for $y_0 < 1$ the energy is transferred to ions, while for larger values it is transferred to electrons.

Just to fix some numbers, we note that a neutral D beam with energy 125 keV, injected to a 5 keV deuterium plasma (with $E_c \cong 148$ keV) delivers to ions a fraction of its energy a factor of two larger than that given to electrons. The heating by the fusion born α-particles (3.5 MeV) is delivered to electron plasma.

In conclusion, NBI exhibits significant advantages, characterized by efficient heating and high deposited power (40 MW on ITER, 24 MW on JET, etc) on the other hand, large beam energies are necessary to penetrate the plasma and heating is not well localized.

In the forthcoming section we will see how *radio-frequency heating* provides a useful complementary tool.

3.4 Radio frequency plasma heating: a few preliminaries

In the previous section we have summarized the physical mechanisms underlying the process of plasma heating by neutral beam injection. We have noted that the process induces heating of both ions and electrons and its role is not different from that played by α-particles.

In this section we describe a further auxiliary heating tool associated with the injection of electromagnetic radiation, which excites plasma waves, eventually damped inside the plasma itself [10].

The working principle is straightforward and according to figure 3.10 we can foresee the following steps:

1. An antenna launches an electromagnetic (e.m.) signal at the plasma edge.
2. The energy of the e.m. wave is transferred to some resonant particles.
3. The acquired energy is lost to plasma via collision redistributions.

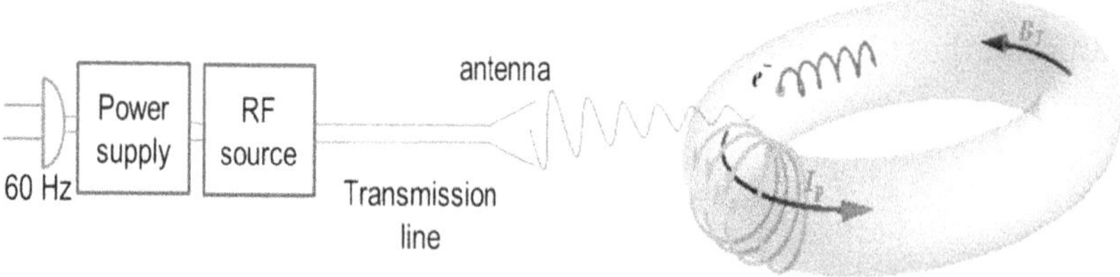

Figure 3.10. Wave launching in plasma and heating.

The tools in our hands to understand the RF heating in plasma are the Maxwell equations for the fields and the Lorentz force equation to study the electrons/ions induced motion.

Before getting into the computational details, it is worth specifying the reference parameters useful to fix the different operating regimes.

The typical parameters of an RF heating wave are given below [11]

$$\text{Electric Field} \cong 20\ \text{kV m}^{-1}$$
$$\text{Magnetic Field} \cong 10^{-3}\ \text{T}$$
$$\text{Frequency} \cong 10 - -200\ \text{GHz}$$

Assuming that inside the chamber the electrons/ions are moving in a magnetic field of few Tesla (say 3 just to fix ideas) it is evident that the perturbation induced on their trajectories by RF electric and magnetic fields is irrelevant.

The wave, passing inside the plasma, accelerates electrons which collide with the environment and dissipate the acquired energy. This mechanism, as we have already learned is not efficient, since it decreases with $T^{-3/2}$. In highly energetic plasma a different mechanism arises, it is the so called resonant absorption, which occurs when the wave matches some plasma frequency. The process is collision-less and therefore is not limited by restriction due to the plasma resistivity reduction with the temperature.

Talking about 'light' absorption implies considerations on the refractive index $n = ck/\omega$ (with k, ω the wave vector and the angular frequency, respectively) the wave behavior inside the plasma depends on its value

$n = 0$ implies wave reflection,

$n < 0$ evanescent propagation (decaying mode),

$n \to \infty$ aborption at resonance.

The prerequisite for an RF heating to be successful is that the wave, proceeding from antenna, enters the plasma without experiencing any appreciable attenuation inside, reaches a region where resonance occurs and there disperses its power.

In view of the fact that the wave produces a tiny perturbation on the plasma environment, the cold plasma approximation can be used, except in presence of the resonance, where higher temperature contributions should be included.

We will therefore treat two distinct cases regarding the regimes of 'cold' and 'hot' magnetized plasma. The adjective cold and hot refer to the electron and ion thermal velocities compared with the phase velocity of the injected wave.

3.4.1 Cold magnetized plasma

A magnetized plasma is said to be cold when electron and thermal velocities are lower than the propagating wave phase velocity.

We need now a little math to proceed, we require therefore a further effort from the reader. The calculations are sketched below.

The starting point of our discussion is the 'rotor' pair of Maxwell equations

$$\vec{\nabla} \times \vec{E} = -\frac{\partial \vec{B}}{\partial t},$$
$$\vec{\nabla} \times \vec{B} = \mu_0 \vec{J} + \frac{1}{c^2}\frac{\partial \vec{E}}{\partial t}. \tag{3.69}$$

If we assume a space–time dependence fixed by the plane wave oscillating behavior,

$$\begin{pmatrix} E \\ B \end{pmatrix} \propto e^{i\left(\vec{k}\cdot\vec{r}-\omega t\right)}, \tag{3.70}$$

after replacing $\vec{\nabla}\times \rightarrow i\vec{k}\times$ and $\frac{\partial}{\partial t} \rightarrow -i\omega$ they can be cast in the form

$$\vec{B} = \frac{\vec{k}}{\omega} \times \vec{E},$$
$$\vec{k} \times \vec{B} = -i\mu_0 \vec{J} - \frac{\omega}{c^2}\vec{E}. \tag{3.71}$$

Considering equations (3.71) as a system with unknown $\vec{E}$ and $\vec{B}$ we write the relevant solution in the slightly unorthodox form

$$\begin{pmatrix} \vec{E} \\ \vec{B} \end{pmatrix} = \begin{pmatrix} \vec{k}\times & -\omega \\ \frac{\omega}{c^2} & \vec{k}\times \end{pmatrix}^{-1} \begin{pmatrix} 0 \\ -i\,\mu_0 \vec{J} \end{pmatrix}. \tag{3.72}$$

We therefore find[3]

$$\left([\vec{k}]^2 + \left(\frac{\omega}{c}\right)^2\right)\vec{E} = -i\omega\mu_0 \vec{J},$$
$$[\vec{k}] = \vec{k}\times, \tag{3.73}$$

and after introducing the conductivity tensor $\bar{\bar{\sigma}}$ as[4]

$$\vec{J} = \bar{\bar{\sigma}} \cdot \vec{E}, \tag{3.74}$$

we obtain the following equation specifying the vector electric field

$$\left([\vec{k}]^2 + \left(\frac{\omega}{c}\right)^2 + i\omega\mu_0\bar{\bar{\sigma}}\right)\vec{E} = 0. \tag{3.75}$$

[3] We recall that the inverse of a $2x2$ matrix

$$A = \begin{pmatrix} ab \\ cd \end{pmatrix}^{-1} \text{ is } A^{-1} = \frac{1}{D}\begin{pmatrix} d & -b \\ -c & a \end{pmatrix}, \text{ where } D = ad - bc.$$

The term in square brackets on the lhs of equation (3.73) is therefore the determinant of the matrix in equation (3.72).

[4] We have used an incorrect notation, employing either tensors and vectors, this is just to speed up the computational procedure. The correctness of the formalism will be restored by the use of the dyadic formalism.

After noting that

$$[\vec{k}]^2\vec{E} = \vec{k}(\vec{k}\cdot\vec{E}) - (\vec{k}\cdot\vec{k})\,\vec{E}, \tag{3.76}$$

and adopting the dyadic notation, we eventually write

$$\left(\bar{\bar{k}}_2 - k^2\bar{\bar{I}} + \left(\frac{\omega}{c}\right)^2\bar{\bar{\varepsilon}}\right)\vec{E} = 0, \tag{3.77}$$

where $\vec{E}$ is understood as a three-component column vector, $\bar{\bar{I}}$ is the unit matrix, $\bar{\bar{k}}_2$ is the dyadic product

$$\bar{\bar{k}}_2 = \begin{pmatrix} k_x \\ k_y \\ k_z \end{pmatrix}\begin{pmatrix} k_x & k_y & k_z \end{pmatrix} = \begin{pmatrix} k_x^2 & k_xk_y & k_xk_z \\ k_xk_y & k_y^2 & k_yk_z \\ k_xk_z & k_yk_z & k_z^2 \end{pmatrix}, \tag{3.78}$$

and $\bar{\bar{\varepsilon}}$ is the dielectric tensor

$$\bar{\bar{\varepsilon}} = \bar{\bar{I}} + i\,\frac{\bar{\bar{\sigma}}}{\omega\varepsilon_0}. \tag{3.79}$$

The missing piece, to draw physical consequences from equation (3.77), is just the conductivity tensor. The relevant evaluation proceeds by determining the electron and ion equations of motion in terms of the electric field and evaluating the associated current density.

The strategy we follow is outlined below:

(a) We write the electron and ion equation of motion in the MHD form

$$\begin{aligned} m_e\left(\frac{\partial}{\partial t}\vec{v}_e + \vec{v}_e\cdot\vec{\nabla}\vec{v}_e\right) &= -e(\vec{E} + \vec{v}_e\times\vec{B}), \\ m_i\left(\frac{\partial}{\partial t}\vec{v}_i + \vec{v}_i\cdot\vec{\nabla}\vec{v}_i\right) &= e(\vec{E} + \vec{v}_i\times\vec{B}). \end{aligned} \tag{3.80}$$

(b) Once the electrons and ions velocities of all the species are determined in terms of $\vec{E}$, we write the current density as

$$\begin{aligned} \vec{J} &= \sum_s n_s\, q_s v_s, \\ s &= \{i_n, e\}, \\ qe &= -e,\ qi_n = e, \end{aligned} \tag{3.81}$$

and the definition of the conductivity tensor follows from (3.74).

In order to solve equations (3.80) we need some approximations to avoid the handling of non-linearities. We linearize around the plasma stationary conditions

$$\begin{aligned} & n_e = n_i \ = n_0, \\ & \vec{v}_e = \vec{v}_i = \vec{0}, \ \vec{E} = \vec{0}, \\ & \vec{B} = B_0\hat{b} = (0, \ 0, \ B_0). \end{aligned} \tag{3.82}$$

The linear form of equation (3.80) reads

$$m \frac{\partial}{\partial t}\vec{v}^{(1)} = q(\vec{E}^{(1)} + \vec{v}^{(1)} \times \vec{B}_0), \tag{3.83}$$

where

$$\begin{aligned} & \vec{v} \approx \vec{v}^{(0)} + \vec{v}^{(1)}, \\ & \vec{E} \approx \vec{E}^{(0)} + \vec{E}^{(1)}, \end{aligned} \tag{3.84}$$

and m, q are the mass and ion/electron charge, respectively. The solution is derived in a form suitable for our purposes, after replacing the derivative operator on the lhs with $-i\omega$ obtaining

$$\begin{pmatrix} -i\omega & -\Omega_c & 0 \\ \Omega_c & -i\omega & 0 \\ 0 & 0 & -i\omega \end{pmatrix} \begin{pmatrix} v_x^{(1)} \\ v_y^{(1)} \\ v_z^{(1)} \end{pmatrix} = \frac{q}{m} \begin{pmatrix} E_x^{(1)} \\ E_y^{(1)} \\ E_z^{(1)} \end{pmatrix}, \tag{3.85}$$

and eventually finding

$$\begin{pmatrix} v_x^{(1)} \\ v_y^{(1)} \\ v_z^{(1)} \end{pmatrix} = \frac{q}{m} \begin{pmatrix} i\dfrac{\omega}{\omega^2 - \Omega_c^2} & -\dfrac{\Omega_c}{\omega^2 - \Omega_c^2} & 0 \\ \dfrac{\Omega_c}{\omega^2 - \Omega_c^2} & i\dfrac{\omega}{\omega^2 - \Omega_c^2} & 0 \\ 0 & 0 & \dfrac{i}{\omega} \end{pmatrix} \begin{pmatrix} E_x^{(1)} \\ E_y^{(1)} \\ E_z^{(1)} \end{pmatrix}, \tag{3.86}$$

where $\Omega_c = q|\vec{B}_0|/m$.

The problem of getting the conductivity tensor is virtually solved. The definition of the current density in equation (3.81) yields

$$\begin{pmatrix} J_x^{(1)} \\ J_y^{(1)} \\ J_z^{(1)} \end{pmatrix} = \begin{pmatrix} \sigma_{x,x} & \sigma_{xy} & 0 \\ \sigma_{yx} & \sigma_{yy} & 0 \\ 0 & 0 & \sigma_{zz} \end{pmatrix} \begin{pmatrix} E_x^{(1)} \\ E_y^{(1)} \\ E_z^{(1)} \end{pmatrix}, \tag{3.87}$$

with the matrix components being specified by

$$\begin{aligned} & \sigma_{xx} = \sigma_{yy} = i\varepsilon_0 \sum_s \frac{\omega\, \Omega_{p,\,s}^2}{\omega^2 - \Omega_{c,\,s}^2}, \\ & \sigma_{xy} = -\, \sigma_{y,x} = \varepsilon_0 \sum_s \frac{\Omega_{p,\,s}^2 \Omega_{c,s}}{\omega^2 - \Omega_{c,\,s}^2}, \\ & \sigma_{zz} = i\varepsilon_0 \sum_s \frac{\Omega_{p,\,s}^2}{\omega}, \end{aligned} \tag{3.88}$$

where the sum is extended to all the species (s) (electron and ions with their own mass and charge).

We are ready to write the dispersion relation in equation (3.77) as [12]

$$\begin{pmatrix} S - n_{\|}^2 & -iD & n_{\|}n_{\perp} \\ iD & S - n^2 & 0 \\ n_{\|}n_{\perp} & 0 & P - n_{\perp}^2 \end{pmatrix} \begin{pmatrix} E_x \\ E_y \\ E_z \end{pmatrix} = 0, \tag{3.89}$$

where

$$\begin{aligned} \vec{n} &= \frac{c\,\vec{k}}{\omega}, \\ \vec{k} &\equiv \begin{pmatrix} k\sin(\vartheta) \\ 0 \\ k\cos(\vartheta) \end{pmatrix} = \begin{pmatrix} k_{\perp} \\ 0 \\ k_{\|} \end{pmatrix}, \\ S &= 1 - \frac{1}{i\,\varepsilon_0\omega}\sigma_{xx}, \\ D &= -\frac{1}{i\varepsilon_0\omega}\sigma_{xy}, \\ P &= 1 - \frac{1}{i\,\varepsilon_0\omega}\sigma_{zz}. \end{aligned} \tag{3.90}$$

The subscripts parallel (||) and orthogonal ($\perp$) refer to the relative direction of the wave vector with the magnetic field.

In the previous discussion we did not mention any involvement of the velocity distribution function f_s and of the associated averages, for the evaluation of the plasma dielectric tensor. Before closing this section, we outline an alternative derivation accomplished employing a perturbative solution of the Vlasov equation (equation (1.92)), which, at the lowest order, reads

$$\frac{\partial f_s^{(1)}}{\partial t} + \vec{v}\cdot\vec{\nabla}_r f_s^{(1)} + \frac{Z_s e}{m_s}(\vec{v}\times\vec{B})\cdot\vec{\nabla}_v f_s^{(1)} = -\frac{Z_s e}{m_s}\left(\vec{E}^{(1)} + \vec{v}\times\vec{B}^{(1)}\right)\cdot\vec{\nabla}_v f_s^{(0)}, \tag{3.91}$$

where

$$f \approx f_s^{(0)} + f_s^{(1)}. \tag{3.92}$$

where the superscript '0' denotes the unperturbed part.

The current density variation associated with $f_s^{(1)}$ is simply given by

$$J^{(1)} = e\sum_s Z_s \int f_s^{(1)}\vec{v}\,d\vec{v}. \tag{3.93}$$

The perturbed term $f_s^{(1)}$ exhibits the same dependence on time and space of the electric and magnetic fields (equation (3.70)) and accordingly

$$\frac{\partial f_s^{(1)}}{\partial t} = -\, i\omega f_s^{(1)},$$
$$\vec{v} \cdot \vec{\nabla}_r f_s^{(1)} = i\vec{k} \cdot v_{t,s} f_s^{(1)}. \tag{3.94}$$

with $v_{t,s}$ the thermal velocity of the species s. In the cold plasma regime, for which $v_{t,s} \ll \omega/k$, the space gradient can be neglected with respect to the time derivative. Equation (3.91) can therefore be written as

$$\left[-i\omega + \frac{Z_s e}{m_s}(\vec{v} \times \vec{B}) \cdot \vec{\nabla}_v\right] f_s^{(1)} = -\frac{Z_s e}{m_s}\left(\vec{E}^{(1)} + \vec{v} \times \vec{B}^{(1)}\right) \cdot \vec{\nabla}_v f_s^{(0)}. \tag{3.95}$$

Furthermore, using the first of equation (3.71) we can write the cross product on the rhs in the form

$$\vec{v} \times \vec{B}^{(1)} = \frac{\vec{v} \times \left(\vec{k} \times \vec{E}^{(1)}\right)}{i\omega} \simeq \frac{iv_{t,s}}{v_p}\left(\hat{v} \times \hat{k} \times \vec{E}^{(1)}\right), \tag{3.96}$$

which can be neglected since in the cold plasma regime $v_{t,s}/v_p \ll 1$.

Therefore, the linearized Vlasov equation reduces to

$$\left[-i\omega + \frac{Z_s e}{m_s}\left(\vec{v} \times \vec{B}\right) \cdot \vec{\nabla}_v\right] f_s^{(1)} = -\frac{Z_s e}{m_s}\vec{E}^{(1)} \cdot \vec{\nabla}_v f_s^{(0)}, \tag{3.97}$$

The previous equation allows the computation of the perturbed part in terms of the $E^{(1)}$ components of the electric field. The identity (3.97) is a non-homogeneous differential equation, which can be solved by the use of different methods.

We first note that

$$\frac{Z_s e}{m_s}\left(\vec{v} \times \vec{B}\right) \cdot \vec{\nabla}_v = \Omega_{c,s}\left(v_y \frac{\partial}{\partial v_x} - v_x \frac{\partial}{\partial v_y}\right), \tag{3.98}$$

which has the transparent physical meaning of an operator yielding a coupling of the transverse velocities[5].

We can therefore write

$$\left[-i\omega + \Omega_{c,s}\left(v_y \frac{\partial}{\partial v_x} - v_x \frac{\partial}{\partial v_y}\right)\right] f_s^{(1)} = -\frac{Z_s e}{m_s}\left(E_x^{(1)} \frac{\partial}{\partial v_x} + E_y^{(1)} \frac{\partial}{\partial v_y} + E_z^{(1)} \frac{\partial}{\partial v_z}\right) f_s^{(0)}, \tag{3.99}$$

which is just a different statement of equation (3.85).

The solution of equation (3.99) can be achieved by standard means (Laplace transform, e.g.), however, for our purposes since we are interested in the integral

[5] It is easily checked that, defining the operators $\hat{L}_+ = v_y \frac{\partial}{\partial v_x}$, $\hat{L}_- = v_x \frac{\partial}{\partial v_y}$, $\hat{L}_3 = v_y \frac{\partial}{\partial v_y} - v_x \frac{\partial}{\partial v_x}$, we obtain the commutation rules of angular momentum like operators $[\hat{L}_+, \hat{L}_-] = \hat{L}_3$, $[\hat{L}_+, \hat{L}_3] = -\hat{L}_+$, $[\hat{L}_-, \hat{L}_3] = \hat{L}_-$.

over the velocities in equation (3.93) it will be sufficient to multiply both sides of equation (3.99) by $\vec{v}$ and get (for the x velocity component)

$$\left[-i\omega v_x + \Omega_{c,s}\left(v_y v_x \frac{\partial}{\partial v_x} - v_x^2 \frac{\partial}{\partial v_y}\right)\right] f_s^{(1)} = -\frac{Z_s e}{m_s}\left(E_x^{(1)} v_x \frac{\partial}{\partial v_x} + E_y^{(1)} v_x \frac{\partial}{\partial v_y} + E_z^{(1)} v_x \frac{\partial}{\partial v_z}\right) f_s^{(0)}. \tag{3.100}$$

After integrating both sides on the velocities and by using the integration by parts, we find

$$\left[-i\omega v_x^{(1)} - \Omega_{c,s} v_y^{(1)}\right] = +\frac{Z_s e}{m_s} E_x^{(1)}, \tag{3.101}$$

with

$$v_x^{(1)} = \int v_x f^{(1)} d^3 v. \tag{3.102}$$

The procedure can be repeated for the other components $v_y v_z$ deriving the matrix in equation (3.85).

A compact notation of the conductivity tensor $\bar{\bar{\sigma}}$ in terms of the dyad has been given in the form [13]

$$\bar{\bar{\sigma}} = i\varepsilon_0 \omega \sum_s \left[\frac{\Omega_{p,s}^2}{\omega^2 - \Omega_{c,s}^2}(\bar{\bar{I}} - \hat{b}\hat{b}) + \frac{\Omega_{p,s}^2}{\omega^2}\hat{b}\hat{b} - \frac{i\Omega_{p,s}^2 \Omega_{c,s}}{\omega(\omega^2 - \Omega_{c,s}^2)}\hat{b} \times \bar{\bar{I}}\right], \tag{3.103}$$

which is easily shown to be equivalent to the equation (3.85) as checked by exploiting the rules reported in appendix A.

In the forthcoming section we use the previously outlined procedure to discuss the hot plasma regime.

3.4.2 Hot magnetized plasma

The hot plasma regime can be treated by handling equation (3.91), without neglecting the spatial gradient and the cross product contributions, thus writing

$$-i\omega f_s^{(1)} + i\vec{k} \cdot \vec{v} f_s^{(1)} + \frac{Z_s e}{m_s}\left(\vec{v} \times \vec{B}\right) \cdot \nabla_v f_s^{(1)} = -\frac{Z_s e}{m_s}\left(\vec{E}^{(1)} + \frac{1}{\omega}\vec{v} \times \left(\vec{k} \times \vec{E}^{(1)}\right)\right) \cdot \nabla_v f_s^{(0)}. \tag{3.104}$$

We can get a more transparent meaning by the use of the orthonormal vector basis ($\hat{x} = \hat{k}_\perp$, $\hat{y} = \hat{b} \times \hat{k}_\perp$, $\hat{b}$) which allows us to express equation (3.104) as

$$\left(-i\omega + ik_{||}v_{||} + \Omega_{c,s}\frac{\partial}{\partial\phi}\right)\left[f_s^{(1)} + \frac{Z_s ei}{m_s\omega}\left(\frac{\partial f_s^{(0)}}{\partial v_{||}} - \frac{v_{||}}{v_\perp}\frac{\partial f_s^{(0)}}{\partial v_\perp}\right)\hat{b}\cdot\vec{E}^{(1)}\right]\exp(i\lambda_s\sin\phi)=$$
$$= \frac{-Z_s e}{m_s}\left[\frac{\partial f_s^{(0)}}{\partial v_\perp} + \frac{k_{||}}{\omega}\left(v_\perp\frac{\partial f_s^{(0)}}{\partial v_{||}} - v_{||}\frac{\partial f_s^{(0)}}{\partial v_\perp}\right)\right]\exp(i\lambda_s\sin\phi) \tag{3.105}$$
$$\left(\frac{\vec{k}_\perp}{k_\perp}cos\phi - \frac{\hat{b}\times\vec{k}_\perp}{k_\perp}sin\phi + \frac{v_{||}}{v_\perp}\hat{b}\right)\cdot\vec{E}^{(1)},$$

where we have introduced the integration factor in the gyrophase ϕ (see figure 3.11)

$$I_\phi(\lambda_s) = \exp(i\lambda_s\sin\phi), \tag{3.106}$$

where $\lambda_s = k_\perp v_\perp/\Omega_{c,s}$.

Using, furthermore, the following Fourier expansion of the gyrophase contribution of the lhs term in equation (3.105)

$$\left[f_s^{(1)} + \frac{Z_s ei}{m_s\omega}\left(\frac{\partial f_s^{(0)}}{\partial v_{||}} - \frac{v_{||}}{v_\perp}\frac{\partial f_s^{(0)}}{\partial v_\perp}\right)\hat{b}\cdot\vec{E}^{(1)}\right]\exp(i\lambda_s\sin\phi) = \sum_{l=\infty}^{\infty}\tilde{F}_{s,l}\exp(il\sin\phi), \tag{3.107}$$

we can get the explicit expression of the Fourier coefficients $\tilde{F}_{s,l}$ employing the Jacobi–Anger expansion

$$\exp(i\lambda_s\sin\phi) = \sum_{n=-\infty}^{\infty}\exp(in\phi)J_n(\lambda_s), \tag{3.108}$$

where J_n is the first kind cylindrical Bessel function of nth order. The use of the orthogonality properties of the circular functions yields

$$\tilde{F}_{s,l} = -\frac{Z_s ei}{m_s}\frac{1}{\omega - k_{||}v_{||} - l\Omega_{c,s}}\left[\frac{\partial f_s^{(0)}}{\partial v_\perp} + \frac{k_{||}}{\omega}\left(v_\perp\frac{\partial f_s^{(0)}}{\partial v_{||}} - v_{||}\frac{\partial f_s^{(0)}}{\partial v_\perp}\right)\right]\vec{u}_l^*\cdot\vec{E}^{(1)}, \tag{3.109}$$

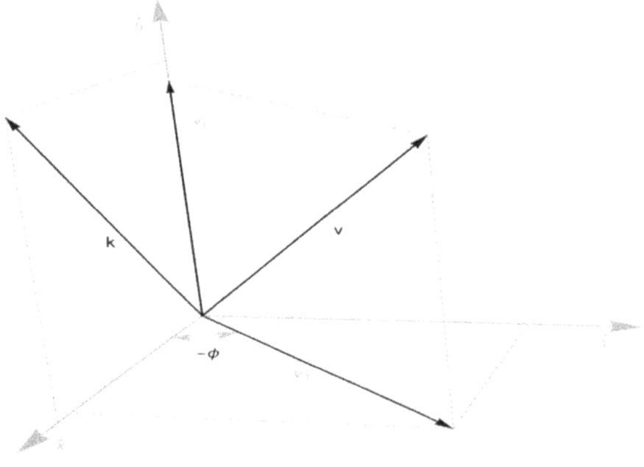

Figure 3.11. Basis for the velocity.

with the complex vector $\vec{u}_l$ given by

$$\vec{u}_l = \frac{J_l(\lambda_s)}{\lambda_s}\hat{k} - iJ_l'(\lambda_s)\hat{b}\times\hat{k} + \frac{v_{\|}}{v_{\perp}}J_l(\lambda_s)\hat{b}, \tag{3.110}$$

and $J_l'(\lambda_s)$ denotes the first derivative of the Bessel function.

The use of the same integration procedure, as before, yields for the dielectric tensor the result reported below

$$\begin{aligned}\bar{\bar{\varepsilon}} = \bar{\bar{I}} + \sum_s \frac{2\pi\Omega_{p,s}^2}{n_s\omega^2}\int_0^\infty dv_{\|}\int_0^\infty dv_{\perp}\Bigg\{\left(v_{\perp}\frac{\partial f_s^{(0)}}{\partial v_{\|}} - v_{\|}\frac{\partial f_s^{(0)}}{\partial v_{\perp}}\right)v_{\|}\hat{b}\hat{b}+ \\ + v_{\perp}^2\left[\frac{\partial f_s^{(0)}}{\partial v_{\perp}} + \frac{k_{\|}}{\omega}\left(v_{\perp}\frac{\partial f_s^{(0)}}{\partial v_{\|}} - v_{\|}\frac{\partial f_s^{(0)}}{\partial v_{\perp}}\right)\right]\sum_{l=-\infty}^{\infty}\frac{\omega\vec{u}_l\vec{u}_l^*}{\omega - k_{\|}v_{\|} - l\Omega_{c,s}}\Bigg\}.\end{aligned} \tag{3.111}$$

The cold plasma approximation is easily regained by noting that in this regime $\lambda_s \ll 1$. Accordingly, keeping the lowest order of the Bessel function expansion

$$J_l(\lambda_s) = \left(\frac{\lambda_s}{2}\right)^l\sum_{r=0}^{\infty}(-1)^r\frac{\left(\frac{\lambda_s}{2}\right)^{2r}}{(l+r)!r!}, \tag{3.112}$$

and retaining the terms u_{-1}, u_0, u_1 we find

$$\begin{aligned}\bar{\bar{\varepsilon}} \simeq \bar{\bar{I}} + \sum_s \frac{2\pi\Omega_{p,s}^2}{n_s\omega^2}\int_0^\infty dv_{\|}\int_0^\infty dv_{\perp} \\ \left\{v_{\perp}v_{\|}\frac{\partial f_s^{(0)}}{\partial v_{\|}}\hat{b}\hat{b} + v_{\perp}^2\frac{\partial f_s^{(0)}}{\partial v_{\perp}}\left(\frac{\omega\vec{u}_{-1}\vec{u}_{-1}^*}{\omega + \Omega_{c,s}} + \frac{\omega\vec{u}_1\vec{u}_1^*}{\omega - \Omega_{c,s}}\right)\right\},\end{aligned} \tag{3.113}$$

which, after using the limits for $\lambda_s \to 0$

$$J_0(\lambda_s) \simeq 1,\ J_{\pm1}(\lambda_s)/(\lambda_s) \simeq \pm1/2,\ J_{\pm1}'(\lambda_s)/(\lambda_s) \simeq \pm1/2, \tag{3.114}$$

yields equation (3.113) in the more compact form

$$\bar{\bar{\varepsilon}} \simeq \varepsilon_{\|}\hat{b}\hat{b} + \varepsilon_{\perp}(\bar{\bar{I}} - \hat{b}\hat{b}) - ig\hat{b}\times\bar{\bar{I}}, \tag{3.115}$$

with

$$\varepsilon_{\perp} = 1 - \sum_s\frac{\Omega_{p,s}^2}{\omega^2 - \Omega_{c,s}^2},\ \varepsilon_{\|} = 1 - \sum_s\frac{\Omega_{p,s}^2}{\omega^2},\ g = -\sum_s\frac{\Omega_{p,s}^2\Omega_{c,s}}{\omega(\omega^2 - \Omega_{c,s}^2)}. \tag{3.116}$$

The use of the equations (3.79), (3.88), and (3.121) and the rules reported in appendix A, allow to establish the full correspondence with the results mentioned in equation (3.87).

In many cases of interest for magnetized plasmas, the velocity distribution for motion perpendicular to $\vec{B}_0$ may be characterized as a non-relativistic Maxwellian; assuming a shifted distribution function for parallel motion we have

$$f_s(v_\perp, v_{||}) = n_s F_\perp^s(v_\perp) F_z^s(v_{||}), \tag{3.117}$$

where

$$\begin{aligned} F_\perp^s(v_{s,\perp}) &= \frac{m_s}{2\pi T_\perp^s} \exp\left(-\frac{m_s v_{s,\perp}^2}{2T_\perp^s}\right), \\ F_{||}^s(v_{||}) &= \left(\frac{m_s}{2\pi T_z^s}\right)^{1/2} \exp\left(-\frac{m_s(v_{s,||} - V_s)^2}{2T_z^s}\right), \end{aligned} \tag{3.118}$$

where $V_s = <v_{s,||}>$.

If we carry out the integration in equation (3.117) with the velocity distributions (3.117)–(3.118), we end up with the following compact expression for the dielectric [14]

$$\bar{\bar{\varepsilon}} = \bar{\bar{I}} + \sum_s \frac{\Omega_{p,s}^2}{\omega^2}\left(\sum_l \left(\zeta_{0,s}\, Z(\zeta_{l,s}) - \left(1 - \frac{1}{\lambda_{T,s}}\right)\left(1 + \zeta_{l,s}\, Z(\zeta_{l,s})\right)\right) e^{-b}\bar{\bar{X}}_{l,s} + 2\eta_{0,s}^2 \lambda_{T,s}\bar{\bar{L}}\right). \tag{3.119}$$

In this, the following definitions have been introduced (see appendix B for more details)

$$\bar{\bar{X}}_{l,s} = \begin{pmatrix} \dfrac{l^2 I_l(b_s)}{b_s} & il[I_l'(b_s) - I_l(b_s)] & -\sqrt{\dfrac{2\lambda_{T,s}}{b_s}}\eta_{l,s} l I_l(b_s) \\ -il[I_l'(b_s) - I_l(b_s)] & \left[\dfrac{l^2}{b_s} + 2b_s\right] I_l(b_s) - 2b_s I_l'(b_s) & i\sqrt{2\lambda_{T,s} b_s}\,\eta_{l,s}[I_l'(b_s) - I_l(b_s)] \\ -\sqrt{\dfrac{2\lambda_{T,s}}{b_s}}\eta_{l,s} l I_l(b_s) & -i\sqrt{2\lambda_{T,s} b_s}\,\eta_{l,s}[I_l'(b_s) - I_l(b_s)] & 2\lambda_{T,s}\eta_{l,s}^2 I_l(b_s) \end{pmatrix}, \tag{3.120}$$

$$\zeta_{l,s} = \frac{\omega - k_{||}V_s + l\Omega_{c,s}}{2^{1/2} k_z v_{T_z^s}},\ \lambda_{T,s} = \frac{T_{z,s}}{T_{\perp,s}},\ b_s = \left(\frac{k_{||} v_{T_{\perp,s}}}{\Omega_{c,s}}\right)^2,$$

$$\eta_{l,s} = \frac{\omega + l\Omega_{c,s}}{2^{1/2} k_z v_{T_z^s}},\ v_{T_{z,s}}^2 = \frac{T_{z,s}}{m_s},\ v_{T_{\perp,s}}^2 = \frac{T_{\perp,s}}{m_s},$$

and $Z(\zeta)$ denoting the plasma dispersion function

$$Z(\zeta) \equiv \frac{1}{\sqrt{\pi}} \int_{-\infty}^{\infty} \frac{\exp(-\beta^2)}{\beta - \zeta} d\beta, \tag{3.121}$$

with $\bar{\bar{L}}$ the matrix

$$\bar{\bar{L}} = \begin{pmatrix} 0 & 0 & 0 \\ 0 & 0 & 0 \\ 0 & 0 & 1 \end{pmatrix}. \tag{3.122}$$

The wave equation for a homogenous plasma is given by equation (3.77) and with the assumption of $n_y = 0$ (see equation (3.90)) we get the following matrix form

$$\begin{pmatrix} \varepsilon_{xx} - n_z^2 & \varepsilon_{xy} & \varepsilon_{xz} n_x n_z \\ \varepsilon_{yx} & \varepsilon_{yy} - n_x^2 - n_z^2 & \varepsilon_{yz} \\ \varepsilon_{zx} + n_z n_x & \varepsilon_{zy} & \varepsilon_{zz} - n_x^2 \end{pmatrix} \begin{pmatrix} E_x \\ E_y \\ E_z \end{pmatrix} = 0. \tag{3.123}$$

The condition that there be non-trivial solutions from this vector equation is that the determinant of the matrix 3 × 3 be zero. This condition provides the dispersion relation for the homogeneous plasma system.

We have so far established the general mathematical framework, which has largely benefited from the use of the dyadic formalism. This mathematical technique has been summarized in appendix A. In the forthcoming sections we will utilize these results to describe more adequately the plasma–wave interaction in the proper physical environment.

3.5 The physics of radio frequency plasma heating

The discussion of the previous section has been devoted to the mathematical description of the wave propagation inside the plasma and the matter we discussed did not include any technological issue associated with the wavelength, power, launching system and so on.

Regarding this last item, we just note that electromagnetic waves are sent inside the plasma from its external contour and move into the plasma towards a fixed target region where they are absorbed by ***collision-less damping***. If the target is located at the plasma core it will produce ***heating***, if, at the outer board, the wave absorption determines ***current drive***. The terms in bold will be discussed in the forthcoming section.

In figure 3.10 we have reported a schematic view of the heating/current drive device, which consists of a power supply, exciting an RF source, which launches the wave to the plasma via a transmission path connected with an antenna.

The heating/drive performance depends on the operating frequency.

It is evident that to be effective the frequencies are chosen to be resonant with the plasma characteristic frequencies, namely $\Omega_{c,e}/(2\pi) \cong 28$ GHz T^{-1} for ECRH (electron cyclotron heating) and $\Omega_{c,i}/(2\pi) \cong 15$ MHz T^{-1} for ion cyclotron heating (ICH).

LHCD (lower hybrid current drive) frequencies, namely a mixture of cyclotron (electron and ion) and plasma frequencies (see below) are located in an intermediate range on the order of few GHz.

The tools corresponding to the quoted frequencies regions are listed below

High–power vacuum tubes	$f < 100$ MHz
klystrons (microwaves)	$f \sim 1$–10 GHz
Gyrotrons (sub–millimeter waves)	$f \sim 10$–200 GHz
CARM (Cyclotron auto–resonance maser)	$f > 200$ GHz

Even though not explicitly stated, the crucial physical mechanisms, underlying the heating processes we are interested in, are the wave propagation inside the plasma and the relevant interaction with the constituting charged particles.

The wave propagation is characterized (among many other things) by the ***phase*** and ***group*** velocities.

The first is understood as the velocity at which an observer should move with the wave in order to 'see' the phase of the wave constant.

The phase wave being

$$\phi = \omega t - \vec{k} \cdot \vec{r}, \tag{3.124}$$

the phase velocity $\vec{v}_p$ is found by setting

$$\frac{d\phi}{dt} = \omega - \vec{k} \cdot \vec{v}_p = 0. \tag{3.125}$$

Which yields for $\vec{v}_p$

$$\vec{v}_p = \frac{\omega}{k}\vec{e}_k, \tag{3.126}$$

accordingly the phase velocity moves along the direction of the wave vector, with a magnitude ω/k.

There is a kind of refrain warning that v_p may be larger than light velocity, without violating relativity and this is due to the fact that 'the phase velocity of a single monochromatic wave does not represent the propagation of any physical quantity like energy'. Notwithstanding, the relevant role in processes like the collision-less damping is crucial, as we will see later in this chapter.

Along with the phase velocity, a further quantity of pivotal importance is the group velocity v_g, associated with the transport of physical quantities and thus of information.

In order to clarify the meaning of this quantity, we consider the wave resulting from the sum of two monochromatic plane waves,

$$u(x, t) = A_0 Re[e^{i(\omega_1 t - k_1 x)} + e^{i(\omega_2 t - k_2 x)}], \tag{3.127}$$

whose square modulus yields

$$|u(x,\, t)|^2 = 2\,|A_0|^2\ \cos\left(\left(\frac{\omega_1 - \omega_2}{2}\right)t - \frac{k_1 - k_2}{2}x\right) = \\ = 2\,|A_0|^2\ \cos\left(\frac{k_1 - k_2}{2}[x - v_g t]\right), \tag{3.128}$$

where $v_g = (\Delta\omega)/(\Delta k)$.

In more general terms, assuming frequencies and wave numbers sufficiently close, we can set

$$v_g = \frac{\partial \omega}{\partial k}, \tag{3.129}$$

and derive the following Rayleigh identity

$$v_g = \left(1 - \lambda\frac{\partial}{\partial \lambda}\right)v_p, \tag{3.130}$$

specifying the relationship between phase and group velocities.

By noting that phase velocity and refractive index are linked by

$$v_p = \frac{c}{n(\omega)}, \tag{3.131}$$

we can also state that

$$v_g = \left(1 + \omega\frac{\partial}{\partial \omega}\right)\frac{c}{n(\omega)} = v_p\left(1 - \omega\frac{n'(\omega)}{n(\omega)}\right). \tag{3.132}$$

In figure 3.12 we have reported the wave in equation (3.128) and specified the relevant form in terms of sine and cosine functions. It consists of an envelope (the one containing the beating terms) and a carrier moving at the phase and group velocity, respectively.

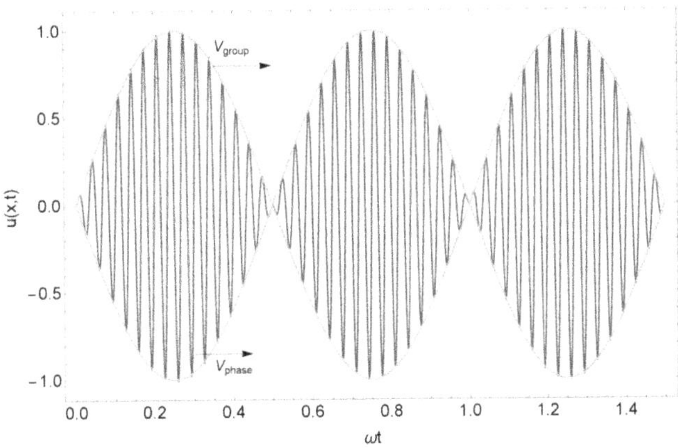

Figure 3.12. Sum of two waves in equation (3.128). The envelope wave with period $T = 2\pi/\omega_{-}$ moves at v_g and the carrier at v_p.

The vector extension of the group velocity is

$$\vec{v}_g = \vec{\nabla}_k \vec{\omega} \tag{3.133}$$

The group velocity always satisfies the condition $v_g < c$, and usually moves along the same direction of the phase counterpart. The two vectors may not be parallel and it also may happen that they propagate in opposite directions. In this case the wave is called ***backward*** .

The next effort is that of putting all the elements together to understand how heating is induced by wave propagation inside the plasma. We therefore go back to what we have tried to convey in generic terms at the beginning of the previous section.

The parameters specifying the wave motion inside the dispersive medium represented by the plasma are:

1. The frequency determined by the e.m. source;
2. The $\vec{k}$ vector (hence the wavelength) determined by the physics of the interaction;
3. The angle of propagation fixed by the antenna.

The wave propagation phenomenology foresees different behaviors, depending on the refractive index (or the phase velocity as well).

We can therefore ask whether $n = 0$ ($v_p \to \infty$) makes sense; these values are appropriate to describe the wave physics at the cutoff, namely the mode with the lowest frequency[6].

At the resonance, namely at the frequency where the radiation is absorbed, $n \to \infty$ ($v_p = 0$), which is consistent with the fact that wave is not propagating anymore.

Within certain limits, an evanescent wave (see figure 3.13) can be characterized by an imaginary refractive index

After these tutorial remarks, we can safely go back to the point of our discussion.

We will exploit the results obtained in equation (3.89), whose solution yields the electric field evolution inside the plasma, in the cold plasma approximation.

The existence of non-trivial solutions requires the vanishing of the determinant of the associated matrix and this condition specifies the dispersion relation, yielding the refractive index as a function of the wave and of the propagating angle. With some algebra and a little patience one finds the bi-quadratic equation

$$\begin{aligned} A(\omega, \vartheta)\, Y^2 - B(\omega, \vartheta)\, Y + C(\omega, \vartheta) = 0, \\ Y = n^2, \end{aligned} \tag{3.134}$$

where

$$\begin{aligned} A &= S \sin^2(\theta) + P \cos^2(\theta), \\ B &= RL \sin^2(\theta) + PS(1 + \cos^2(\theta)), \\ C &= PRL, \end{aligned} \tag{3.135}$$

[6] Note that where $k = \omega / v_p$ in correspondence to the cutoff frequency ω_c we have $k_c = 0$.

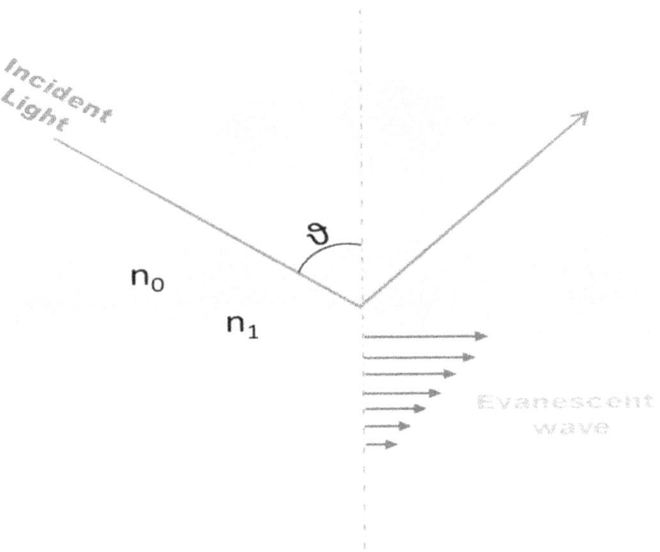

Figure 3.13. Wave reflected and transmitted above the critical angle. The transmitted field is said to be evanescent, because characterized by a decaying amplitude when it moves away from the interface region.

We have denoted by R, L the right and left handed terms, defined as

$$R = S + D = 1 - \sum_s \frac{\Omega_{p,s}^2}{\omega(\omega + \Omega_{c,s})},$$
$$L = S - D = 1 - \sum_s \frac{\Omega_{p,s}^2}{\omega(\omega - \Omega_{c,s})}, \tag{3.136}$$

and satisfying the identity $S^2 - D^2 = RL$ exploited to derive equations (3.134) and (3.135).

The nature of the equation is such that in correspondence of any ω we have two solutions $n^2 = n_{+,-}^2$ and thus two different modes of propagation.

In plasma heating the choice of wave propagation along the perpendicular direction to the confining field is commonly adopted. We are therefore left with $n_{\|} = 0$.

In a simplified model, describing purely perpendicular propagation, it is possible to calculate the frequency resonances using the condition

$$A\left(\omega, \vartheta = \frac{\pi}{2}\right) = 0 \rightarrow S = 0 \rightarrow 1 - \sum_s \frac{\Omega_{p,s}^2}{\omega^2 - \Omega_{c,s}^2} = 0, \tag{3.137}$$

which can be cast in the form

$$\sum_i \frac{\Omega_{p,i}^2}{\omega^2 - \Omega_{c,i}^2} + \frac{\Omega_{p,e}^2}{\omega^2 - \Omega_{c,e}^2} = 1. \tag{3.138}$$

We consider the range of frequencies close to the electron cyclotron interval. In this region $f_{p,e} \cong f_{c,e} \gg f_{p,i}$, it is evident that in this interval the first term associated with ion frequencies can be neglected, thus finding

$$\frac{\Omega_{p,\,e}^2}{\omega^2 - \Omega_{c,\,e}^2} = 1 \to \omega^2 = \Omega_{p,\,e}^2 + \Omega_{c,\,e}^2 = \omega_{\mathrm{UH}}^2. \tag{3.139}$$

The subscript UH indicates upper-hybrid, because it involves both plasma and electron cyclotron frequencies. By recalling that $f_{c,e}[\mathrm{GHz}] \cong 28|\vec{B}[\mathrm{T}]|$ and that in the core of fusion plasma it is comparable with plasma frequency, we end up with the conclusion that the suitable sources candidates are gyrotron and CARM devices.

In the case of ITER ***24-Gyrotrons*** at 170 GHz will deliver 1 MW power per unit.

Suppose that the plasma is composed by two ions species, we can therefore write the condition (3.137) as

$$\frac{\Omega_{p,\,1}^2}{\omega^2 - \Omega_{c,\,1}^2} + \frac{\Omega_{p,\,2}^2}{\omega^2 - \Omega_{c,\,2}^2} = 1 - \frac{\Omega_{p,\,e}^2}{\omega^2 - \Omega_{c,\,e}^2}, \tag{3.140}$$

and consider frequencies such that $\omega \cong \Omega_{c,\,i} \ll \Omega_{c,e}$, the lhs contributions are dominant with respect to the rhs so that the resonant frequencies are specified by

$$\omega_{\mathrm{IIH}}^2 = \frac{\Omega_{p,\,1}^2\Omega_{c,\,2}^2 + \Omega_{p,\,2}^2\Omega_{c,\,1}^2}{\Omega_{p,\,1}^2 + \Omega_{p,\,2}^2}, \tag{3.141}$$

where IIH stands for ion–ion hybrid, the relevant frequency range lies in the region (20–100) MHz (for further comments see the next concluding section).

We eventually consider the ***lower hybrid*** condition, which attains the region $f_{c,i} \ll f \ll f_{c,e}$, which allows us to write (3.138) as

$$\sum_i \frac{\Omega_{p,\,i}^2}{\omega^2} + \frac{\Omega_{p,\,e}^2}{\Omega_{c,\,e}^2} = 1, \tag{3.142}$$

which yields

$$\omega_{\mathrm{LH}}^2 = \frac{\sum_i \Omega_{p,\,i}^2}{1 + \left(\dfrac{\Omega_{p,e}}{\Omega_{c,e}}\right)^2}, \tag{3.143}$$

with typical values ranging from 1 to 8 GHz.

In the forthcoming section we will provide further elements regarding heating/current drive in the spectral regions we have just mentioned.

3.6 Generalities on beam plasma energy transfer

This section is aimed at specifying, mainly at a qualitative level and with the minimum amount of math, some aspects of the wave-plasma energy transfer. The forthcoming concluding sections will be more detailed and will go deeper into the relevant physics.

The discussion developed in the first part of the chapter has fixed the 'numbers' characterizing the e.m. heating process in plasma. Quoting again the electron cyclotron frequency ($\Omega_{c,e}$[GHz] = 28 B [T]) we note that for a magnetic field on the order of a few Tesla, the relevant heating frequencies should range within the region of hundreds of GHz (few millimeters wavelength), where they can be launched inside plasma from vacuum, in the form of well-focused high energy densities beams. According to the few notions we have acquired, the plasma–wave interaction is resonant, therefore, the power deposition is well localized.

In conclusion, ECRH is characterized by unique features due to a combination of heat local deposition from adjustable and well collimated beams.

We have also mentioned the importance of a physical quantity like the refractive index, which, for the cold plasma regime (see equation (3.134)), is written in the so called Appleton–Hartree form [11, 15]

$$n = 1 - 2\frac{A - B + C}{2A - B \pm F}, \tag{3.144}$$

where F is the discriminant of equation (3.134) and reads

$$F^2 = (RL - PS)^2 \sin(\vartheta)^4 + (2PD\cos(\vartheta))^2. \tag{3.145}$$

The $\pm$ sign appearing in the denominator of equation (3.144) is not insignificant and accounts for two modes of propagation, known as the extraordinary (X-mode) and ordinary (O-mode), respectively (see below).

The previous equations can be conveniently reshuffled in a more physically transparent form, yielding the propagation angle in terms of the refractive index, namely

$$\tan(\vartheta)^2 = \frac{P(n^2 - R)(n^2 - L)}{(Sn^2 - RL)(n^2 - P)}, \tag{3.146}$$

which allows us to get straightforwardly the regions of interest associated with the conditions given below for the different direction of propagation

$$\begin{aligned} &\text{Parallel} \quad \vartheta = 0 \rightarrow \quad P = 0,\ n^2 = R,\ n^2 = L, \\ &\text{Perpendicular} \quad \vartheta = \frac{\pi}{2} \rightarrow \quad n^2 = P\left(\text{O–Mode}\right),\ n^2 = \frac{RL}{S}\left(\text{X–Mode}\right). \end{aligned} \tag{3.147}$$

In order to understand the importance of the previous identities, we provide the refractive index for right/left (R/L) handed wave in parallel propagation and single ion species using the equations in equation (3.136)

$$n_{R/L}^2 = 1 - \frac{\Omega_{p,e}^2}{\omega(\omega \pm \Omega_{c,e})} - \frac{\Omega_{p,i}^2}{\omega(\omega \pm \Omega_{c,i})}, \tag{3.148}$$

which yields a resonance frequency at $\omega = \Omega_{c,e}$ for right-handed wave and $\omega = \Omega_{c,i}$ for left-handed wave and a cutoff ($n = 0$) at

$$\omega_{R/L} = \frac{\pm(|\Omega_{c,e}| - \Omega_{c,i})}{2} + \left[\left(\frac{|\Omega_{c,e}| + \Omega_{c,i}}{2}\right)^2 + \Omega_{p,\,e}^2 + \Omega_{p,\,i}^2\right]^{\frac{1}{2}}, \tag{3.149}$$

the '+' sign in front of the square root ensures that $\omega_{R/L} > 0$.

The plot of n_R^2 versus ω is given in figure 3.14. The relevant behavior near the origin ($\omega \approx 0$) can be determined as follows.

Equation (3.148) can be cast in the form

$$n_{R/L}^2 = 1 - \frac{\Omega_{p,\,e}^2 + \Omega_{p,\,i}^2}{(\omega \pm \Omega_{c,e})(\omega \pm \Omega_{c,i})}, \tag{3.150}$$

where

$$\Omega_{p,\,e}^2\Omega_{c,i} + \Omega_{p,\,i}^2\Omega_{c,e} = 0, \tag{3.151}$$

assuming that the background plasma satisfies the quasi-neutrality condition ($n_i = n_e$).

With $\omega_{R/L}$ being the zeros of the refractive index, we can eventually write

$$n_{R/L}^2 = \frac{(\omega \mp \omega_R)(\omega \pm \omega_L)}{(\omega \mp |\Omega_{c,e}|)(\omega \pm \Omega_{c,i})}, \tag{3.152}$$

for vanishing ω we end up with

$$n_{R/L}^2 = \frac{\omega_R\omega_L}{|\Omega_e|\Omega_i} = 1 + \frac{\Omega_{p,\,e}^2 + \Omega_{p,\,i}^2}{|\Omega_e|\Omega_i}, \tag{3.153}$$

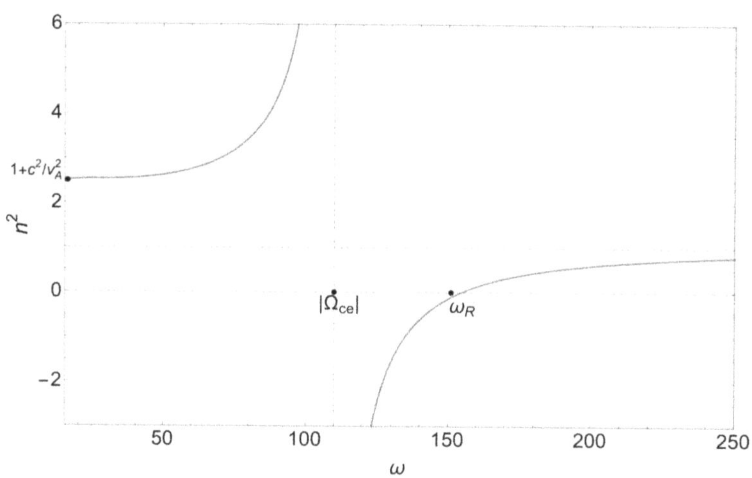

Figure 3.14. Dispersion relation for right-handed wave.

where from equation (3.149)

$$\omega_R\omega_L = |\Omega_e|\Omega_i + \Omega_{p,\,e}^2 + \Omega_{p,\,i}^2. \tag{3.154}$$

Equation (3.153) can be cast in terms of more appealing physical quantities as, where v_A is the so called Alfvén velocity as follows from the following identities,

$$n_{R,\,L}^2|_{\omega\to 0} = 1 + \frac{c^2}{v_A^2}, \tag{3.155}$$

where

$$\frac{\Omega_{p,\,e}^2 + \Omega_{p,\,i}^2}{|\Omega_e|\Omega_i} \approx \frac{\Omega_{p,\,e}^2}{|\Omega_e|\Omega_i} = \frac{n_e e^2/(\varepsilon_0 m_e)}{(eB_0/m_e)(eB_0/m_i)} = \frac{\mu_0 n_e m_i}{B_0^2}c^2 = \frac{c^2}{v_A^2}, \tag{3.156}$$

where v_A is given by[7]

$$v_A = \frac{B_0}{\sqrt{\mu_0 n m_e}}. \tag{3.157}$$

A more complicate behavior is provided by the analysis of the conditions for perpendicular wave propagation, namely $n^2 = RL/S$. The associated analytical forms albeit awkward are useful to get the localization of the resonances. If we impose the condition $S = 0$, corresponding to $n \to \infty$, we can evaluate the resonance specified by the relationship

$$\omega^2 = \frac{\Omega_{p,\,e}^2 + \Omega_{c,\,e}^2 + \Omega_{p,\,i}^2 + \Omega_{c,\,i}^2}{2} \pm \sqrt{\left(\frac{\Omega_{p,\,e}^2 + \Omega_{c,\,e}^2 + \Omega_{p,\,i}^2 + \Omega_{c,\,i}^2}{2}\right)^2 + \Omega_{p,\,i}^2\Omega_{p,\,e}^2}, \tag{3.158}$$

which yields a slightly more general expression than that discussed in the previous section. If we neglect, for example, terms $\Omega_{c,i}/\Omega_{c,e} \ll 1$, the previous identity yields the already derived upper and lower hybrid resonance conditions and the dispersion relation can be cast in a useful expression

$$n^2 = \frac{(\omega - \omega_R)(\omega + \omega_L)(\omega + \omega_R)(\omega - \omega_L)}{(\omega - |\Omega_{c,e}|)(\omega + \Omega_{c,i})(\omega + |\Omega_{c,e}|)(\omega - \Omega_{c,i})}\frac{(\omega_{c,\,e}^2 - \omega^2)(\omega_{c,\,i}^2 - \omega^2)}{(\omega^2 - \omega_{\rm LH}^2)(\omega^2 - \omega_{\rm UH}^2)} = \frac{(\omega^2 - \omega_R^2)(\omega^2 - \omega_L^2)}{(\omega^2 - \omega_{\rm LH}^2)(\omega^2 - \omega_{\rm UH}^2)}. \tag{3.159}$$

[7] A simple explanation of the Alfvén velocity is provided by the following hydrodynamic analogy, equating magnetic pressure and kinetic pressure, we get with $\rho v_A^2/2 = B^2/(2\mu_0)$ where $\rho = m_i n_i + m_e n_e = n(m_i + m_e) \approx n m_i$.

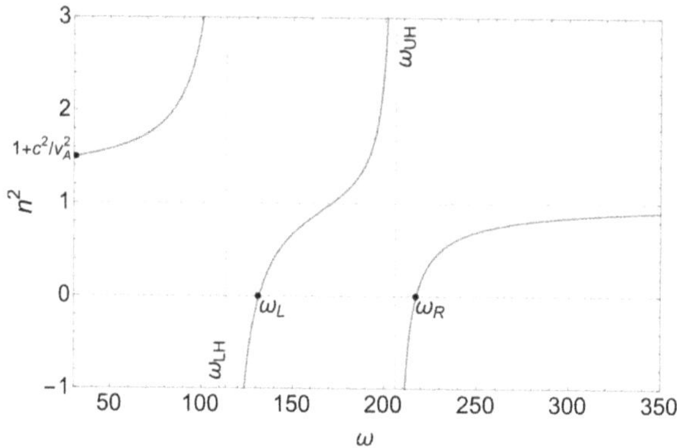

Figure 3.15. Dispersion relation for extraordinary wave.

In figure 3.15 we summarize in a (n^2, ω) plot the various regions of interest, with specific marks about resonance and cutoff frequencies.

The discussion should be completed with the radial dependence of the confining field in terms of the radius R, which is a further element allowing a local wave deposition, as discussed in the second part of this book.

Even though not explicitly stated yet, we recall that the resonance mechanism is the physical process allowing the energy transfer from wave to plasma. This absorption ensures the heating of a given ensemble of particles, with the kinematic variables matching the resonance condition. Plasma and wave parameters should therefore be properly adjusted to allow the most efficient power transfer, thus allowing energy exchange with the largest number of particles.

Before concluding the chapter we touch on two topics we have just mentioned and which will be more carefully discussed in the second part of the book.

Regarding the collisionless damping, the first condition to be satisfied is the resonance matching, resonant particles move at velocity

$$v_{\parallel} \cong \frac{\omega}{k_{\parallel}} = v_p. \tag{3.160}$$

This is easily understood because $v_{\parallel}$ is the phase velocity and, in their own reference frame these particles see essentially a static wave. Consider now particles with slightly lower velocities, they result as accelerated for a large period of time, thus producing a net energy gain of the particles. Conversely, those with larger velocity are decelerated for a similar amount of time, therefore, particles experience a loss of energy.

Suppose now that the particle velocity distribution is such that $\frac{\partial f(v_{\parallel})}{\partial v_{\parallel}} < 0$, namely that there are more particles with $v_{\parallel} < v_p$, if so the wave is damped.

We will reconsider this (appealing) physical picture in the forthcoming chapter within the context of a more appropriate mathematical framework.

The next topic is current drive, or better, non-inductive current drive. To distinguish it from the transformer-like induced part.

The previously quoted heating effects can be combined with the driving of current and lead to longer Tokamak pulses. Furthermore, the use of waves may locally modify the plasma phase space distribution and create the conditions to cure or prevent instabilities.

An example, of how the mechanism occurs, is provided by the collision-less damping argument, which creates a local modification of the current profile. This is just an abrupt description, the associated phenomenology is rather intriguing and we prefer to discuss it, on the basis of specific examples, after having explored the conditions of particle wave energy transfer.

3.7 The mechanism of radio frequency–plasma interaction

The launched wave, propagating inside the plasma, accelerates the charged particles, which in turn experience collisions, thus losing the acquired energy. The energy is therefore dissipated inside the environment and the wave results to be damped, because of the drag induced by the multiple collisions. We have already discussed this effect and we have noted that the collision, hence the plasma resistivity, decreases with the temperature. It is therefore evident that the collision mechanism is not efficient to transfer energy in hot plasma, where the temperature values are approximately $T_i \approx T_e \approx 5\,\text{keV}$ and the density value is $n = 5.0 \cdot 10^{-19}\text{m}^{-3}$. The derived collision frequency v (1.83) is of the order of a few kHz

$$v = 2.9 \cdot 10^{-12} n ln\Lambda T^{-3/2} \approx 20\text{kHz}. \tag{3.161}$$

In the electromagnetic theory the ratio $v/(2\pi f)$ is characteristic of the importance of dissipative effects due to the collisions with respect to the wave oscillation. For smaller values of v/ω, the motion is almost dissipation-less. Huge fields and large perturbations in the particle motion are necessary if any significant amount of energy is to be damped in the plasma.

In a plasma, with the previous characteristics, non-collisional resonant mechanisms at the frequencies in the MHz range, or higher, play the major role in absorption. Under resonance conditions, a small excitation determines either a huge response in the particle's motion (wave-particle resonance where only few particles fulfill resonance conditions) or large wave field build up (wave resonance for which a collective effect involves all particles meeting the resonance conditions).

The absence (or better the inefficiency) of colliding mechanisms, yielding the dissipation load, makes the wave–particle interaction, in a plasma, similar to that of a particle oscillating in vacuum under the action of a wave field. It is well known that, under these conditions (Woodward–Lawson theorem[8]), any net energy transfer from the wave to the particle is not possible. Other mechanisms should therefore be

[8] It can be shown from Maxwell equations in vacuum that energy gain from optical fields is not possible under the following assumptions: (a) the region of interaction is infinite, (b) the laser fields are in vacuum with no walls or boundaries present, (c) the electron is highly relativistic, ($v \approx c$) along the acceleration path, (d) no static electric or magnetic fields are present, and (e) nonlinear effects due to ponderomotive, $\vec{v} \times \vec{B}$, and radiation reaction forces are neglected. One or more of the assumptions of the Woodward–Lawson theorem must be violated in order to achieve a nonzero net energy gain.

invoked to ensure wave plasma heating. If we were capable of transforming the field carried by a wave into a static component, the problem would be naturally solved. Regarding the wave–particle interaction, two different mechanisms determine the frequency and wavelength ranges of interest.

The first mechanism is based on the Landau damping which, as already discussed in the previous section, is one of the key tools ensuring the wave–plasma power transfer. It is indeed effective when the particle, with an initial velocity of magnitude v, matches the wave phase velocity, thus seeing, in its own reference frame, a constant field.

In a steady magnetic field $\vec{B}_0$, the component of the particle's velocity along $\vec{B}_0$ is unaffected by the magnetic field, the Landau resonance follows from the condition $\omega - k_{||}v_{||} = 0$, with $v_{||} = \omega/k_{||}$ and $k_{||} = \mathbf{k} \cdot \vec{B}_0/|\vec{B}_0|$. This condition allows two different mechanisms, determining the wave energy transfer to the particle:

- Electrostatic force effect ($F_{||} = qE_{||}$), whenever the wave has a component of the electric field and a component of the wave vector both parallel to $\vec{B}_0$, the force allowing the wave–beam energy exchange is simply given by the electrostatic force;
- Magnetic field gradient driving force $\left(\vec{F}_{||} = \frac{1}{2}\left(\frac{mv_\perp^2}{B_0}\frac{d\vec{B}_z^W}{dz}\right)\right)$, due to the gradient of the wave magnetic field (B_z^W) component interacting with the magnetic moment of the particle (the z direction defined by the static magnetic field $\vec{B}_0$); in fact, for any slow variation of the magnetic field given, for example, by $\vec{B}^W = (0, 0, B(y))$ in the y direction the Lorentz force F, averaged on the Larmor period acting on the particle, assumes the expression $<F> = -\frac{1}{2}mv_\perp^2 B'(y)/B_0\mathbf{e}_y$. This mechanism is usually referred to as transit time magnetic pumping (**TTMP**).

The consequences of the **Landau damping** mechanism can be better appreciated if we describe the relevant physics in a coordinate system moving at the wave phase velocity. In this reference frame the electric field of the wave and the associated electrical potential will be fixed in time. The particle, having total energy less than the electrostatic energy potential of the wave field, will be trapped by the wave and oscillates within the bucket between two consecutive maxima of the electrostatic potential (A, B in figure 3.16).

In this frame a particle, having velocity less than the wave phase velocity, moves in the opposite direction of the propagating wave, till it reaches the point A (in figure 3.16) where it will be reflected (absorbing energy from the wave), while a particle with a velocity greater than the wave phase velocity will be reflected at point B (giving energy to the wave).

This process must be averaged over the distribution function of the particles velocities. The distribution, we consider, is centered around $v_{\rm ph}$, and the number of particles with $v < v_{\rm ph}$ is larger than those with $v > v_{\rm ph}$. In these conditions the mechanism of energy absorption from the wave dominates the process and the final result is a damping of the wave itself.

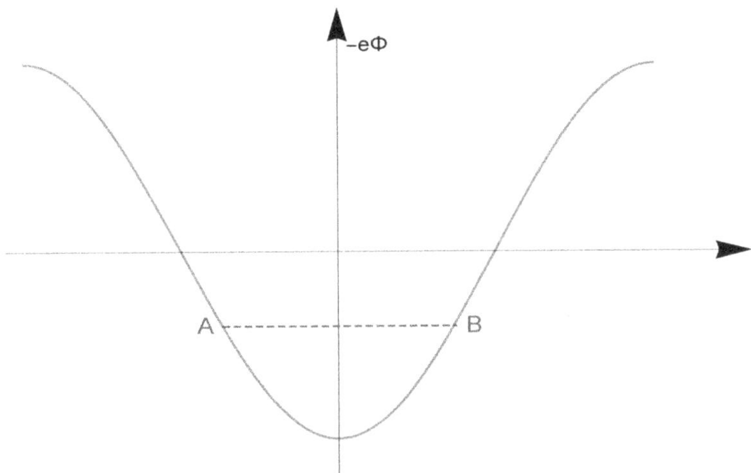

Figure 3.16. The potential energy $-e\Phi$ versus the particle motion direction in the reference frame of the wave.

The previous physical picture is amenable for a toy model interpretation, accounting for the interaction between the charged particle and a one-dimensional electrostatic (longitudinal) wave with $\vec{k}||\vec{E}$, in the absence of the magnetic field. Taking $\vec{v} = v\hat{z}$ and $\vec{E} = E\cos(kz - \omega t)\hat{z}$, the single particle equation of motion can be written as

$$m\frac{dv}{dt} = qE\cos(kz - \omega t). \tag{3.162}$$

The 0th order solution (for $E = 0$) $z = v_0 t + z_0$ can be substituted into the first order equation to give

$$m\frac{dv_1}{dt} = qE\cos(kz_0 + kv_0 t - \omega t), \tag{3.163}$$

whose solution, for the initial condition $v_1 = 0$ at $t = 0$, is given by

$$v_1 = \frac{qE}{m}\frac{\sin(kz_0 + kv_0 t - \omega t) - \sin(kz_0)}{kv_0 - \omega}. \tag{3.164}$$

The time rate of the kinetic energy change, averaged over initial position z_0, is

$$\left\langle \frac{d}{dt}\frac{mv^2}{2} \right\rangle_{z_0} = \frac{q^2E^2}{2m}\left[-\frac{\omega\sin(\alpha t)}{\alpha^2} + t\cos(\alpha t) + \frac{\omega t\cos(\alpha t)}{\alpha}\right], \tag{3.165}$$

where $\alpha = kv_0 - \omega$ and averaging over the distribution $f(v_0)$ of initial velocity v_0

$$f(v_0) = f(\frac{\alpha + \omega}{k}) = g(\alpha), \tag{3.166}$$

we get

$$\left\langle \frac{d}{dt} \frac{mv^2}{2} \right\rangle_{z_0,v_0} = -\frac{\omega q^2 E^2}{|k|2m} P \int_{-\infty}^{\infty} \frac{g(\alpha)\sin(\alpha t)}{\alpha^2} d\alpha, \tag{3.167}$$

P being the Cauchy principal value[9].

The expansion of the distribution function in the neighborhood of $\alpha = 0$

$$g(\alpha) = g(0) + \alpha g'(0) + \frac{\alpha^2}{2} g'' + \ldots \tag{3.168}$$

gives

$$\left\langle \frac{d}{dt} \frac{mv^2}{2} \right\rangle_{z_0,v_0} \approx -\frac{\pi \omega q^2 E^2}{k|k|2m} \left[\frac{df(v_0)}{dv_0} \right]_{v_0=\omega/k}. \tag{3.169}$$

This expression yields the transfer energy from the wave to the resonant particles, with velocity close to the wave phase velocity.

The linear regime of the Landau damping can be applied for a short time with respect to the bounce period inside the potential bucket (see figure 3.16). A comparison of the bounce period relative to the plasma frequency period is given through the argument presented below.

In the frame moving at velocity close to the phase velocity, the electric field is essentially static, namely $E = E_0 \sin(kx)$. The potential, ruling the dynamics of an electron inside the wave, is therefore, reported in figure 3.16 and the relevant equations of motion near the minimum are reduced to

$$m\ddot{x} = -eE = -eE_0 \sin(kx) \approx -eE_0\, kx, \tag{3.170}$$

which is the equation of a harmonic oscillator with a bounce frequency given by

$$\omega_b = \sqrt{eE_0\, k/m} = k\sqrt{e\Phi_0/m}. \tag{3.171}$$

Accordingly, for $\omega \gg \omega_{pe}$ we find that the ratio between the bounce and electron plasma wave periods, satisfies the condition

$$\frac{\tau_b}{\tau_{pe}} \approx \frac{\omega_{pe} c_s}{\omega\sqrt{e\Phi_0/m}} = \frac{\omega_{pe}}{\omega} \sqrt{\frac{k_B T}{e\Phi_0/m}} \ll 1. \tag{3.172}$$

The relation (3.172) shows that the approximation of the linear Landau damping regime is verified for a few periods of the plasma wave.

[9] The Cauchy principal value of a function ($f(x)$, $x \in \mathbb{R}$) which is integrable on the complement of one point x_0, if it exists, is the limit of the integrals of the function over subsets in the complement of this point as these integration domains tend to that point symmetrically from all sides:

$$PV \int f(x)dx := \lim_{\varepsilon \to 0} \int_{\mathbb{R}\setminus(x_0-\varepsilon, x_0+\varepsilon)} f(x)dx$$

We are now going to analyze the effect induced by the magnetic field gradient of the wave ($\vec{B}_z^w$) in the direction of the external confining magnetic field $\vec{B}_z^0$. The equation of the charged particle motion due to the force induced by the wave magnetic gradient ($\vec{\nabla}\vec{B}_z^w$) yields

$$m\frac{dv_{||}}{dt} = -\mu\hat{b}\cdot\nabla|\vec{B}^w(r,t)|, \tag{3.173}$$

μ being the magnetic momentum of the charged particle and $\hat{b} = \vec{B}_0/|\vec{B}_0|$ the unit vector. This equation describes the already defined **TTMP damping** which is the magnetic analog of Landau damping, whose dynamical roots can be traced back to the equation of motion

$$m\frac{dv_{||}}{dt} = -q\hat{b}\cdot\nabla\phi, \tag{3.174}$$

obtained by replacing the magnetic moment μ in equation (3.173) with the charge q and the magnetic field wave module $|\vec{B}_z^w|$ with the electrostatic potential $\nabla\phi$.

TTMP and the Landau damping are both resonating for $v_{||} = v_{\rm ph}$ and the two process are said to be coherent. The net result of the total interaction depends on the relative phase of the fields $E_{||}$ and the perturbed $B_{||}^{(1)}$. We cannot calculate them separately and simply add them together, hence cross term contributions must be considered.

As an example, we can consider the case of the compressional Alfvén[10] wave, propagating in the direction of the magnetic field with $|k_{||}v_{\rm th||}^e| \simeq \omega \ll \Omega_{c,i}$. The last assumption, along with the conditions $T_\perp = T_{||}$ and $V = 0$, yields the following results at the lowest order in b_s [16, 17] (see appendix B for the explicit derivation)

$$\varepsilon_{xx} = 1 + \sum_s\sum_{l=-\infty}^{\infty}\frac{\Omega_{p,s}^2}{\omega|k_{||}|v_{t,s}}\frac{l^2 I_l(b_s)}{b_s}\exp(-b_s)Z(\zeta_{s,l}) \approx 1 + \sum_s\frac{\Omega_{p,s}^2}{\Omega_{c,s}^2} = \varepsilon_\perp, \tag{3.175}$$

$$\begin{aligned}\varepsilon_{yy} = 1 + \sum_s\sum_{l=-\infty}^{\infty}\frac{\Omega_{p,s}^2}{\omega|k_{||}|v_{t,s}}\left[\frac{l^2 I_l(b_s)}{b_s} + 2b_s(I_l(b_s) - I_l'(b_s))\right]\\ \exp(-b_s)Z(\zeta_{s,l}) \approx \varepsilon_\perp + i\Delta\varepsilon_\perp,\end{aligned} \tag{3.176}$$

where

$$\Delta\varepsilon_\perp = \sum_s 2b_s\delta_s,\ \delta_s = \frac{\Omega_{p,s}^2}{\omega|k_{||}|v_{t,s}}\sqrt{\pi}\exp\left(-\zeta_{s,0}^2\right), \tag{3.177}$$

[10] They are the waves described by the ideal MHD stability theory, for which the magnetic field plays a central role. The magnetic field perturbation due to the mass motion induces a magnetic stress which tends to restore the equilibrium so that oscillatory modes appear. The plasma can support three different types of stable waves. Firstly there is the shear Alfvén. This is an incompressible wave and it is therefore unaffected by the plasma pressure. The other two involve compression and their wave velocities involve both the Alfvén speed and the sound speed.

$$\varepsilon_{zz} = 1 + \sum_s \sum_{l=-\infty}^{\infty} 2\frac{\Omega_{p,s}^2}{\omega|k_{||}|v_{t,s}} I_l(b_s)\exp(-b_s)\zeta_{s,l}[1 + \zeta_{s,l}Z(\zeta_{s,l})] \approx \varepsilon_{||} + i\Delta\varepsilon_{||}, \quad (3.178)$$

with

$$\varepsilon_{||} = -\sum_s \left(\frac{\Omega_{p,s}}{|k_{||}|v_{t,s}}\right)^2 Re(Z'), \ \Delta\varepsilon_{||} = \sum_s 2\zeta_{s,0}^2\delta_s, \quad (3.179)$$

$$\begin{aligned}\varepsilon_{yz} = -\sum_s \sum_{l=-\infty}^{\infty} \frac{\Omega_{p,s}^2}{\omega|k_{||}|v_{t,s}} \frac{Z_s}{|Z_s|}\sqrt{2b_s}[I_l'(b_s) - I_l(b_s)] \\ \exp(-b_s)[1 + \zeta_{s,l}Z(\zeta_{s,l})] \approx \varepsilon_m + i\Delta\varepsilon_m,\end{aligned} \quad (3.180)$$

where

$$\varepsilon_m = \sum_s -\frac{1}{2}\frac{n_\perp}{n_{||}}\frac{Z_s}{|Z_s|}\frac{\Omega_{p,s}^2}{\omega\Omega_{c,s}} Re(Z'), \ \Delta\varepsilon_m = \sum_s \frac{Z_s}{|Z_s|}\frac{n_\perp}{n_{||}}\frac{\omega}{\Omega_{c,s}}\delta_s, \quad (3.181)$$

$$\begin{aligned}\varepsilon_{xy} &= \sum_s \sum_{l=-\infty}^{\infty} \frac{\Omega_{p,s}^2}{\omega|k_{||}|v_{t,s}} l[I_l'(b_s) - I_l(b_s)]\exp(-b_s)Z(\zeta_{s,l}) \approx \\ &\approx \sum_s \sum_{l=-\infty}^{\infty} \frac{\Omega_{p,s}^2}{\omega|k_{||}|v_{t,s}} l[I_l'(b_s) - I_l(b_s)](1 - b_s)\frac{|k_{||}|v_{t,s}}{l\Omega_s} \approx 0,\end{aligned} \quad (3.182)$$

$$\varepsilon_{xz} = \sum_s \sum_{l=-\infty}^{\infty} \frac{\Omega_{p,s}^2}{\omega|k_{||}|v_{t,s}} \frac{Z_s}{|Z_s|}\sqrt{\frac{2}{b_s}}\, lI_l(b_s)\exp(-b_s)[1 + \zeta_{s,l}Z(\zeta_{s,l})] \approx 0, \quad (3.183)$$

which is a consequence of

$$\zeta_{s,l}Z(\zeta_{s,l}) = -1 + O\left(\frac{1}{\zeta_{s,l}^2}\right) \text{ for } l \neq 0. \quad (3.184)$$

Now using equation (3.77) and the second in equation (3.90) we get

$$\begin{pmatrix} \varepsilon_\perp - n_{||}^2 & 0 & n_{||}n_\perp \\ 0 & \varepsilon_\perp + i\Delta\varepsilon_\perp - n^2 & i(\varepsilon_m + i\Delta\varepsilon_m) \\ n_{||}n_\perp & -i(\varepsilon_m + i\Delta\varepsilon_m) & \varepsilon_{||} + i\Delta\varepsilon_{||} - n_\perp^2 \end{pmatrix}\begin{pmatrix} E_x \\ E_y \\ E_z \end{pmatrix} = 0. \quad (3.185)$$

Damping is described by the anti-Hermitian parts of the wave equation which occur only in the lower right 2×2 part of the matrix, so the relevant wave equation is

$$\underbrace{\begin{pmatrix} \epsilon_\perp + i\Delta\epsilon_\perp - n^2 & i(\epsilon_m + i\Delta\epsilon_m) \\ -i(\epsilon_m + i\Delta\epsilon_m) & \epsilon_{||} + i\Delta\epsilon_{||} \end{pmatrix}}_{A}\begin{pmatrix} E_y \\ E_z \end{pmatrix} = 0, \quad (3.186)$$

where we have taken into account that $\varepsilon_{||}$ is larger compared to the other dielectric term in particular to $n_{\perp}^2$.

Assuming $\omega = \omega_i + i\omega_r$ with $\omega_i \ll \omega_r$ the dispersion relation ($D(\omega_i, \omega_r) = \mathrm{Det}(A) = 0$) expansion reads

$$\begin{aligned} D(\omega_i, \omega_r) &= D_r(\omega_r, \omega_i) + iD_i(\omega_r, \omega_i) \\ &\approx D_r(\omega, 0) + iD_i(\omega, 0) + \omega_i \frac{\partial D_r(\omega, 0)}{\partial \omega}\bigg|_{\omega=\omega_r} + i\omega_i \frac{\partial D_i(\omega, 0)}{\partial \omega}\bigg|_{\omega=\omega_r} \\ &= D_r(\omega, 0) + \omega_i \frac{\partial D_r(\omega, 0)}{\partial \omega}\bigg|_{\omega=\omega_r} + i\left(D_i(\omega, 0) + \omega_i \frac{\partial D_i(\omega, 0)}{\partial \omega}\bigg|_{\omega=\omega_r}\right) = 0. \end{aligned} \tag{3.187}$$

According to the use of the Cauchy–Riemann[11] relations, providing a link between real and imaginary parts of the dispersion function, we can cast equation (3.187) as

$$\begin{aligned} &D_i(\omega, 0) - \omega_i \frac{\partial D_r(\omega, 0)}{\partial \omega}\bigg|_{\omega=\omega_i} = \\ &= -\omega_i \frac{\partial}{\partial \omega}\left[(\varepsilon_\perp - n^2)\varepsilon_{||} - \Delta\varepsilon_\perp \Delta\varepsilon_{||} + (\Delta\varepsilon_m)^2 - \varepsilon_m^2\right] + \\ &+ \varepsilon_\perp \Delta\varepsilon_{||} + \varepsilon_{||}\Delta\varepsilon_\perp - n^2 \Delta\varepsilon_{||} - 2\varepsilon_m \Delta\varepsilon_m = 0. \end{aligned} \tag{3.188}$$

We are allowed to make some approximations to simplify the previous relations. We note in particular that, $\varepsilon_{||}$ being larger than other components of the dielectric tensor, we find from equation (3.185) for the compressional Alfvén waves, the following result for the relevant refractive index

$$n^2 = n_\perp + n_{||} = \varepsilon_\perp - \frac{\varepsilon_m^2}{\varepsilon_{||}} \approx \varepsilon_\perp. \tag{3.189}$$

Equation (3.188), using equation (3.189), at the first order in δ reduces to

$$\omega_i \frac{\partial}{\partial \omega}\left[(\varepsilon_\perp - n^2)\varepsilon_{||} - \varepsilon_m^2\right] + \varepsilon_{||}\Delta\varepsilon_\perp + \frac{\varepsilon_m^2}{\varepsilon_{||}}\Delta\varepsilon_{||} - 2\varepsilon_m \Delta\varepsilon_m \approx 0. \tag{3.190}$$

The last three terms in the equation (3.188) represent TTPM damping, Landau damping and the cross term, respectively.

The power density absorption under the assumption that $|\omega_i| \ll |\omega_r|$ (for $\omega = \omega_r + i\omega_i$), can be calculated once both the electric field $\vec{E}$ and the susceptibility tensor $\bar{\bar{\chi}}$ are given.

[11] The harmonic dispersive function $D(\omega_r, \omega_i) = D_r(\omega_r, \omega_i) + iD_i(\omega_r, \omega_i)$ must be holomorphic which means D_r, D_i have first partial derivative with respect to ω_i, ω_r and satisfy the Cauchy–Riemann equation:

$$\frac{\partial D_r}{\partial \omega_r} = \frac{\partial D_i}{\partial \omega_i};\ \frac{\partial D_r}{\partial \omega_i} = -\frac{\partial D_i}{\partial \omega_r}$$

From the generalized equation of the energy

$$\mathcal{E} = \frac{1}{2}\varepsilon_0 \vec{E}^* \cdot \bar{\bar{\chi}}^a \cdot \vec{E}, \tag{3.191}$$

where

$$\bar{\bar{\chi}}^a = \frac{1}{2i}(\bar{\bar{\chi}} - \bar{\bar{\chi}}^*). \tag{3.192}$$

Multiplying by ω equation (3.191), the power density absorbed by the species s is written as

$$P_s = \frac{\varepsilon_0 \omega}{2}\vec{E}^* \cdot \bar{\bar{\chi}}_s^a \cdot \vec{E}. \tag{3.193}$$

The absorption power due to the Landau damping, with $\vec{E} = \hat{z}|\vec{E}|$ and compressional Alfvén wave approximation, is determined by the imaginary part of the ε_{zz}^s.

From equation (3.120) assuming $T_{||} = T_{\perp}$, $V \neq 0$ the ε_{zz}^s coefficient is written as

$$\begin{aligned}
\mathrm{Im}[\varepsilon_{zz}^s] &= \frac{\Omega_{p,s}^2}{\omega^2}\sum_l \zeta_{0,s}\mathrm{Im}[Z(\zeta_{l,s})]e^{-b_s}2\eta_{l,s}^2 I_l(b_s) \\
&\approx \frac{\Omega_{p,s}^2}{\omega^2}2\zeta_{0,s}\mathrm{Im}[Z(\zeta_{0,s})]e^{-b_s}\eta_{0,s}^2 I_0(b_s) = \\
&= \frac{n_s q_s^2}{\sqrt{2}k_{||}^2 k_B T \varepsilon_0}\frac{\omega - k_{||}V_s}{k_{||}v_{\mathrm{th}||}}e^{-b_s}I_0(b_s)\exp\left[-\left(\frac{\omega - k_{||}V_s}{k_{||}v_{\mathrm{th}||}}\right)^2\right].
\end{aligned} \tag{3.194}$$

from which absorption power assumes the following expression [17]

$$\begin{aligned}
P_s &= \frac{\omega\varepsilon_0}{2}\mathrm{Im}[\varepsilon_{zz}^s]|E_{||}|^2 \\
&= \frac{n_s q_s^2}{\sqrt{2}k_{||}^2 k_B T \varepsilon_0}\frac{\omega|E_{||}|^2\varepsilon_0}{2}\frac{(\omega - k_{||}V_s)}{k_{||}v_{\mathrm{th}||}}e^{-b_s}I_0(b_s)\exp\left[-\left(\frac{\omega - k_{||}V_s}{k_{||}v_{\mathrm{th}||}}\right)^2\right].
\end{aligned} \tag{3.195}$$

Regarding the case of TTMP damping with $\vec{E}_{\perp} = \hat{x}|E_x| + \hat{y}|E_y|$, $E_{||} = 0$ the contribution to the power dispersion due to the Alfvén wave is given by the imaginary part of the ε_{yy} coefficient which is written as

$$\begin{aligned}
\mathrm{Im}[\varepsilon_{yy}^s] &= \frac{\Omega_{p,s}^2}{\omega^2}\sum_l(\zeta_{0,s})\mathrm{Im}[Z(\zeta_{l,s})]e^{-b_s}\left[\left(\frac{l^2}{b_s} + 2b_s\right)I_l(b_s) - 2b_s I_l'(b_s)\right] \approx \\
&\approx \frac{\Omega_{p,s}^2}{\omega^2}\frac{(\omega - k_{||}V_s)}{k_{||}\sqrt{2}v_{\mathrm{th}||}}e^{-b_s}2b_s[I_l(b_s) - I_l'(b_s)]\sqrt{\pi}\exp\left[-\left(\frac{\omega - k_{||}V_s}{k_{||}v_{\mathrm{th}||}}\right)^2\right].
\end{aligned} \tag{3.196}$$

Using the previous relation and equation (3.71) we find, for the absorbed energy [17]

$$\begin{aligned} P_s = \frac{\omega\varepsilon_0}{2}\mathrm{Im}[\varepsilon^{s}_{yy}]|E_y|^2 &= 2b_s e^{-b_s}[I_0(b_s) - I_0'(b_s)]\frac{\omega\varepsilon_0|E_y|^2}{2}\frac{\Omega^2_{p,\,s}}{\omega^2}\frac{(\omega - k_{\|}V_s)\sqrt{\pi}}{|k_{\|}|\sqrt{2}\,v_{\mathrm{th}\|}} \\ &\exp\left[-\left(\frac{\omega - k_{\|}V_s}{k_{\|}v_{\mathrm{th}\|}}\right)^2\right] = \\ &= \frac{\cancel{2}k_\perp^2\cancel{m_s}k_BT_s}{\cancel{q^2}B^2}e^{-b_s}[I_0(b_s) - I_0'(b_s)]\frac{\omega\cancel{\varepsilon_0}|E_y|^2}{\cancel{2}}\frac{n_s\cancel{q^2}\sqrt{\pi}(\omega - k_{\|}V_s)}{\cancel{m_s}\cancel{\varepsilon_0}\omega^2|k_{\|}|v_{\mathrm{th}\|}} \\ &\exp\left[-\left(\frac{\omega - k_{\|}V_s}{k_{\|}v_{\mathrm{th}\|}}\right)^2\right] = \\ &= \frac{n_sk_BT_s}{B^2}e^{-b_s}[I_0(b_s) - I_0'(b_s)]\omega|B_{\|}|^2\frac{(\omega - k_{\|}V_s)\sqrt{\pi}}{|k_{\|}|v_{\mathrm{th}\|}}\exp\left[-\left(\frac{\omega - k_{\|}V_s}{k_{\|}v_{\mathrm{th}\|}}\right)^2\right]. \end{aligned} \tag{3.197}$$

If $B_{\|}^{(1)} \neq 0$ and $E_{\|} \neq 0$, Landau and TTPM dumping will appear together and cross term ($\sim E_{\|}^{*}B_{\|}^{(1)}$) contributions from ε_{yz} and ε_{zy} must be included for the calculation of the total power absorbtion.

The second mechanism, due to the **cyclotron damping**, occurs when the wave electric field has a component perpendicular to $\vec{B}_0$. In this case the cyclotron resonance can increase the energy of the particle if the electric field vector has a circularly polarized component, rotating at the same frequency, with induces charged particles gyration around $\vec{B}_0$ with the same handedness. The rotation angular frequency is given by $\Omega_c = qB_0/m$ (with q and m the charge and the mass of the particle, respectively) and the resonance condition for resonance is:

$$\omega - k_{\|}v_{\|} = \Omega_c. \tag{3.198}$$

The physical picture of the cyclotron damping can be described in the following way.

Consider a beam of particles traveling along a uniform magnetic field $\vec{B}_0 = B_0\hat{z}$ at velocity $\vec{v} = V\hat{z}$, with no velocity spread (zero temperature). Assume a wave traveling along the magnetic field $\vec{k} = k\hat{z}$ ($k = k_{\|}$) with the electrical field perpendicular to the magnetic field ($E_{\|} = 0$). The linearized equation of motion is

$$m\frac{\partial\vec{v}^{(1)}}{\partial t} + mV\frac{\partial\vec{v}^{(1)}}{\partial z} = q\left(\vec{E}^{(1)} + \vec{v}^{(1)} \times \hat{z}B_0 + V\hat{z} \times \vec{B}^{(1)}\right), \tag{3.199}$$

assuming

$$\begin{aligned} \vec{v} &\approx V\hat{z} + \vec{v}^{(1)}, \\ \vec{B} &\approx \vec{B}_0 + \vec{B}^{(1)}. \end{aligned} \tag{3.200}$$

The physical structure of equation (3.199) is such that the relevant solution should satisfy the following conditions:

(a) The transverse velocity components are coupled and the longitudinal component translates under the action of the electric field;

(b) The temporal evolution of the transverse component is ruled by the equation given below which combines the effect of the electric field and the rotation due to the magnetic counterpart.

Namely, we can cast equation (3.199) in the following

$$\begin{aligned}\left(\frac{\partial}{\partial t}+\frac{\partial}{\partial z}\right)\begin{pmatrix}v_x^{(1)}\\ v_y^{(1)}\end{pmatrix} &= i(kV-\omega)\begin{pmatrix}v_x^{(1)}\\ v_y^{(1)}\end{pmatrix}-\\ &-\frac{1}{2}A(kz-\omega t)e^{i(\omega-kV)t}\begin{pmatrix}\cos(\Omega_c t) & \sin(\Omega_c t)\\ -\sin(\Omega_c t) & \cos(\Omega_c t)\end{pmatrix}\begin{pmatrix}E_x^{(1)}\\ -E_y^{(1)}\end{pmatrix},\end{aligned} \tag{3.201}$$

with

$$A(kz-\omega t)=q\frac{\omega-kV}{\omega}e^{i(kz-\omega t)}. \tag{3.202}$$

The structure of the above equation suggests the use of rotating coordinates and we cast accordingly the solution in the form

$$v_{\pm}^{(1)}=iA(kz-\omega t)\frac{1-e^{i\delta_{\mp}t}}{\delta_{\mp}}E_{\pm}^{(1)}, \tag{3.203}$$

with

$$\begin{aligned}E_{\pm}^{(1)} &= \frac{E_x^{(1)}\pm iE_y^{(1)}}{2},\\ \delta_{\mp} &= \omega-kV\mp\Omega_c.\end{aligned} \tag{3.204}$$

A different form of equations (3.199) or (3.201) is obtained by considering the time dependent derivative only, namely

$$\begin{pmatrix}\dfrac{dv_x^{(1)}}{dt}\\ \dfrac{dv_y^{(1)}}{dt}\end{pmatrix}=\underbrace{\begin{pmatrix}-ikV & \Omega_c\\ -\Omega_c & -ikV\end{pmatrix}}_{A}\begin{pmatrix}v_x^{(1)}\\ v_y^{(1)}\end{pmatrix}+\frac{q}{m}\frac{\omega-kV}{\omega}\underbrace{\begin{pmatrix}E_x e^{ikz-i\omega t}\\ E_y e^{ikz-i\omega t}\end{pmatrix}}_{F(t)}, \tag{3.205}$$

the present form is that of a first order non-homogeneous equation, which can be solved by standard means, namely

$$\vec{V}=e^{At}\int_0^t e^{-At'}F(t')dt'. \tag{3.206}$$

The use of the following matrix exponentiation identities

$$e^{At} = e^{-ikVt}\begin{pmatrix} \cos(\Omega_c t) & \sin(\Omega_c t) \\ -\sin(\Omega_c t) & \cos(\Omega_c t) \end{pmatrix}, \tag{3.207}$$

and

$$\int_0^t e^{-At'}F(t')dt' = \frac{\omega - kV}{\omega}e^{ikz}. \begin{pmatrix} \frac{e^{irt}(ir\cos(\Omega_c t) + \Omega_c \sin(\Omega_c t)) - ir}{(\Omega_c - r)(\Omega_c + r)}E_x - \frac{\Omega_c + e^{irt}(ir\sin(\Omega_c t) - \Omega_c\cos(\Omega_c t))}{(\Omega_c - r)(\Omega_c + r)}E_y \\ \frac{\Omega_c + e^{irt}(ir\sin(\Omega_c t) - \Omega_c\cos(\Omega_c t))}{(\Omega_c - r)(\Omega_c + r)}Ex + \frac{e^{irt}(ir\cos(\Omega_c t) + \Omega_c\sin(\Omega_c t)) - ir}{(\Omega_c - r)(\Omega_c + r)}E_y \end{pmatrix}, \tag{3.208}$$

where $r = (kV - \omega)$, eventually leads to the solution

$$\begin{pmatrix} v_x^{(1)} \\ v_y^{(1)} \end{pmatrix} = e^{At}\int_0^t e^{-At'}F(t')dt' = \frac{\omega - kV}{\omega}e^{ikz}. \begin{pmatrix} \frac{e^{-ikvt}\left(e^{irt}(\Omega_c E_y + iE_x r) + \sin(\Omega_c t)(\Omega_c E_x - iE_y r) - \cos(\Omega_c t)(\Omega_c E_y + iE_x r)\right)}{(\Omega_c - r)(\Omega_c + r)} \\ \frac{e^{-ikvt}\left(e^{irt}(-\Omega_c E_x + iE_y r) + \sin(\Omega_c t)(\Omega_c E_y + iE_x r) + \cos(\Omega_c t)(\Omega_c E_x - iE_y r)\right)}{(\Omega_c - r)(\Omega_c + r)} \end{pmatrix}, \tag{3.209}$$

from which, in the rotational coordinate, we get a more simple expression

$$v^{+}=\frac{1}{2}(v_x^{(1)} + iv_y^{(1)}) = \frac{iE_{+}e^{ikz-i\omega t}(1 - e^{i(\omega - kV - \Omega_c)t})}{\omega - kV - \Omega_c}, \tag{3.210}$$

and

$$v^{-}=\frac{1}{2}(v_x^{(1)} - iv_y^{(1)}) = \frac{iE_{-}e^{ikz-i\omega t}(1 - e^{i(\omega - kV + \Omega_c)t})}{\omega - kV + \Omega_c}, \tag{3.211}$$

or in the compact form given in equation (3.203).

The velocities $v^{\pm}$ grow linearly with time, when the particle feels the wave electric field oscillating at the cyclotron frequency, namely when it is resonant at the fundamental harmonic cyclotron resonance ($\omega - kV = \pm\Omega_c$).

Including in the treatment a velocity distribution $f(V)$, the average of the previous quantities, over these distributions yields

$$\langle \vec{v}_{\perp}(z, t)\rangle = Re\left[\frac{iqe^{ikz-i\omega t}}{2m}[(c_{+} + c_{-})\vec{E}_{\perp} + i(c_{+} - c_{-})\vec{E}_{\perp} \times \hat{z}]\right], \tag{3.212}$$

where $c^{\pm}(t) = \alpha^{\pm}(t) - i\beta^{\pm}(t)$, being

$$\begin{aligned}\alpha^{\pm}(t) &= \int_{-\infty}^{\infty} dV \frac{f(V)(1 - kV/\omega)[1 - \cos(\omega t - kVt \mp \Omega_c t)]}{\omega - kV \mp \Omega_c},\\ \beta^{\pm}(t) &= \int_{-\infty}^{\infty} dV \frac{f(V)(1 - kV/\omega)\sin(\omega t - kVt \mp \Omega_c t)}{\omega - kV \mp \Omega_c}.\end{aligned} \tag{3.213}$$

For large value of t, these integrals approach asymptotic values independent of t

$$\begin{aligned}\alpha^{\pm} &\rightarrow P\int_{-\infty}^{\infty} \frac{f(V)(1 - kV/\omega)}{\omega - kV \mp \Omega_c} dV\\ \beta^{\pm} &\rightarrow \pm\frac{\pi\Omega}{\omega|k|} f\left(\frac{\omega \mp \Omega_c}{k}\right).\end{aligned} \tag{3.214}$$

According to the previous relations, in the Doppler shifted wave frame, the charged particle sees an electric field in phase with its gyro-motion, thus leading to a constant absorption of energy (see figure 3.17(a)).

The situation with higher harmonics is slightly more complicated. For instance, a Doppler shifted frequency corresponding to $\omega - k_{||}v_{||} = 2\Omega_c$ does not resonate with the gyro motion. However, when $k_{\perp} \neq 0$ the particles are not driven by a spatially uniform sine wave in time because of the finite perpendicular wavelength. In fact, when the perpendicular wavelength is comparable to the gyro radius, at a given instant of time, the electric field reverses sign across the orbit, as shown in

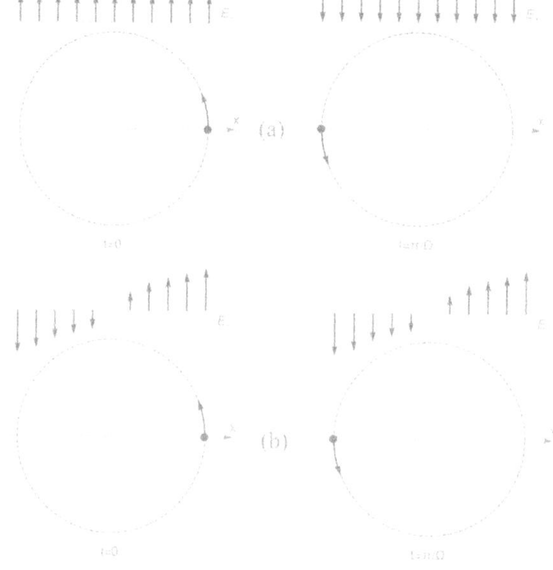

Figure 3.17. Cyclotron resonance mechanism at the fundamental (a) and at second harmonics (b).

figure 3.17(b). This spatially induced change in sign again brings the particle into resonance. The effect can be seen explicitly in the equation describing the evolution of v_+,due to the superposition of the cyclotron motion component and the effect of the second harmonic in Ω_c, namely

$$\frac{dv_+}{dt} + i\Omega_c v_+ = \frac{eE_+}{m} e^{-i(2\Omega_c t - k_\perp x)}, \tag{3.215}$$

by noting that

$$x = r_L \sin(\Omega_c t) = \frac{v_\perp}{\Omega_c} \sin(\Omega_c t), \tag{3.216}$$

and, after using the dipole approximation $k_\perp r_L \ll 1$, we write equation (3.215) as

$$\frac{dv_+}{dt} + i\Omega_c v_+ \approx \frac{eE_+}{m} e^{-i2\Omega_c t}\left[1 + \frac{k_\perp v_\perp}{2\Omega_c}(e^{i\Omega_c t} - e^{-i\Omega_c t})\right]. \tag{3.217}$$

The spatial term proportional to $(k_\perp v_\perp)/(2\Omega_c)e^{i\Omega_c t}$ beats with the second harmonic driving frequency producing a finite gyro radius induced resonant interaction. It is by this mechanism that energy is absorbed at higher cyclotron harmonics.

3.8 X-mode and O-mode transfer power

The previous (cumbersome) discussion has been devoted to establishing a sound analytical environment to frame the problems of wave absorption in magnetized plasma. A more pragmatic approach is now in order. The wave–particle power transfer calculation can be performed by a difference of the power (P_{in}), flowing into the plasma from the launcher in the perpendicular direction with respect to the external magnetic field (B_0) at plasma boundary ($x = -a$), and the exit power P_{out} in the opposite direction ($x = a$). The difference $P_{\text{in}} - P_{\text{out}}$ represents the power absorbed by the plasma. This implies that the heating efficiency η_h, which is defined as the fraction of the power absorbed, is given by

$$\eta_h = \frac{P_{\text{in}} - P_{\text{out}}}{P_{\text{in}}}. \tag{3.218}$$

In the cold plasma limit (without other dissipative effects) the relation between P_{in} and P_{out} suggested by the Lambert–Beer law is the following

$$P_{\text{out}} = P_{\text{in}} \exp\left(-\int_{-a}^{a} \alpha_\omega dx\right), \tag{3.219}$$

from which for equation (3.218) related to the heating efficiency we get

$$\eta_h = 1 - e^{-\lambda}, \tag{3.220}$$

where

$$\lambda = \int_{-\infty}^{\infty} \alpha_\omega dx. \tag{3.221}$$

The change in the integration limits from finite to infinite thickness, is due to the fact that the absorption coefficient rate depends on the frequency band, which is very narrow in space. The advantage being that the approximation leads to negligible errors and the integral can be evaluated in analytical terms (see below).

The α_ω coefficient will be derived starting from the particle motion ruled by the electromagnetic wave field and, as we have seen before, deriving the variation of the energy transfer dW/dt and averaging it to any given initial velocity

$$\frac{d\bar{W}}{dt} = \frac{k_{||}}{2\pi}\int_0^{2\pi/k_{||}} \frac{dW}{dt} dz_i, \tag{3.222}$$

z_i being the initial position of an electron.

Finally, the power gained per unit volume (G_ω) will be derived multiplying equation (3.222) by the particle distribution $f_0(v_{||}, v_\perp)$ and integrating over all velocities ($v_{||}, v_\perp$).

The last step is to derive the damping rate by means of a simpler power balance between the power loss and the power gain by the resonant particles.

In general, the total energy U stored in the wave averaged over one oscillation period in time, consists of the sum of the electric, magnetic, and plasma kinetic energy densities. In the limit of the energy slowly decaying in time with $\omega_i \ll \omega_r$, without any other dissipative phenomena the total energy is written as [11]

$$U(t) = \frac{1}{2}\left(\frac{\varepsilon_0|E|^2}{2} + \frac{|B|^2}{2\mu_0} + \sum_s\sum_j \frac{n_0 m_j|u_{j,s}|^2}{2}\right)e^{2\omega_i t} = U_0 e^{2\omega_i t}, \tag{3.223}$$

s being the particle species in the plasma, and for the energy conservation we get

$$dU(t)/dt = G_\omega e^{2\omega_i t}, \tag{3.224}$$

or

$$(2\omega_i U_0 + G_\omega)e^{2\omega_i t} = 0, \tag{3.225}$$

from which we get the damping rate

$$\omega_i = -G_\omega/(2U_0), \tag{3.226}$$

where the U_0 will be evaluated from the cold plasma analysis.

We have noted that the dimensions of the absorption α coefficient are those of the inverse of a length. It can, therefore, be viewed as the wave vector associated with the damped wave, through its phase velocity, namely

$$\alpha_\omega = k_{\perp i} = \frac{\omega_i}{V_{g\perp}}. \tag{3.227}$$

Furthermore, with the power density (per surface unit) lost by the wave given by (see for more details [11])

$$P_\perp = V_{g\perp} U_0, \tag{3.228}$$

we finally end up with (see equation (3.226))

$$\alpha_\omega = -G_\omega/(2P_\perp). \tag{3.229}$$

These are general considerations, aimed at fixing the physical meaning of the absorption coefficient.

In order to calculate the power absorption for the O-mode and X-mode an expression for the α_ω is necessary.

The spatial rate (α_ω) of the **O-mode absorption** from which only $E_{||} \neq 0$ is written as

$$\alpha_\omega = k_{\perp i} = \frac{\omega}{c}\frac{\mathrm{Im}[\varepsilon_{zz}]}{n_\perp}. \tag{3.230}$$

From equation (3.119) with $V = 0$ and $l = 1$, the $\mathrm{Im}[\varepsilon_{zz}]$, neglecting the contribution due to the Landau damping, assumes the expression

$$\mathrm{Im}[\varepsilon_{zz}] = \sqrt{\pi}\frac{\Omega_{p,e}}{\omega k_{||}v_{\mathrm{th}||}}e^{-b}I_1(b)\zeta^2 e^{-\zeta^2}, \tag{3.231}$$

where $\zeta = \omega - \Omega_{c,e}/(k_{||}v_{\mathrm{th}||})$.

Approximating the product $e^{-b}I_1(b)$ at the first order in b we get

$$e^{-b}I_1(b) \approx \frac{n_\perp^2}{4}\frac{v_{\mathrm{th}||}^2}{c^2}\frac{\omega^2}{\Omega_{c,\,e}^2}, \tag{3.232}$$

from which with the accessibility of the O-mode condition $n_\perp = (1 - \omega_{\mathrm{pe}}^2/\omega^2)^{1/2}$ (see equation (3.147)) we get

$$\alpha_\omega = \frac{\sqrt{\pi}}{4}\frac{v_{\mathrm{th}||}\Omega_{p,\,e}^2}{k_{||}c^3}\left(1 - \frac{\Omega_{p,\,e}^2}{\Omega_{c,\,e}^2}\right)^{1/2}\zeta^2 e^{-\zeta^2}. \tag{3.233}$$

Finally, the improper integral of the equation (3.233) gives for λ the expression

$$\lambda = \frac{\pi}{4}\frac{v_{\mathrm{th}||}^2}{c^2}\frac{\Omega_{c,e}R_0}{c}\frac{\Omega_{p,\,e}^2}{\Omega_{c,\,e}^2}\left(1 - \frac{\Omega_{p,\,e}^2}{\Omega_{c,\,e}^2}\right)^{1/2}. \tag{3.234}$$

We note that the optimum density to maximize the absorption is given by $\Omega_{p,\,e}^2/\Omega_{c,\,e}^2 = 2/3$ which yields for λ

$$\lambda = \frac{\pi}{6\sqrt{3}}\frac{v_{\mathrm{th}||}^2}{c^2}\frac{\Omega_{c,e}R_0}{c}, \tag{3.235}$$

namely, λ is proportional to the plasma temperature, which means that, starting from a reasonable coefficient much greater than 1 for the usually Tokamak parameters, we can achieve $\eta_h \approx 1$ (all the injected power can be absorbed). The only problem is that the density optimization value is slightly below the maximum density limit. This means that the wave does not reach the plasma core and the heating may be inefficient.

It is worth noting that, albeit we assumed $k_{||}$ finite (even tough small), the maximized value in equation (3.234) is independent of $k_{||}$. Therefore, keeping $k_{||} \to 0$ should be consistent with the results, obtained so far. It happens however that in this limit $\alpha_w \to 0$, and therefore the heating efficiency is zero. To resolve the contradiction it must be underlined that in this limit causality is violated and a correct treatment requires a calculation involving *ab initio* the relativistic invariance. We just mention this delicate point and address the reader to the specialized literature.

The spatial rate (α_ω) for the **X-mode absorption** needs some preliminary specifications in terms of the mode accessibility. In the cold plasma approximation form the dispersion relation we get, for the perpendicular component of the diffractive index, the expression

$$n_\perp = \frac{(\omega^2 + \omega\Omega_{c,e} - \Omega_{p,\,e}^2)(\omega^2 - \omega\Omega_{c,e} - \Omega_{p,\,e}^2)}{\omega^2(\omega^2 - \Omega_{c,\,e}^2 - \Omega_{p,\,e}^2)}, \tag{3.236}$$

derived under the assumption that $\omega \gg \Omega_{c,i}$ and $\omega \gg \Omega_{p,i}$.

If we evaluate equation (3.236) at fundamental harmonics ($\omega = \Omega_{c,e}$), considering the toroidal variation of $\Omega_{c,e} = \Omega_{c0,e}R_0/R$, we observe that the accessibility to the plasma center is more easily achieved with a launcher from high field side and is substantially reduced for a low field side launcher due to the cut-off density and the upper hybrid resonance (figure 3.18).

Even if recently some configurations have been tested for high field side launcher [18] nowadays is still preferable to work with the second harmonics of the X-mode. Since in this case for $\omega = 2\Omega_{c,e}$ equation (3.236) reduces to

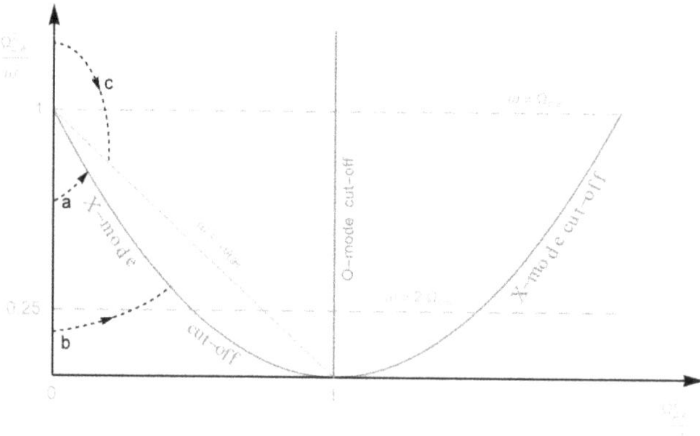

Figure 3.18. CMA diagram showing the X-mode and O-mode cut-off and the accessibility for three different launching X-modes: (a) the fundamental from the low field side which does not reach any resonance due the density cut-off, (b) at second harmonic range which is able to reach the second harmonic resonance, (c) the fundamental from the high field side which can propagate to the fundamental and reach the upper hybrid resonance.

$$n_{\perp}^{2} = \frac{(6 - \Omega_{p,e}/\Omega_{c,e}^{2})(2 - \Omega_{p,e}^{2}/\Omega_{c,e}^{2})}{4(3 - \Omega_{p,e}^{2}/\Omega_{c,e}^{2})}, \tag{3.237}$$

having a cut-off for $\Omega_{p,e}^{2}/\Omega_{c,e}^{2} = 2$ and the upper hybrid resonance for $\Omega_{p,e}^{2}/\Omega_{c,e}^{2} = 3$. It follows that the second harmonic of the X-mode has good accessibility to the plasma core and with a density limit twice that provided by the O-mode.

The use of the same procedure adopted for the O-mode along with equation (3.237) for $n_{\perp}$ allows us to cast the α_{ω} coefficient as

$$\alpha_{\omega} = 2\sqrt{\pi}\frac{v_{\text{th}||}\Omega_{c,e}^{2}}{c^{3}k_{||}}G(\alpha)e^{-\zeta^{2}}, \tag{3.238}$$

where $\alpha = \Omega_{p,e}^{2}/\Omega_{c,e}^{2}$ and

$$G(\alpha) = \frac{\alpha(2 - \alpha)^{1/2}(6 - \alpha)^{5/2}}{32(3 - \alpha)^{5}/2}, \tag{3.239}$$

and with ζ specified in the limit of a thin resonant layer

$$\zeta^{2} \approx \left(\frac{2\Omega_{c,e}}{k_{||}v_{\text{th}||}}\right)^{2}\frac{x^{2}}{R_{0}^{2}}. \tag{3.240}$$

From equation (3.238) we get the absorbing coefficient

$$\lambda = 2\int_{-\infty}^{\infty}\alpha_{\omega}dx = 2\pi G(\alpha)\frac{v_{\text{th}||}^{2}}{c^{2}}\frac{\Omega_{c,e}R_{0}}{c}. \tag{3.241}$$

The maximum value for $G(\alpha)$ occurs for $\alpha = 1.75$ which eventually yields

$$\lambda = 3.66\frac{v_{\text{th}||}^{2}}{c^{2}}\frac{\Omega_{c,e}R_{0}}{c}. \tag{3.242}$$

It exhibits the same scaling versus $\Omega_{p,e}/\Omega_{c,e}$ as for the O-mode, except for a multiplicative constant ten times larger, which enhances the X-mode absorption (see figure 3.19).

3.9 Practical formulae for plasma physics and fusion devices

This and the forthcoming section close the first part of the book. We have gone through the major aspects of magnetized plasma physics and we have discussed some elements of Tokamak design and the physics of external heating devices. We have often mentioned the scaling formulae, which are frequently exploited to provide a quick understanding of the feasibility of Tokamaks and of their performances in terms of 'output' power.

The scaling formulae discussed here and those regarding the design of the heating sources reported in the forthcoming part of the book, trace back to a common

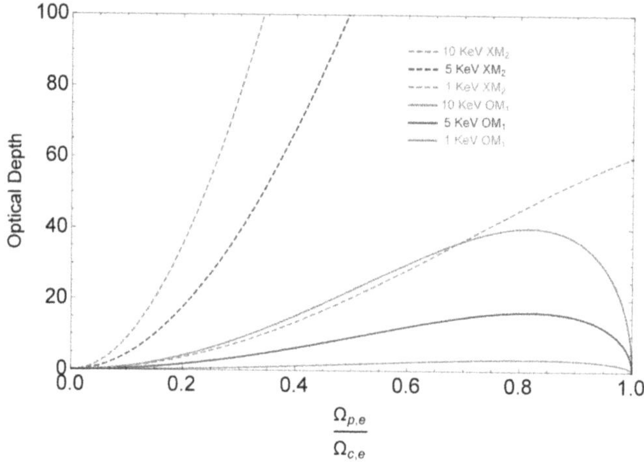

Figure 3.19. The optical thickness of the O-mode and X-mode versus the plasma density for a 140 GHz RF wave with Tokamak major radius $R = 6$ m at different plasma temperature values.

origin. They are the result of the overlapping and synthesis of different experiences (theoretical, numerical, experimental, etc) all combined in a wise fashion, in order to identify global quantities capable of accounting for the phenomenology one is interested in [19].

We provide an idea of the usefulness of these procedures, when applied to the design of a magnetic confinement device and of the relevant limits as well.

Before starting, it is worth fixing the terms of the forthcoming discussion by recalling that the goal of any fusion program is that of providing fusion power P_{fus} with the minimum auxiliary external heating, thus maximizing the ratio $Q = P_{\text{fus}}/P_{\text{heat}}$. Put in even simpler terms, the problem is that of fixing a given fusion power and then specifying key parameters, like the size and the magnetic field of a reactor. Such orientation, emerging from the previous discussion, is not an easy task.

The 'laws' we are going to describe are called empirical (a term not fully appropriate) and deal with relationships among collective variables, merging physical, geometrical and engineering quantities. The intent is that of putting on a common ground physical and engineering parameters and checking the mutual consistency, which is not easily achieved because it is hard to reconcile the requirements from plasma and engineering parameters (see the forthcoming section).

The two sets concur in defining a number of dimensionless parameters, like the collision frequency (multiplied by the transit time) and the beta coefficient defined as the ratio between the outward (plasma) inner (magnetic) pressure.

Although there are a number of parameters characterizing the magnetic fusion devices, we will concentrate on four critical quantities (the plasma density, the thermal energy, the torus major radius and the magnetic field). They can be further reduced, after assuming that the fusion cross section exhibits a quadratic dependence on the temperature.

There is no doubt that one of the most significant quantities (or figures of merit) of any Tokamak is the foreseen output power, which is a quantity depending on different parameters, characterizing the device itself.

According to the discussions of the first two chapters we can say that the fusion power scales as (see chapter 2, equations (2.1–2.6)

$$P_{\text{fus}} \propto \frac{R^3}{A^2} B_e^4, \tag{3.243}$$

where A is the Tokamak aspect ratio, namely the ratio of the torus major to the minor radius.

In order to better specify the geometrical environment we are referring to, figure 3.20 is reported for convenience

The previous equation is a scaling relation. It states that the fusion power can be increased by designing devices with larger major radius, larger confining magnetic field and lower aspect ratio.

The largest existing Tokamak (JET), characterized by $R = 3$ m, $B = 4$ T, has been designed to operate near the break-even. The next-step device ITER (R = 6.2 m, B = 5.2 T), presently under construction, aims at providing a fusion power larger than a factor 10. An economically attractive fusion power plant requires even larger values and has to fulfill several other conditions, such as long pulses or steady-state operation (effectively allowing continuous electricity production).

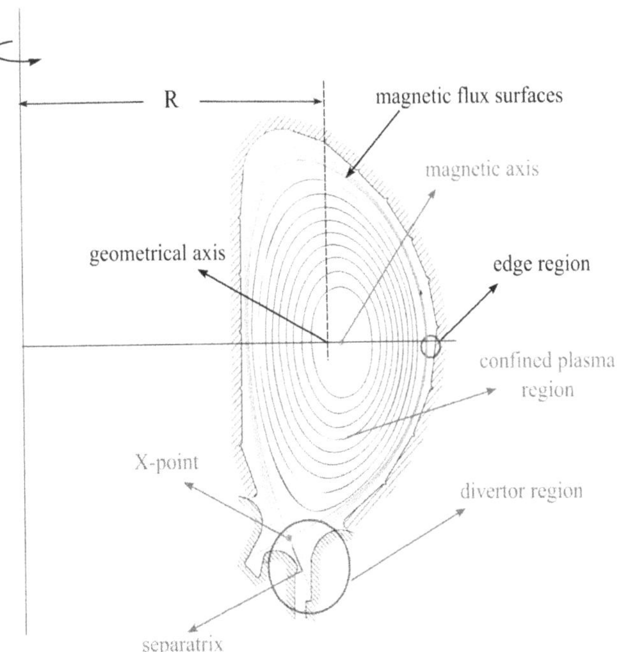

Figure 3.20. Tokamak poloidal torus section: magnetic surfaces, separatrix point, divertor and axes. Taken from [20] with some modifications by the authors. Copyright 2017 EURATOM. CC BY 3.0.

It is evident that to transform equation (3.243) into something useful for design purposes it is necessary to plug the proportionality factor in, which should be independent of the specific device under study and provide, within the real of magnetic fusion devices, a kind of 'universal' constant.

In the introductory section of the previous chapter, we have addressed the problem of determining the Tokamak characteristics starting from elementary considerations involving the definition of 'beta' parameters (see equations 2.1 and 2.6); we concluded that a commercially attractive confining machine would require large beta values. We just assumed a working value around 5% and got quantities of interest (confining time, Tokamak size, wall power load, etc) without justifying the difficulties with the beta stability limit.

We address this section to the derivation of equation (3.243) including the proportionality factor.

In order to estimate the fusion power we should consider parameters like the plasma density, the thermal energy and the plasma current.

We follow the notation proposed in [21] according to which the quantities denoted with a superimposed ^ are defined as

$$\begin{aligned} &\hat{n} = 10^{19} n \, [\mathrm{m}^{-3}], \\ &\hat{T} \equiv \text{Temperature, Energy in keV}, \\ &\hat{I}_p \equiv \text{Plasma current in MA}, \end{aligned} \tag{3.244}$$

those without anything superimposed are expressed in SI units.

The strategy we follow is that of starting from the definition of P_{fus} (see chapter 1, equation (1.10)) and express the variables therein in terms of the machine parameters.

The plasma density is fixed invoking the so-called Greenwald limit [22], according to which it cannot exceed the value

$$\begin{aligned} \hat{n}_G &= C_n \frac{A^2 \hat{I}_p}{R^2}, \\ C_n &= \frac{10}{\pi}. \end{aligned} \tag{3.245}$$

By recalling that the plasma current is linked to the poloidal magnetic field by the Ampère law, namely

$$\begin{aligned} &\mu_o I_p = L_c B_p, \\ &L_c = 2\pi a \, F, \\ &F \equiv \text{plasma silhohuette coefficient}, \end{aligned} \tag{3.246}$$

employing the definition of the q-safety factor (see equation (1.114))

$$q = \frac{1}{A} \frac{B_t}{B_p}, \tag{3.247}$$

we find

$$\mu_0 I_p = \frac{2\pi a}{q}\frac{F}{A}B_t = F\frac{2\pi a^2}{qR}B_t = F\frac{2\pi}{qA^2}R\,B_t, \tag{3.248}$$

and eventually

$$\hat{I}_p = C_I\frac{RB_e}{A^2 q}, \tag{3.249}$$

with

$$C_I = \frac{2\pi\ 10^{-6}}{\mu_0}, \tag{3.250}$$

after noting that $\hat{I}_p = 10^{-6} I_p$ and that $B_e \cong B_t \gg B_p$.

The plasma Greenwald density can accordingly be written as

$$\hat{n}_G = C_n C_I \frac{B}{qR}. \tag{3.251}$$

With $\hat{n}_G$ being an upper limit, we introduce the fraction coefficient $n_N = \hat{n}/\hat{n}_G$ and write the actual plasma density according to

$$\hat{n} = C_n C_I n_N \frac{B}{qR}. \tag{3.252}$$

The next step is the inclusion of the temperature, which can be associated with the magnetic field through the 'beta' dimensionless parameters (see chapter 1.1, equations (1.84–1.86)

$$\begin{aligned} \beta &= 2\mu_0\frac{\langle p\rangle}{B_e^2}, \\ \langle p\rangle &= 2n\ KT. \end{aligned} \tag{3.253}$$

The factor 2 in the definition of the plasma kinetic pressure comes from the fact that both electrons and ions are assumed to be at the same temperature.

Since the value of beta is given in % we can convert the previous relationship into our practical units, thus getting (see the previous discussion on units)

$$\begin{aligned} \beta[\%] &= C_\beta\frac{\hat{n}\hat{T}}{B_e^2}, \\ C_\beta &= 4\cdot 10^2\cdot\mu_0\cdot 10^{19}\cdot 10^3\cdot e \cong 0.805. \end{aligned} \tag{3.254}$$

We have already noted that β is subject to limitations due to plasma instabilities; it comes out that the maximum value scales with $1/(qA)$, which, from the previous identities, comes out to be

$$(qA)^{-1} = \frac{\mu_0 I_p}{2\pi a\ B_e}. \tag{3.255}$$

We take advantage from the above identity by imposing that

$$\beta_{\%} = \beta_N \frac{\hat{I}_p}{a\ B_e}, \tag{3.256}$$

where β_N is called '***normalized beta***' and is subject to certain constraints to be specified later.

Putting everything together we find that the density-temperature product satisfies the identity

$$\hat{n}\ \hat{T} = \frac{\beta_N}{C_\beta}\frac{A\hat{I}_p B}{R} = \frac{C_l}{C_\beta}\frac{\beta_N B_e^2}{q\ A}. \tag{3.257}$$

We are almost done, the next effort is to obtain a suitable 'practical' formula for the fusion power. This can be achieved by following the prescription given below

(i) Consider a D–T mixture with $n_\mathrm{D} = n_\mathrm{T} = n/2$ and write the fusion power as

$$P_f = \frac{n^2}{4}\langle\sigma v\rangle_{\mathrm{DT}} E_{\mathrm{DT}} V, \\ V = 2\pi^2\ \kappa\ R^2 a, \tag{3.258}$$

or

$$P_f = \pi^2\frac{n^2}{2}\langle\sigma v\rangle_{\mathrm{DT}} E_{\mathrm{DT}}\ \kappa\ A^{-2}R^3. \tag{3.259}$$

(ii) Use the following approximation for the D–T reactivity [21]

$$\langle\sigma v\rangle_{\mathrm{DT}} \cong 1.18 \cdot 10^{-24}\hat{T}^2 \mathrm{m}^3\ \mathrm{s}^{-1}, \tag{3.260}$$

and putting everything together we get

$$\hat{P}_f = C_f \kappa\ A^{-2}R^3\hat{n}^2\hat{T}^2. \tag{3.261}$$

(iii) The coefficient C_f is obtained after a little arithmetic of units, namely if P_f is expressed in MW and recalling that the fusion energy released in the D–T reaction is 17.6 MeV[12],

$$C_f \cong 17.6 \times e \times 1.18 \cdot 10^{-24} \times 10^{2\cdot 19} \times \frac{\pi^2}{2} \cong 1.64 \cdot 10^{-3}. \tag{3.262}$$

(iv) Use equation (3.257) to express the power in terms of β_N, namely

$$\hat{P}_f = d\frac{C_f C_l^2}{C_\beta^2}\kappa\ R^3\frac{\beta_N^2 B_e^4}{q^2 A^4}. \tag{3.263}$$

[12] The reaction energetic balance is D + T → ^{4}He(3.56 MeV) + n(14.03 MeV), with $E_n/E_\alpha \cong 4$.

The previous equation yields the power in terms of the engineering machine parameters, it is evident that it becomes a powerful tool to get a quick idea of the feasibility of the magnetic fusion device itself.

What next?

We have fixed the fusion power, which is an important reference value, but not exhaustive. It should be complemented by deriving analogous results for the fusion gain, defined as

$$Q = \frac{P_f}{P_{a,h}}, \tag{3.264}$$

where $P_{a,h}$ is the auxiliary heating power[13].

The fusion power is distributed between neutrons and α-particles, assuming that

$$P_n = (\lambda - 1)P_\alpha, \tag{3.265}$$

we can write

$$P_f = \lambda\, P_\alpha, \tag{3.266}$$

and it is experimentally found that $\lambda \cong 4.94$ (see footnote [12]).

It is not difficult to realize that the net power is just provided by (a more appropriate discussion is reported in [19, 21])

$$P_{\text{net}} = \gamma_r(P_\alpha + P_{\text{ha}}) = \gamma_r P_f\left(\frac{1}{\lambda} + \frac{1}{Q}\right), \tag{3.267}$$

where $\gamma_r(0 < \gamma_r < 1)$ is a coefficient associated with the radiation losses (bremsstrahlung and synchrotron see [21]).

In conclusion we find

$$\hat{P}_{\text{net}} = \gamma_r\, C_f\kappa\, A^{-2}R^3\hat{n}^2\hat{T}^2\left(\frac{1 + \frac{Q}{\lambda}}{Q}\right). \tag{3.268}$$

We have so far sketched a strategy aiming at developing a 'formulary' sufficiently elaborated to guarantee a synthetic view to a Tokamak dimensioning. It is not yet a designing tool—how to accomplish such a goal will be shown in the forthcoming section.

3.9.1 Scaling

In the previous section we have embedded general and well established formulae in plasma physics and written them in an easily amenable form for quick computation.

[13] Equation (3.264), as it stands, is not fully representative of the fusion gain since it does not contain the power necessary to feed other ancillary quantities, like those necessary to pump the coolant in the blankets or the efficiency to convert the thermal into electric power. This definition looks, however, appropriate, since it refers to fusion balance efficiency only.

What we are looking for is a set of formulae allowing a calculation protocol according to which fusion power and gain Q are, e.g., specified and corresponding machine parameters are eventually predicted [23–25].

In the previous section we did not specify the confinement time which is one of the key parameters of any fusion device and is not associated with simple formulae associating it with the plasma and machine parameters.

The best one can do is therefore to rely on scaling relations derived experimentally on different operating machines. Without entering into more specific details, which can be [26, 27], we note that the confining time is parameterized as

$$\tau_E = C_{\mathrm{SL}} M^{\alpha_M} \kappa^{\alpha_\kappa} \varepsilon^{\alpha_\varepsilon} \hat{n}^{\alpha_n} \hat{I}_p^{\alpha_I} R^{\alpha_R} B^{\alpha_B} \hat{P}_{\mathrm{net}}^{\alpha_p}, \tag{3.269}$$

where the parameters are obtained via a fitting procedure of the experimental data. They are reported in table 3.1 and further specific details can be obtained from the references therein.

The confinement time is defined as the ratio between the internal energy and the transported power, namely

$$\tau_E = \frac{\hat{W}}{\hat{P}_{\mathrm{tr}}}, \tag{3.270}$$

where $\hat{W}$ (the internal energy in MJ) is specified by the relation

$$\hat{W} = \frac{3}{2}\hat{n}(\hat{T}_e + \hat{T}_i)V. \tag{3.271}$$

After keeping $\hat{T}_e = \hat{T}_i$, from the previous equation we get

$$\hat{W} = C_{\mathrm{tr}} \kappa \frac{\hat{n}\hat{T}}{A^2} R^3, \tag{3.272}$$

Table 3.1. Numerical values for the exponents, appearing in the scaling relation (3.269).

Name Ref.	IPB(98)(y,2) [[26], equation (20)]	DS03 [27]	L-mode [[26], equation (24)]
C_{SL}	0.056 2	0.028	0.023
α_M	0.19	0.14	0.20
α_k	0.78	0.75	0.64
α_ε	0.58	0.30	−0.06
α_n	0.41	0.49	0.40
α_I	0.93	0.83	0.96
α_R	1.97	2.11	1.78
α_B	0.15	0.07	0.03
α_P	−0.69	−0.55	−0.73

with

$$C_{\rm tr} = 6\,\pi^2 10^{19} \times e \times 10^{-3} \cong 0.095, \tag{3.273}$$

where, from (3.270),

$$\hat{P}_{\rm tr} = C_{\rm tr}\kappa \frac{\hat{n}\hat{T}}{A^2 \tau_E} R^3 = \hat{P}_{\rm net}. \tag{3.274}$$

Finally, for the triple product (3.268) we end up with the following expression

$$\hat{n}\hat{T}\tau_E = \frac{C_{\rm tr}}{\gamma_r C_f}\left(\frac{1+\dfrac{Q}{\lambda}}{Q}\right)^{-1}. \tag{3.275}$$

On the other side, using equations (3.267) and (3.269) we also find

$$\hat{n}\;\hat{T}\tau_E = C_{\rm SL} M^{\alpha_M} \kappa^{\alpha_\kappa} \varepsilon^{\alpha_\varepsilon} \hat{n}^{\alpha_n+1} \hat{T}\; \hat{I}_p^{\alpha_I} R^{\alpha_R} B^{\alpha_B} \left[\gamma_r P_f \left(\frac{1+\dfrac{Q}{\lambda}}{Q}\right)\right]^{\alpha_p}. \tag{3.276}$$

Before proceeding further we note that according to equations (3.275)–(3.257) the confining time can be expressed in the form

$$\tau_E = \frac{C_{\rm tr} C_\beta}{\gamma_r C_f C_I} \frac{q\;A}{\beta_N B_e^2} \frac{Q}{\left(1+\dfrac{Q}{\lambda}\right)}. \tag{3.277}$$

In terms of scaling coefficients, after embedding equations (3.275)–(3.276) we end up with

$$\left[\frac{Q}{\gamma_r\left(1+\dfrac{Q}{\lambda}\right)}\right]^{1+\alpha_p} = C_{\rm sl} C_n^{\alpha_n} C_I^{\gamma_I} \frac{C_f}{C_\beta C_{\rm tr}} \hat{P}_f^{\alpha_p} M^{\alpha_M} \kappa^{\alpha_\kappa} \varepsilon^{\gamma_\varepsilon} q^{-\gamma_I} n_N^{\alpha_n} \beta_N R^{\gamma_R} B^{\gamma_B}, \tag{3.278}$$

with

$$\begin{aligned} \gamma_I &= 1 + \alpha_n + \alpha_I, \quad & \gamma_\varepsilon &= 1 + \alpha_\varepsilon + 2\alpha_I, \\ \gamma_R &= \alpha_R + \alpha_I - \alpha_n, \quad & \gamma_B &= \alpha_B + \alpha_n + \alpha_I + 2. \end{aligned} \tag{3.279}$$

Equations (3.263) and (3.278) are the pivotal equations of our analysis, since they express fusion power and gain in terms of machine constructive parameters. The way they are exploited is clear, one prescribes P_f, Q and find the values of Tokamak radius, magnetic field compatible with these values (figure 3.21).

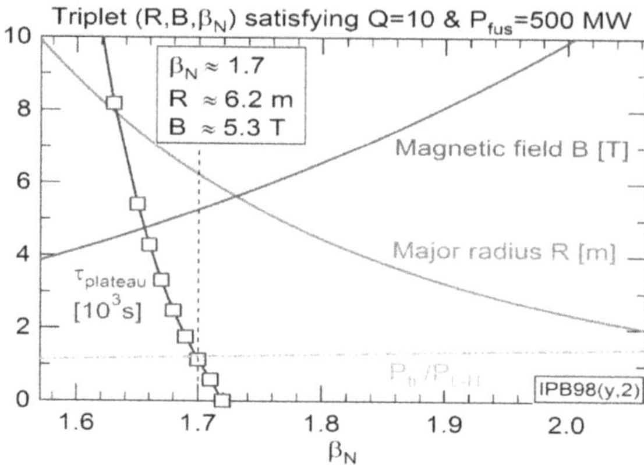

Figure 3.21. Tokamak design parameters versus β_n as a result of the scaling formulae optimization (reproduced from [21], copyright 2019 IAEA, Vienna).

It is worth noting that we have provided the scaling relations in terms of β_N and n_N whose constraints are defined in the previous section.

References

[1] Swesson J 2011 *Tokamaks* (Oxford: Oxford Science Publication)

[2] Bigot B 2019 Progress toward ITER's first plasma *Nucl. Fusion* **59** 1–11

[3] Koch R 2010 The ion cyclotron, lower hybrid and Alfven wave heating methods *Fusion Sci. Technol.* **57** 206–13

[4] Anderson H 2000 Neutral beam stopping and emission in fusion plasmas I: deuterium beams *Plasma Phys. Control. Fusion* **42** 781–806

[5] Hanada M 2001 *Development of Multi-megawatt Negative Ion Sources and Accelerators for Neutral Beam Injectors* (Vienna: IAEA) https://www-pub.iaea.org/MTCD/publications/PDF/csp_008c/fec1998/pdf/ftp_20.pdf

[6] Oikawa T *et al* 2000 Heating and non-inductive current drive by negative ion based NBI in JT-60U *Nucl. Fusion* **40** 435

[7] Vincenzi P 2016 *Interaction between neutral beam fast particles and plasma in fusion experiments* PhD Thesis Joint Research Doctorate in Fusion Science and Engineering, Cycle XXVIII

[8] Kazakov Y, Van Eester D and Ongena J 2015 Plasma heating in present-day and future fusion machines *12th Carolus Magnus Summer School on Plasma and Fusion Energy Physics, Leuven (Belgium), 24 Aug–4 Sep* **vol 298** 290–7 (Jülich: Forschungszentrum Jülich GmbH Zentralbibliothek, Verlag) pp 290–7 https://juser.fz-juelich.de/record/283655

[9] Stix T H 1972 Heating of toroidal plasmas by neutral injection *Plasma Phys.* **14** 367–84

[10] Kock R 2008 The coupling of electromagnetic power to plasmas *Fusion Sci. Technol.* **53** 184–93

[11] Freidberg J 2007 *Plasma Physics and Fusion Energy* (Cambridge: Cambridge University Press)

[12] Hutchinson I H 2004 *Introduction to Plasma Physics I–Courses in Nuclear Science and Engineering* http://www.ismolindell.com/publications/monographs/pdf/Aftis.pdf

[13] Parra Felix I 2019 *Collisionless Plasma Physics* http://www-thphys.physics.ox.ac.uk/people/FelixParra/CollisionlessPlasmaPhysics/notes/
[14] Miyamoto K 2007 *Controlled Fusion and Plasma Physics* (Series in Plasma Physics) (London: Taylor & Francis)
[15] Boyd T J M and Sanderson J J 2003 *the Physics of Plasma* (Cambridge: Cambridge University Press)
[16] Stix T H and Nierenberg W A 1962 *The Theory of Plasma Wave* (New York: McGraw-Hill Book Company)
[17] Stix T H 1992 *Waves in Plasmas* (New York: Springer)
[18] Elserafy H 2019 HFS injection of x-mode for EBW conversion in quest *Plasma Fusion Res.* **14** 1205038
[19] Freidberg J P, Mangiarotti F J and Minervini J 2015 Designing a tokamak fusion reactor. How does plasma physics fit in? *Phys. Plasmas* **22** 070901
[20] Federici G, Biel W, Gilbert M R, Kemp R, Taylor N and Wenninger R 2017 European demo design strategy and consequences for materials *Nucl. Fusion* **57** 092002
[21] Sarazin Y, Hillairet J, Duchateau J-L, Gaudimont K, Varennes R and Garbet X 2019 Impact of scaling laws on tokamak reactor dimensioning *Nucl. Fusion* **60** 016010
[22] Greenwald M 2002 Density limits in toroidal plasmas *Plasma Phys. Control. Fusion* **44** R27–53
[23] Coda S 2019 Physics research on the TCV tokamak facility: from conventional to alternative scenarios and beyond *Nucl. Fusion* **59** 112023
[24] Crisanti F 2017 The Divertor Tokamak Test facility proposal: physical requirements and reference design *Nucl. Mater. Energy* **12** 1330–5
[25] Green Book 2019 *Divertor Tokamak Test Facility Interim Design Report* https://http://www.dtt-project.it/index.php/dtt-green-book.html
[26] ITER Physics Expert Group on Confinement and Transport *et al* 1999 *Nucl. Fusion* **39** 2175
[27] Slips A C C 2018 Assessment of the baseline scenario at $q_{95} \sim 3$ for ITER *Nucl. Fusion* **58** 126010

Part II

External additional heating sources

G Dattoli, E Di Palma, S P Sabchevski and I P Spassovsky

Chapter 4

Undulator based free electron laser

4.1 Introduction

This chapter opens the second part of the book. We will focus here on the tools foreseeable for the production of the electromagnetic radiation, to be exploited to heat the plasma.

According to the general discussion of the previous chapters, a heating source should satisfy a number of requirements:

(a) To operate within a specific range of frequencies.
(b) To have large enough power.
(c) To have large efficiency.

Whilst (a) and (b) are clear, the third needs a few words of comments.

Suppose that the heating source is produced through a process in which power is subtracted to an electron beam (e-beam) and is transferred to an electromagnetic field oscillating at a given frequency. The relevant efficiency is, within this context, defined as the fractional change of e-beam power. Just to fix ideas, we fix such a value in excess of 30%.

This represents a critical requirement to develop the future electron cyclotron systems for next generation fusion devices.

We have already explored the mechanisms, ensuring the plasma heating, through millimeter wave power absorption. The wave supplies the power to the plasma electrons, which in turn transmit it to the surrounding particle 'soup'.

The candidate sources for this type of heating process are systems which we grossly ascribe to the family of free electron laser (FEL) generators.

These tools are characterized by an e-beam of a given energy, a system allowing the resonant emission and a mechanism providing the coupling and the amplification of the corresponding electromagnetic (em) field, either in a waveguide or in vacuum.

doi:10.1088/978-0-7503-2464-9ch4

The frequency of the emitted radiation depends on the beam energy (and on other quantities to be specified later). It is accordingly evident that the devices we will discuss are specified by different processes all characterized by their own efficiency.

E-beam acceleration, which may occur via electrostatic or radio-frequency (RF) means, e-beam power transfer from the electron to the field. The wall plug efficiency is defined by the combination of the single efficiency.

In the following we will deal with different devices which can be framed within the context of FEL or vacuum tube devices. The specific definition depends on the field of provenence of the user.

Apart from the specific names we recall that, in the past it was said that a 'power grid tube is a device using the flow of free electrons in a vacuum to produce useful work'. Such a definition is well appropriate for any FEL-like device, for which the emission of radiation is ensured by bremsstrahlung.

In the following we will discuss undulator FEL (U-FEL), gyrotron and Cyclotron Auto Resonance Maser (CARM). U-FELs will be described not for their use in magnetic fusion plasma but because they can be exploited as a paradigmatic tool to introduce the relevant physics. The gyrotron is a kind of FEM (M stands for maser) working in the GHz region of mm waves from 20 to 500 GHz. They exhibit a significant breakdown of efficiency above 200 GHz, the region in which CARM may play a significant role.

Before concluding this introductory section, we would like to stress that the present state of the art sees that the most reliable and mature technology for the electron cyclotron resonance heating (ECRH) and electron cyclotron current drive (ECCD) is provided by the gyrotron system. In the ITER installation, twenty four 170 GHz gyrotron systems with 1 MW microwave power each has been released with the collaboration of Japan, Russia and EU. Studies for self-sustained DEMO operation claims for the ECCD system an efficiency exceeding 30% for large continuous-wave power ($\approx$ 1 MW) at high frequency ($\approx$ 250 GHz) [1]. Nowadays, the scientific community is devoting a special effort to extending above 200 GHz the operational frequency range of the gyrotron. A notable result has been recently achieved with the demonstration of a 300 GHz source with 0.5 MW output power and an efficiency of 20% [2]. The main problems with the gyrotron appear when we need to increase the output power using a moderately relativistic beam as the resonance condition, at which the beam–particle energy exchange takes place, will be affected by the relativistic factor with a considerable reduction of the system efficiency. Conversely, the lose resonance during the gyrotron interaction can be balanced considering an RF system tuned at the Doppler-shift interaction (CARM), as will be discussed in chapter 6.

4.2 Undulator based FEL, generalities

The previous three chapters have dealt with generalities on plasma and Tokamak physics, on the relevant design issues and on the problems associated with the additional heating. In this respect we have covered a number of items, which can be exploited for this purpose, we have discussed the importance of electromagnetic sources, but we did not specify any device conceived to accomplish this task. The forthcoming chapters deal with these aspects.

Free electron coherent devices, like gyrotrons, are currently exploited to launch electromagnetic radiation inside the plasma. The radiation is characterized in terms of intensity, frequency, polarization etc by the requirements discussed in chapter 3. Furthermore, being based on a mechanism of power transfer from an e-beam to a narrow band electromagnetic field, they should be designed to guarantee high transfer efficiency.

This chapter is intended as introduction to FELs, which is a term describing a fairly broad family of devices, including different sources that provide coherent radiation through a beam of free electrons, where 'free' should be understood as 'not bound' in an atomic or molecular system.

One such system is the klystron, a known workhorse feeding the Linac RF cavities and realizing the conditions to bring the electrons at relativistic energies [3].

This chapter is devoted to framing FEL-type sources within a unitary context, keeping the U-FELs. To accomplish this task we follow the line developed in previous publications by some of the present authors [4, 5].

U-FELs have become powerful tools for a variety of applications in physics. Since the first successful experiment during the second half of the 1970s [6, 7], the next forty years yielded successful operation in different configurations, like oscillators and single passage devices [3, 5, 8].

The U-FEL technology was developed in parallel with the synchrotron light sources and merged at the beginning of this century. The advances in the technology of high brightness Linacs have allowed the production of x-ray beams with un-precedented characteristics in terms of coherence, flux and brightness. The physical mechanisms underlying the U-FEL operation are generally well understood and will be reviewed herein. Subsequently, the underlying mechanism of U-FEL will be used as a pivot to establish a common thread with other free electron coherent sources.

We can define the FELs (U, but not exclusively) in a non-technical, but (hopefully) effective, language, as devices 'stealing' power from an e-beam and transforming it into em radiation [9].

The e-beam power is defined as the product between the current it bears and its energy, namely, expressing the current in Ampère and the energy in MeV, the power in MW is written as[1]

$$\hat{P}_E[\mathrm{MW}] = \hat{I}[\mathrm{A}]\,\mathcal{E}\,[\mathrm{MeV}]. \tag{4.1}$$

In the case of electron bunch of charge Q, displaying a Gaussian distribution with rms time duration σ_τ the current is expressed as

$$\hat{I} = \frac{Q}{\sqrt{2\,\pi}\sigma_\tau}. \tag{4.2}$$

[1] The superimposed 'hat' denotes peak current.

In order to extract energy from electrons and transform it into electromagnetic energy, it is necessary:

1. To find a suitable device allowing the coupling of the electrons with an external electromagnetic field.
2. To select a suitable resonance mechanism.

In figure 4.1 we have reported the tool ensuring the previous two conditions, namely an undulator magnet [10, 11], with an alternating magnetic field directed along the vertical direction, namely

$$\begin{aligned}
&\vec{B} = B_0\left(0,\ \sin\left(2\,\pi\frac{z}{\lambda_u}\right),\ 0\right),\\
&B_0 \equiv \text{on axis field},\\
&\lambda_u \equiv \text{undulator period},\\
&L_u = N\lambda_u \equiv \text{undulator length}.
\end{aligned} \tag{4.3}$$

When the e-beam (usually with ultra-relativistic energies) enters the magnet, the Lorentz force induces a transverse component of motion (see figure 4.1), which ensures the conditions for the coupling of the beam to a co-propagating electromagnetic wave.

The electrons execute transverse oscillations around the vertical direction and emit a flash of bremsstrahlung radiation, at each undulator period.

Positive interference occurs at the wavelength fixed by the phase difference between electrons and photons after each period

$$\begin{aligned}
&\delta = (1 - \beta_z)\,\lambda_u = \lambda,\\
&\beta_z = \sqrt{1 - \frac{1}{\gamma_z^2}}.
\end{aligned} \tag{4.4}$$

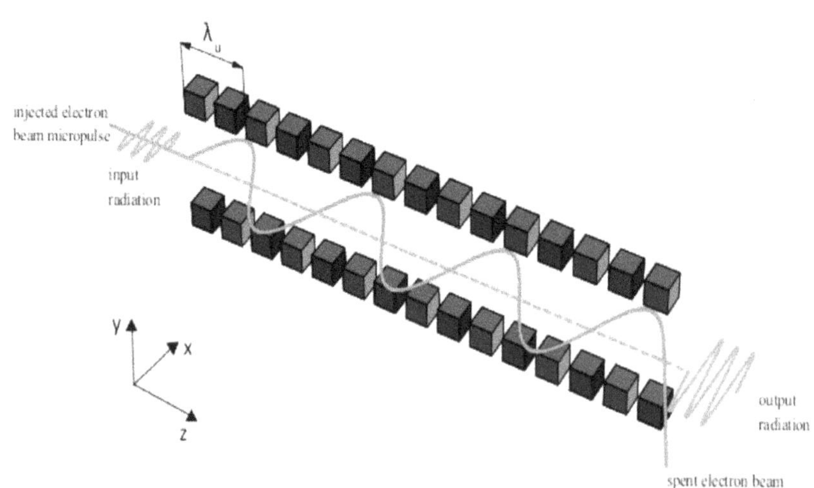

Figure 4.1. Electrons moving inside an undulator along with the photon flashes induced by bremsstrahlung.

Taking into account the relativistic nature of the electron motion, we can expand the longitudinal velocity at the lowest order in the relativistic factor, namely

$$\beta_z \cong 1 - \frac{1}{2\,\gamma_z^2}, \tag{4.5}$$

and finally getting for the wavelength

$$\lambda \cong \frac{\lambda_u}{2\,\gamma_z^2}. \tag{4.6}$$

The longitudinal Lorentz factor contains the modulation induced by the presence of the transverse component of the motion inside the undulator, in terms of the kinematical variables β_z reads

$$\beta_z^2 = 1 - \frac{1 + (\gamma\,\beta_\perp)^2}{\gamma^2}. \tag{4.7}$$

The transverse reduced velocity $\beta_\perp$, induced by the Lorentz force and averaged on the undulator period, written as

$$\begin{aligned} \beta_\perp &\cong \frac{K}{\gamma\sqrt{2}}, \\ K &= \frac{eB_0\lambda_u}{2\pi m_e c}, \end{aligned} \tag{4.8}$$

where K, usually referred to as undulator strength parameter, is a measure of the electron deviation from the undulator axis. Accordingly, the average angular deviation from the axis is

$$\langle\vartheta\rangle \cong \frac{K}{\sqrt{2}\gamma}. \tag{4.9}$$

The central emission frequency can therefore be written as

$$\lambda \cong \frac{\lambda_u}{2\,\gamma^2}\left(1 + \frac{K^2}{2}\right). \tag{4.10}$$

According to the previous discussion, the prerequisite for an FEL operation (we omit the prefix U and add it whenever necessary) is due to the dynamics impressed on the electron motion by the undulator magnetic field.

The induced transverse motion and its periodicity determine the conditions for the emission at a fixed frequency. The first part of the process is sometimes referred to as '***spontaneous emission***' to underline the absence of any stimulating field, determining the lasing process. We can include this element in the game by noting that the electrons may re-interact with the radiation they produced and become periodically accelerated and decelerated, while absorbing and emitting radiation. At the microscopic level, the net emission process is therefore the result of a kind of

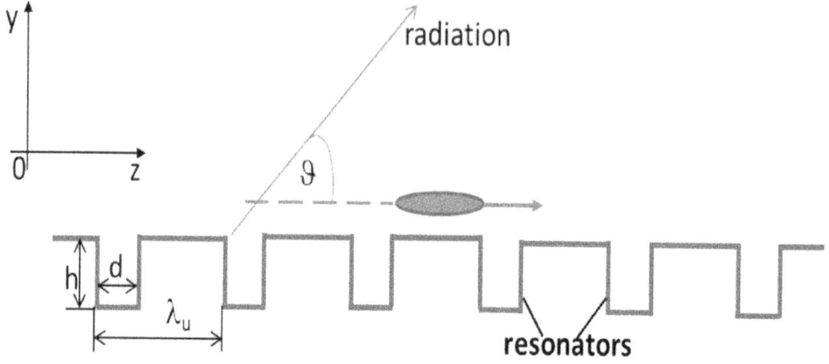

Figure 4.2. Geometry of Smith–Purcell device.

balance between the absorption and emission of the individual electrons. At the macroscopic scale, the net emission is based on the average on the whole beam.

Leaving the microscopic view, we can describe the mechanism underlying the FEL process, as summarized in figure 4.2.

We recognize four different phases:

(a) Beam energy modulation (induced by external seed and/or noise).
(b) Bunching.
(c) Coherent emission.
(d) Saturation.

This description in different phases is rather sloppy and it is just a crude exemplification. In the following we will see that the process is more intriguing, there is not such a sharp distinction and the path to saturation is accompanied by a rich phenomenology, including the emission at higher harmonics and a degradation of the beam qualities, which become more and more dispersed in energy.

According to this description, it seems that any effect of an elementary nature, involving the electron–photon interaction, does not play any significant role in FEL physics.

The nature of the mechanism triggering the U-FEL was strongly debated at the beginning of FEL physics and for some time the importance of quantum effects was emphasized, but over the course of time, one opinion prevailed. Even though quantum effects were necessary to trigger the FEL operation itself, they are hardly observed in practice and therefore FEL is a classical device. The relevant lasing process has accordingly been reduced to the mechanisms governing the amplification and the oscillation in traveling wave tubes (TWTs), discussed in the forthcoming chapters.

4.3 U-FEL and other free electron sources of coherent electromagnetic radiation

The process of emission by a charge, moving along a given trajectory, requires that specific kinematical conditions, allowing the coupling to an external electromagnetic field, be satisfied. The emission may occur either in vacuum or in a waveguide.

A further mechanism, defining the resonance condition, should be added to specify the wavelength selection. In the case of radiation by a relativistic charge moving inside an undulator, it is specified by equation (4.4), representing the conditions of positive interference for the field radiated at each undulator period. It can also be written as

$$\omega\,(1-\beta_z)=\omega_u \rightarrow \omega_0=\frac{\omega_u}{1-\beta_z}\cong\frac{2\,\gamma^2}{1+\dfrac{K^2}{2}}\omega_u, \tag{4.11}$$

where $\omega_u = 2\,\pi\;c/\lambda_u$ is a pseudo frequency associated with the charge oscillations, induced by the magnet itself. The last expression is interesting, because it resembles the Compton backscattering of pseudo-photons with wavelength $2\,\lambda_u$ and the comparison with the resonance condition of other processes, implying the emission by free charges.

We are entitled to use this point of view to understand the emission by charges moving in other periodic structures. In figure 4.2 we have sketched one such device, known as Smith–Purcell [12, 13]. In which an electron passing in the proximity of a metal grating emits radiation with wavelength fixed by

$$\lambda_0=\lambda_u(\beta_z^{-1}-\cos(\Theta)), \tag{4.12}$$

which is just a different way of expressing the previous 'resonance condition'.

It is well known that electrons in vacuum do not radiate, because such a process would be a violation of energy and momentum [14]. The condition underlying the Cerenkov emission reflects the previous statement and the emission angle between the direction of the electron motion and the emitted photon is

$$\cos(\Theta)=\frac{1}{\beta_2\sqrt{\varepsilon_r}}\left[1-\frac{\varepsilon_r-1}{2}\sqrt{1-\beta_2^2}\left(\frac{h\omega}{m_e c^2}\right)\right], \tag{4.13}$$

where the relative permittivity ε_r summarizes the interaction of the charge with the medium.

If we neglect the term containing the reduced Planck constant and consider small deviation of ε_r from unit we find

$$\Theta^2\cong\frac{1}{\beta_z}\delta_\varepsilon, \tag{4.14}$$

which, once confronted with equation (4.9), allows the identification

$$\delta_\varepsilon\cong\frac{K^2}{2\,\gamma^2}. \tag{4.15}$$

In order to further pursue the analogy, we consider the Cyclotron Auto Resonance Maser (CARM), which will be thoroughly discussed in the forthcoming chapter 6.

It can be viewed as a different flavor of FEL device. It exploits a moderately relativistic e-beam, injected into the waveguide of a resonant cavity, where an axial static magnetic field constrains the electrons, traveling with relativistic factor γ, along a helical path characterized by the cyclotron frequency

$$\Omega_0 = \frac{eB}{m_e}, \tag{4.16}$$

and by the helix period (see figure 4.3)

$$\begin{aligned} \Lambda &= 2\pi\frac{c}{\Omega}, \\ \Omega &= \frac{\Omega_0}{\gamma}. \end{aligned} \tag{4.17}$$

The interaction of the electrons, with the cavity environment, produces the loss of power towards a cavity mode, provided that the appropriate resonance conditions are satisfied. The interaction process can be reduced to what is sketched in figure 4.3, namely an electron, with longitudinal velocity v_z, propagating in a longitudinal magnetic field in the presence of a co-propagating electromagnetic field characterized by a wave-vector k_z, linked to the wave phase velocity v_p

$$k_z = \frac{\omega}{v_p}. \tag{4.18}$$

The kinematics characterizing the CARM operation is fully analogous to that of U-FEL and indeed we can write

$$\begin{aligned} \beta_z^2 + \beta_\perp^2 &= 1 - \gamma^{-2}, \\ \beta_{z,\perp} = \frac{v_{z,\perp}}{c}&, \ \alpha = \frac{v_\perp}{v_z}, \end{aligned} \tag{4.19}$$

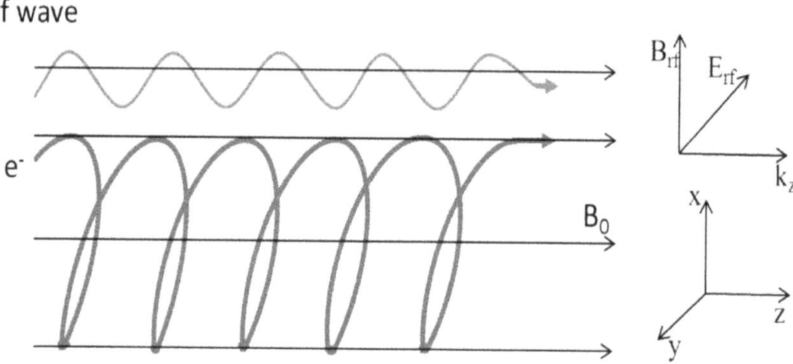

Figure 4.3. Geometry of the CARM interaction.

where α, the so called pitch factor, plays a role of pivotal importance, as seen below. The previous discussion allows the following correspondence between U-FEL and CARM devices

$$\begin{aligned} \lambda_u &\leftrightarrow \Lambda, \\ K &\leftrightarrow \gamma\beta_\perp. \end{aligned} \tag{4.20}$$

It is important to note that the transverse velocity component, associated with the pitch factor, plays the same role of the undulator strength parameter and is responsible for the coupling with the cavity field.

We can accordingly derive the CARM resonance condition, in a U-FEL fashion; we can invoke the same synchronism condition in equation (4.4) and write

$$\lambda \cong (v_p - v_e)\,\frac{\Lambda}{c}. \tag{4.21}$$

We expect indeed that, the electron and radiation velocities being different, positive interference requires that at each helix path, the radiation overcomes the electron by a wave length.

In the forthcoming chapter we will provide a more sound argument, involving the cavity dispersion relation, which imposes further constraints to the resonance condition.

This section has been devoted to frame different electron coherent sources within a unifying context. The forthcoming part of the chapter will be devoted to specific U-FEL aspects.

4.4 Free electron laser phenomenology and gain

The physics of U-FEL is very well understood, therefore the relevant theoretical treatment can be afforded using a very general argument, grounded essentially on first principles.

We have already noted that the laser process occurs through different steps, going from ‘spontaneous’ emission, exponential growth and saturation.

Regarding the first step, we note that having defined the central emission frequency through equation (4.11), we should complete the picture by defining the emission line shape and the relevant width. The radiation pulse at the end of the undulator can be viewed as a window function, with a length $N\delta$ and a time duration $\Delta t = N\delta/c$. The Fourier transform of this function is a Sinc function and its square modulus yields the associated spectral amplitude (SA) (see figure 4.4). The Parseval identity, specifies the spectral width, namely

$$\Delta\,\omega \cong \frac{\pi}{\Delta t} = \pi\frac{c}{N\,\delta}. \tag{4.22}$$

Let us now define a more appropriate set of dimensionless variables to characterize the SA. In equation (4.11) we have defined the central emission angular frequency, which is used to introduce the dimensionless variable, usually called ‘detuning parameter’, as

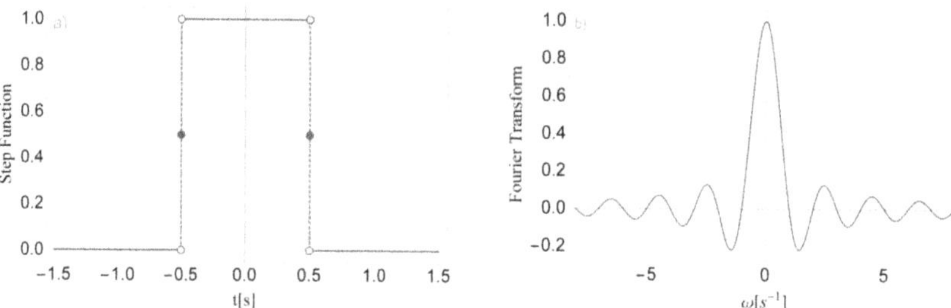

Figure 4.4. Temporal (a) and frequency (b) shapes of the bremsstrahlung emission process in the undulator.

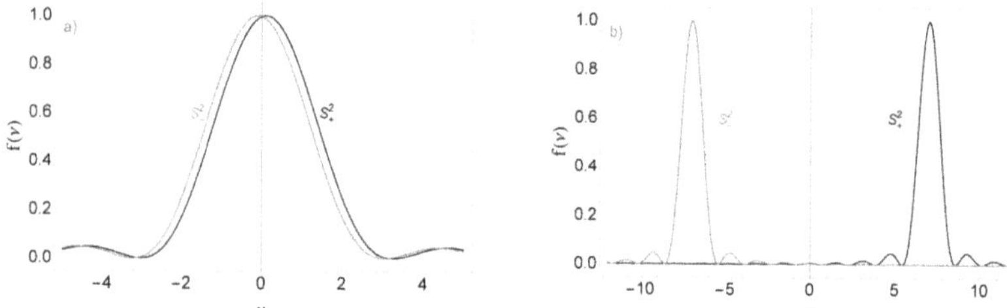

Figure 4.5. (a) Spontaneous emission line shape. (b) Shifted spontaneous emission curve.

$$\nu = 2\pi N \frac{\omega_0 - \omega}{\omega_0}. \tag{4.23}$$

The spectral line shape, given by the square of the Fourier transform, has the sinc^2 form, reported below

$$f(\nu) \propto \pi \left[\frac{\text{sinc}\left(\frac{\nu}{2}\right)}{\frac{\nu}{2}} \right]^2, \tag{4.24}$$

which is in turn provided by the integral transform

$$f(\nu) \propto 2\pi Re \int_0^1 (1 - \xi)\, e^{-i\nu\xi} d\xi. \tag{4.25}$$

In figure 4.5(d) we have reported the line shape, we should now consider a further element. The FEL process is more complicated than the simple spontaneous emission, it can be imagined as the balance between the emission and the absorption of a photon. The emission of a photon corresponds to the loss of energy by the emitting electron, the absorption to an energy increase. Assuming that the two mechanisms leave unaltered the line shape and produce only a negative or positive shift of the detuning, we can therefore write that the line shape regulating the emission/absorption process is

$$f(\nu, \delta_\nu) = f(\nu - \delta_\nu) - f(\nu + \delta_\nu), \tag{4.26}$$

where $\mp\delta_\nu$ is associated with the energy loss of the electrons in the process of emission/absorption of a photon. Such a quantity being small, compared to ν, we can write at the lowest order in δ_ν,

$$g(\nu) \cong -2\delta_\nu\, \partial_\nu f(\nu) \propto 2\pi Re\ i \int_0^1 \xi(1 - \xi)\, e^{-i\nu\,\xi} d\xi, \tag{4.27}$$

we can now develop a simple argument.

The amplitude is constructed inside the undulator as a continuous emission process, if we define the dimensionless time variable

$$\tau = \frac{z}{L_u},\ 0 \leqslant \tau \leqslant 1, \tag{4.28}$$

we define the complex amplitude inside the undulator

$$a(\tau, \nu) \propto 2\pi \int_0^\tau \xi(\tau - \xi)\, e^{-i\nu\,\xi} d\xi = 2\pi\ i \int_0^\tau d\tau' \int_0^{\tau'} d\tau'' \tau'' e^{-i\nu\,\tau''}. \tag{4.29}$$

By keeping the derivative of both sides of the previous identity with respect to τ we find

$$\partial_\tau a(\tau, \nu) \propto 2\pi \int_0^\tau d\tau' \tau' e^{-i\nu\,\tau'}, \tag{4.30}$$

which is the differential equation, yielding the complex laser field evolution inside the undulator. Equation (4.30), as it stands, is insufficient to describe the FEL field growth, because it describes a one-body process and it does not include the memory of the field variation along the undulator itself.

The price to be paid to include all these contributions is to write (we drop the explicit dependence on the detuning parameter)

$$\partial_\tau a(\tau) = i\ \pi\ g_0 \int_0^\tau \tau' e^{-i\nu\,\tau'} a(\tau - \tau') d\tau', \tag{4.31}$$

where the small signal gain coefficient g_0 will be specified below. Equation (4.31) is the FEL integral equation. It is sufficiently general to account for the field growth inside the undulator and the 'low gain' regime, characterized by keeping the field constant in the rhs of equation (4.29), namely

$$a(\tau - \tau') \cong a(\tau) \cong a(0). \tag{4.32}$$

In the forthcoming section we will be more specific by specifying the meaning of the complex amplitude $a(\tau)$, whose square modulus is associated with the field intensity, and of the small signal gain coefficient, which are crucial quantities to understand the FEL phenomenology.

4.5 FEL low and high gain regimes

Equation (4.31) describes the 'linear' growth of the FEL amplitude due to the interaction with an e-beam inside an undulator. The gain coefficient g_0 should therefore include the beam current, the undulator parameters and other quantities characterizing the dynamics and the geometry of the interaction itself.

We report below the explicit expression for the coefficient g_0, along with the definition of small signal gain, which is associated with the relative variation of the square modulus of the dimensionless amplitude a

$$\begin{aligned}
g_0 &= \frac{16\pi}{\gamma}\frac{|J|}{I_0}\lambda_0 L N^2 \xi f_b^2 = \\
&= 2\pi\frac{|J|}{I_0}\left(\frac{N}{\gamma}\right)^3 (\lambda_u K f_b)^2, \text{ small signal gain FEL coefficient,} \\
G &= \frac{|a|^2 - |a_0|^2}{|a_0|^2}\text{FEL–Gain,} \\
J &\equiv \text{electron beam density,} \\
I_0 &\equiv \text{Alfvén} - \textit{current}, \\
f_b(K) &= J_0(\xi) - J_1(\xi),\ \xi = \frac{1}{4}\frac{K^2}{1 + \frac{K^2}{2}}.
\end{aligned} \tag{4.33}$$

Regarding the FEL gain, defined as the relative variation of $|a|^2$, we attempt to clarify this crucial concept, by keeping equation (4.31) and making the low gain approximation and writing

$$\partial_\tau a(\tau) = i\,\pi\, g_0 a_0 \int_0^\tau \tau' e^{-i\nu\tau'} d\tau', \tag{4.34}$$

with solution

$$a(\tau) = a_0\left[1 + i\,\pi\, g_0 \int_0^\tau d\tau' \int_0^{\tau'} \tau'' e^{-i\nu\tau''} d\tau''\right], \tag{4.35}$$

namely, the same as equation (4.29).

If we evaluate the square modulus of $a(\tau)$ at $\tau = 1$ (this means at the end of the undulator, see below for further comments) neglecting the second order contribution in g_0 (this is due to the assumption of low gain regime), we eventually obtain

$$G = \frac{2\pi}{\nu^3} g_0[2\,(1 - \cos(\nu)) - \nu\sin(\nu)] = -\pi\, g_0 \partial_\nu \left[\frac{\sin\left(\frac{\nu}{2}\right)}{\frac{\nu}{2}}\right]^2. \tag{4.36}$$

Namely, the celebrated asymmetric gain curve (see figure 4.6), measured for the first time by Madey and coworkers in 1976 during the FEL amplification experiment [6].

The small signal gain coefficient is usually given in %, the neglecting of higher order in g_0 is justified for values not larger than 30%, but to better appreciate this point we should clarify what values can be achieved in an actual device.

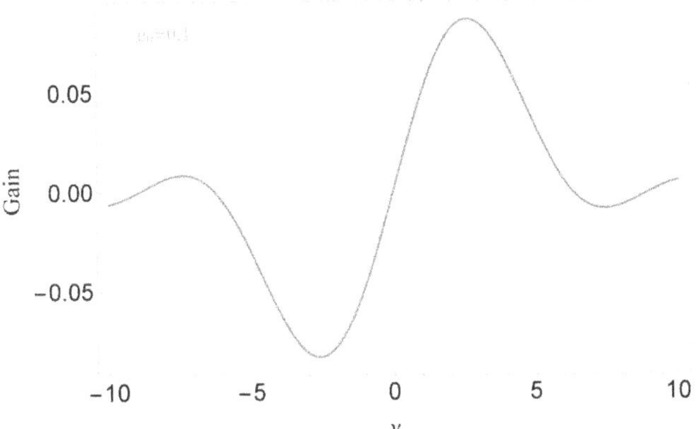

Figure 4.6. Low gain FEL curve, $g0 = 0.1$.

We consider therefore the quantity

$$g_0 = 2\pi \frac{|J|}{I_0} \left(\frac{N}{\gamma} \right)^3 (\lambda_u K f_b)^2, \tag{4.37}$$

and recall that, in practical units, the undulator strength reads

$$K = \frac{\lambda_u[cm] B_0[KG]}{10.71}. \tag{4.38}$$

Furthermore, noting that

$$\gamma \cong \sqrt{\frac{\lambda_u}{2\lambda}\left(1 + \frac{K^2}{2}\right)}, \tag{4.39}$$

we can write the small signal gain coefficient as

$$g_0 = 2\pi \frac{I_e}{I_0} \left(\frac{N\sqrt{2\lambda}}{\sqrt{\lambda_u \left(1 + \frac{K^2}{2}\right)}} \right)^3 \left(\frac{\lambda_u K f_b}{\sqrt{\Sigma_e}} \right)^2. \tag{4.40}$$

After writing the current density as

$$|J| = \frac{I_e}{\Sigma_e}, \tag{4.41}$$

where Σ_e is the e-beam transverse section.

We consider an undulator characterized by the following values $N = 70$, $\lambda_u = 3$ cm, $L_u = 3$ m, $K = \sqrt{2}$ and limit ourselves to wavelength up to UV, namely

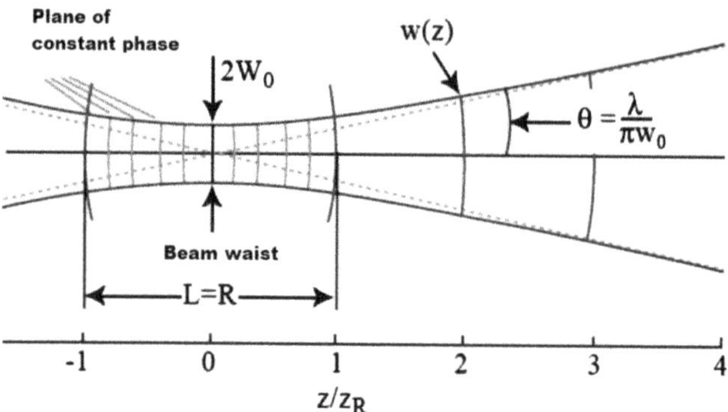

Figure 4.7. Optical cavity and relevant geometry.

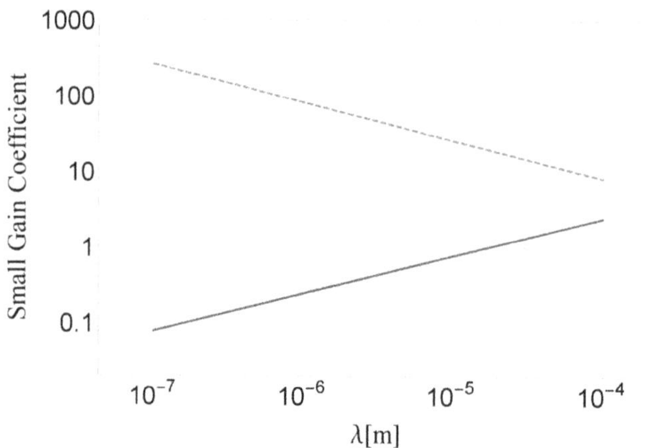

Figure 4.8. Small signal gain coefficient (continuous-blue) and beam energy (dashed-red) versus λ[m]. $N = 70 \lambda_u = 3$ cm, $K = \sqrt{2}$, $I_e \cong 10$ A, $L_c = 5$ m, $I_e = 20$ A.

λ not larger than 100 nm. Considering an FEL operating in the oscillator configuration and assuming the undulator matches the beam cavity waist (see figure 4.7)

$$\Sigma_e \cong \pi \frac{\lambda L_c}{2}, \qquad L_c \equiv \text{cavity–length}. \tag{4.42}$$

In figure 4.8 we have reported g_0 versus λ, in the interval ranging from 100 nm to 100 μm (UV-FIR). To make the picture more complete, we have also added the beam energy versus wavelength; for the chosen parameters, the energy range spans from hundreds to tens of MeV.

We have chosen a beam current of 20 A and the small signal gain coefficient is larger than 30% in the region above the mid-IR (medium infrared).

The parameters we have chosen are typical of oscillator FEL configurations and show that, also within this framework, the low gain regime condition might be violated. What should be expected if we include in the solution of our integral equation higher order corrections in the small signal gain coefficient, is easily understood through a perturbative analysis (namely a naïve expansion in terms of the coefficient g_0), which yields

$$\begin{aligned} G(g_0, \nu) &\cong \sum_{s=1}^{3} g_0^s g_s(\nu), \\ g_1(\nu) &= \frac{2\pi}{\nu^3}[2(1 - \cos(\nu)) - \nu \sin(\nu)], \\ g_2(\nu) &= \frac{\pi^2}{3\nu^6}[84(1 - \cos(\nu)) - 60\nu \sin(\nu) + 3\nu^2 + 15\nu^2 \cos(\nu) + \nu^3 \sin(\nu)], \\ g_3(\nu) &= \frac{\pi^3}{60\nu^9}[11520(1 - \cos(\nu)) - 9000\nu \sin(\nu) + 360\nu^2 + 2880\nu^2 \cos(\nu) + \\ &\quad + 480\nu^3 \sin(\nu) - 20\nu^4(1 + 2\cos(\nu)) - \nu^5 \sin(\nu)]. \end{aligned} \tag{4.43}$$

In figure 4.9 we have reported the gain including higher order corrections, for $g0 = \{1, 10\}$, which evidently show gain curves which lose their anti-symmetric shape.

The maximum gain is furthermore given by

$$G_M \cong 0.85 g_0 + 0.192 g_0^2 + 4.23 \cdot 10^{-3} g_0^3, \tag{4.44}$$

and it is easily checked that deviation from the low gain regime start to be evident above 20% (see figure 4.10).

The parameters we have discussed are suitable for FEL operating in the FEL configuration, which works whenever there are cavities available to confine the FEL radiation.

The problems arise for shorter wavelengths (VUV-X) where there are not efficient mirrors available to provide sufficient feedback to contain the radiation growth, up to the saturation.

In this case totally different parameters should be adopted.

Before going further, let us note that the dimensionless variables we have exploited so far are better suited for the oscillator regime than the operation in

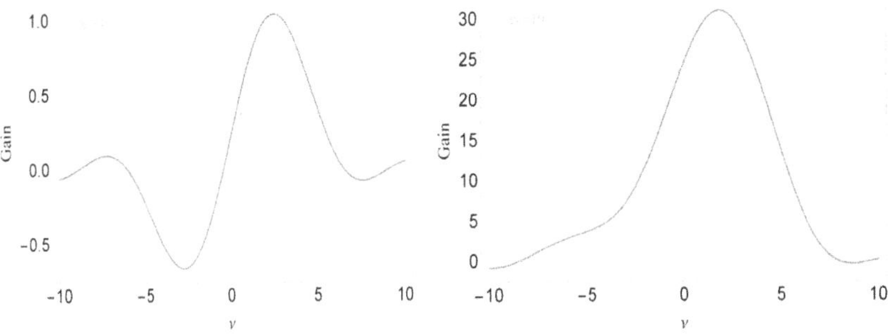

Figure 4.9. FEL high gain curves including the non-linear term in g_0.

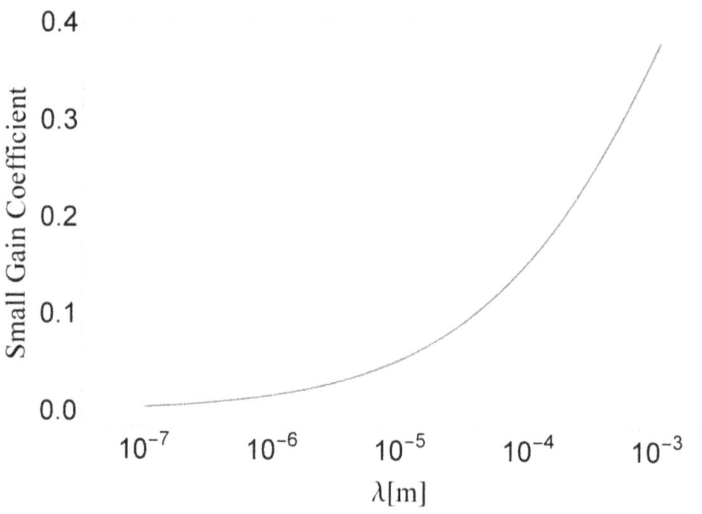

Figure 4.10. Small signal gain coefficient versus λ[m].

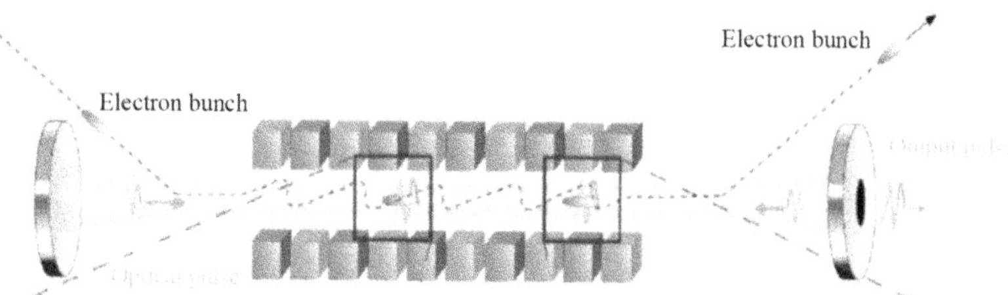

Figure 4.11. FEL oscillator.

single passage at high gain regime. The drawback of the previous set of variables is that either the small signal gain coefficient, the dimensionless time and the detuning parameter are explicitly depending on the number of undulator period. A quantity is predetermined in an oscillator FEL device, for the laser field growth inside the optical cavity, after many bounces back and forth and interaction with the 'active medium' inside the resonator (see figure 4.11).

In the high gain regime the situation is different (see figure 4.12), the field growth is determined by the beam shot noise and the coherent part of the bremsstrahlung radiation and becomes self-amplified.

In this last case it is convenient to write the FEL integral equation (4.31) after replacing τ with z

$$\partial_z a(z) = i\,\frac{\pi\, g_0}{(N\,\lambda_u)^3}\int_0^{\tau} z' e^{-i\nu\frac{z'}{N\lambda_u}}a(z-z')dz', \tag{4.45}$$

where it is worth noting that

$$\tilde{g}^3 = \frac{\pi\, g_0}{N^3} = 2\pi^2\frac{|J|}{I_0}\left(\frac{1}{\gamma}\right)^3(\lambda_u K f_b)^2 \tag{4.46}$$

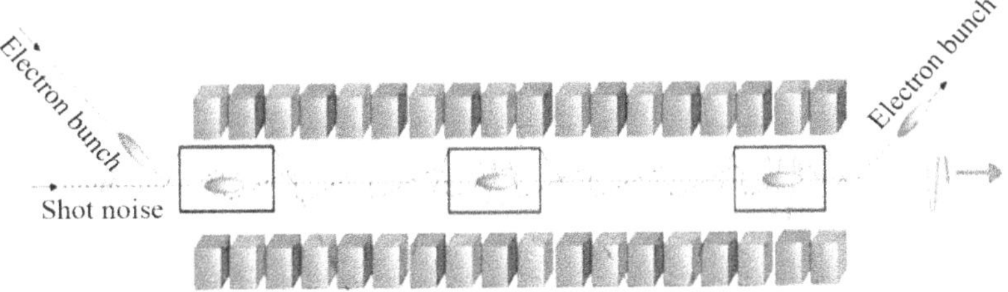

Figure 4.12. Self-amplified spontaneous emission (SASE) process: the field is 'seeded' by the shot noise (see below) and then goes through the canonical phases (energy modulation, bunching etc).

is no more explicitly depending on N. The same holds for the term containing the detuning in the oscillating term inside the undulator, which can be replaced by

$$\tilde{\nu} = \frac{1}{4\pi\sqrt{3}\rho}\frac{\omega_0 - \omega}{\omega_0}, \tag{4.47}$$

thus writing equation (4.45) as

$$\partial_z a(z) = i\left(\frac{1}{\tilde{\lambda}_u}\right)^3 \int_0^z z' e^{-i\tilde{\nu} z'} a(z - z')dz' \tag{4.48}$$

with

$$\tilde{\lambda}_u = \frac{\lambda_u}{\tilde{g}}. \tag{4.49}$$

The message contained in the above restyling is that the field growth is ruled by the parameter $\tilde{\lambda}_u$, which measures the growth per a specific length, to be defined below.

We introduce the following normalized longitudinal coordinate

$$\begin{gathered} \tilde{z} = \frac{z}{L_g},\ L_g = \frac{\lambda_u}{4\pi\sqrt{3}\rho}, \\ \rho = \frac{\tilde{g}}{4\pi} \end{gathered} \tag{4.50}$$

in which L_g is the gain length.

Using equations (4.50), equation (4.48) can eventually end up with

$$\begin{gathered} \partial_{\tilde{z}} a(\tilde{z}) = \frac{i}{3\sqrt{3}} \int_0^z \tilde{z}' e^{-i\tilde{\nu} z'} a(\tilde{z} - \tilde{z}')d\tilde{z}', \\ \tilde{\nu} = \frac{1}{2\sqrt{3}\rho}\frac{\omega_0 - \omega}{\omega_0}. \end{gathered} \tag{4.51}$$

It is evident that the reference quantity for the high gain regime is no longer g_0, but ρ commonly referred to as Pierce parameter. To get an idea of the values it can take, we make reference to figure 4.17, where we have

$$(\pi g_0)^{\frac{1}{3}} = \frac{1}{4\pi}\left(\frac{\sqrt{2\lambda}}{\sqrt{\lambda_u\left(1+\frac{K^2}{2}\right)}}\right)\left[2\pi^2, \frac{I_e}{I_0}\left(\frac{\lambda_u K f_b}{\sqrt{\Sigma_e}}\right)^2\right]^{\frac{1}{3}}, \tag{4.52}$$

where Σ_e is replaced by

$$\Sigma_e = 2\pi\sigma_x\sigma_y, \tag{4.53}$$

representing the beam cross area (we will better specify the definition of the transverse length σ in the forthcoming sections). The corresponding values of the Pierce parameter are given in figure 4.13 for high gain devices operating in the X-VUV regions.

The gain length L_g measures the growth rate, the relevant values in correspondence of ρ are given in figure 4.14.

The intensity signal growth (see appendix C) is obtained by solving equation (4.45) thus eventually obtaining (see figure 4.15 to have an idea of the relevant growth along the undulator axis)

$$I(\tilde{z}) = \frac{I_0}{9}\left[3 + 2\cosh\left(\frac{z}{L_g}\right) + 4\cos\left(\frac{\sqrt{3}}{2}\frac{z}{L_g}\right)\cosh\left(\frac{z}{2L_g}\right)\right] \tag{4.54}$$

with I_0 being the intensity of the seed field, if the device works as an amplifier.

In figure 4.15 we have selected some points, and we have marked the associated gain and the relevant transition from low gain to high gain regimes (see next section for further comments).

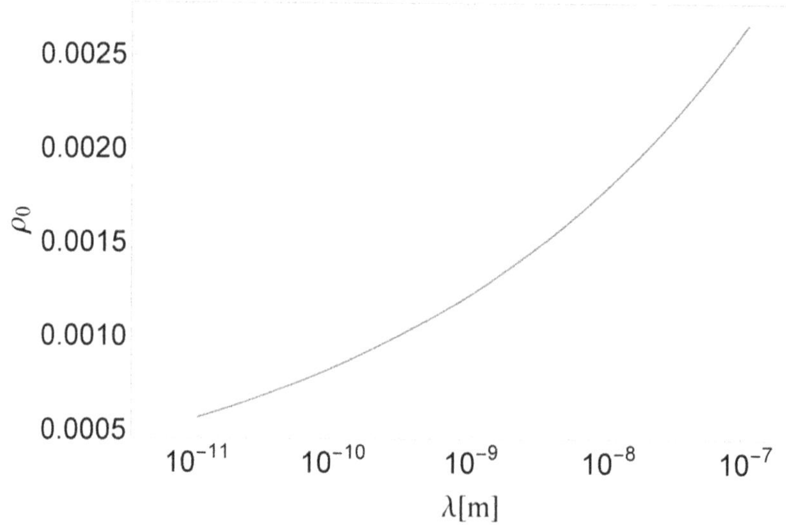

Figure 4.13. Pierce parameter versus wavelength for $I = 20$ A, $K = \sqrt{2}$, $\lambda_u = 2$ cm, $N = 70$.

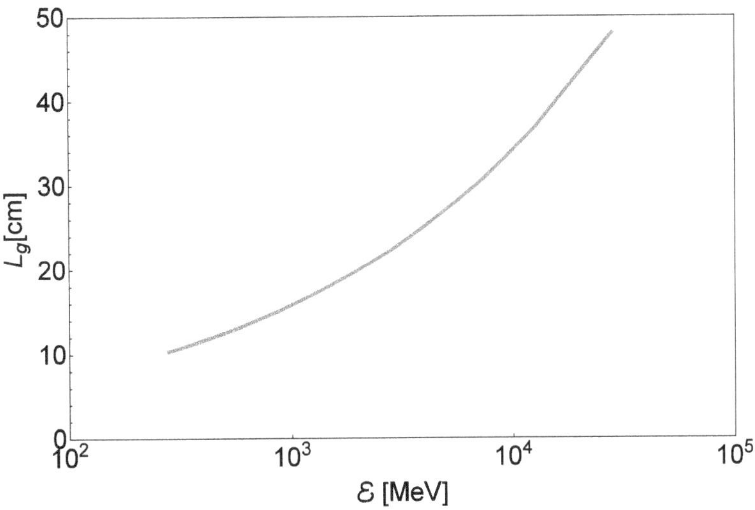

Figure 4.14. Gain length versus energy [MeV] with $I = 1000$ A, $K = \sqrt{2}$, $\lambda_u = 2$ cm.

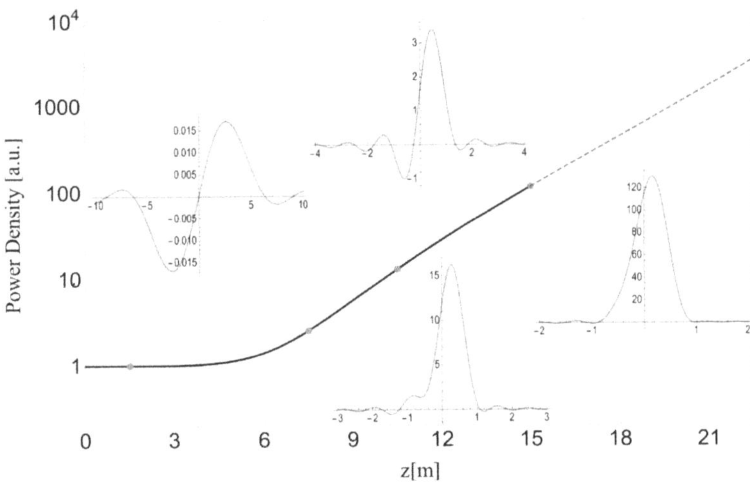

Figure 4.15. Power Density [a.u.] versus z and gain function 'measured' at different points inside the undulator. Above the last point, where the dotted curve starts, the small signal gain is not properly defined because the small signal approximation does not strictly apply. The curves have been derived for the parameters $\rho_0 = 6.8 \cdot 10^{-4}$ and $\lambda_u = 5$ cm.

The previous analysis concerns the field evolution in the absence of any feedback mechanism inducing saturation effects. The discussion of the associated issues is presented in the forthcoming section.

4.6 Non-linear regime and saturation

We have closed the previous section with an intensity growth curve and some points marking the gain regime. The underlying physical meaning can be understood as follows:

(a) Equation (4.54) yields the amplification of an input seed and the gain, at each specified point inside the undulator, is defined as the relative variation of the intensity, namely

$$G_i(\nu) = \frac{|a_i|^2 - |a_0|^2}{|a_0|^2},\qquad |a_i|^2 \propto I_i. \tag{4.55}$$

(b) The intensity, after a mild lethargic behavior (which lasts $Z_l \cong 2.1L_g$) follows an exponential pattern not counteracted by any non-linear mechanism inducing saturation, namely the reduction of the gain and the achievement of a stationary condition.

Point (a) is clear, the region where the gain curve exhibits the anti-symmetric shape is that corresponding to undulators with a short number of periods, as in the first experiments employing the oscillator configuration.

The second point regards more general considerations, concerning the very physical nature of the FEL interaction.

The physical conditions we have so far considered (small signal regime) assume that neither the energy of electrons nor their distribution remain constant during the interaction. The energy conservation imposes, however, that, while giving power to the field, the electrons loose energy. The resonance condition is no longer satisfied, where the relative energy loss linked to the detuning variation by the condition is

$$\Delta\nu = 4\pi N \frac{\Delta\mathcal{E}}{\mathcal{E}}, \tag{4.56}$$

assuming $\Delta\nu \approx 2 \cdot 2.6$ (see figure 4.16) we obtain for efficiency

$$\eta = \frac{\Delta\mathcal{E}}{\mathcal{E}} \approx \frac{2 \cdot 2.6}{4\pi N} \approx \frac{1.3}{\pi N}. \tag{4.57}$$

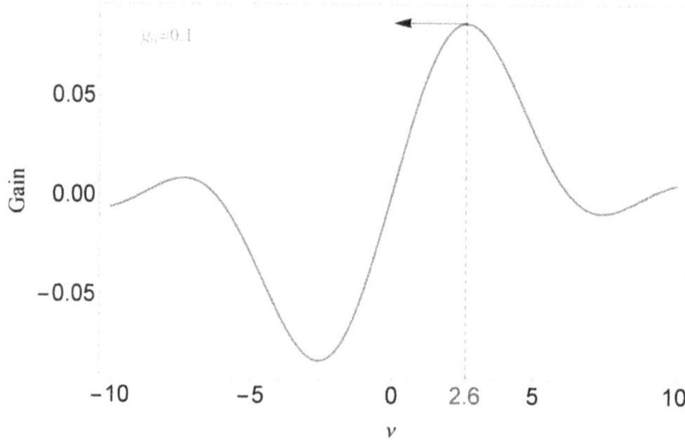

Figure 4.16. Low gain small signal gain versus ν. The maximum gain occurs at $\nu^* \cong 2.6$, when the beam loses energy the emission occurs at lower frequencies and the peak shifts towards the absorption region.

This is only a part of the story; when the field grows the induced energy spread increases and the gain is reduced. A more reliable conclusion is that the low gain efficiency is given by

$$\frac{\Delta\mathcal{E}}{\mathcal{E}} \approx \frac{1}{4N}. \tag{4.58}$$

All the dynamics requires, therefore, a more appropriate mathematical treatment.

Without entering a specific derivation, which can, however, be found in other books, we note that the variable ν can be defined as[2]

$$\begin{aligned}
&\nu = \frac{d}{d\tau}\zeta,\\
&\zeta = (k_u + k_r)\int_0^t \bar{v}_z(t')\,dt' - \omega t,\\
&k_u = \frac{2\pi}{\lambda_u} \ll k_r,\\
&\bar{v}_z(t) \equiv \text{longitudinal velocity averaged over the undulator period.}
\end{aligned} \tag{4.59}$$

The extra variable we should include is that of the field amplitude; considering ν, ζ as a pair of conjugate variables, we can fit in the FEL dynamics in the following Hamiltonian

$$\begin{aligned}
H &= \frac{1}{2}\nu^2 - [ae^{-i\zeta} + a^*e^{i\zeta}],\\
a &= |a|\, e^{i\phi}.
\end{aligned} \tag{4.60}$$

The associated equations of motion are therefore

$$\begin{aligned}
\frac{d}{d\tau}\zeta &= \frac{\partial H}{\partial \nu} = \nu,\\
\frac{d}{d\tau}\nu &= -\frac{\partial H}{\partial \zeta} = |a|\,\cos(\zeta + \phi),
\end{aligned} \tag{4.61}$$

thus ending up with the 'pendulum equation'

$$\frac{d\zeta}{d\tau^2} = |a|\,\cos(\zeta + \phi). \tag{4.62}$$

To make the system self-consistent we should couple to the last equation that regarding the field evolution; it follows from the Maxwell equations and in the slowly varying amplitude approximation reads

$$\frac{da}{d\tau} = -2\,\pi\,g_0\langle e^{-i\zeta}\rangle, \tag{4.63}$$

[2] Note indeed that $\frac{d}{d\tau}\zeta = (k_u + k_r)\int_0^t \bar{v}_z(t')\,dt' - \omega\,t = N\,((k_u + k_r)\,c\bar{\beta}_z - \omega) \cong 2\,\pi N\frac{\omega r - \omega}{\omega r}$.

where the brackets $\langle ... \rangle$ denote averages on the coordinate ζ.

The definition of the dimensionless field amplitude in terms of physical quantities is given below

$$|a| = \frac{k_s}{\gamma^4}\frac{K}{\sqrt{2}}(N\lambda_u)^2\left(1 + \frac{K^2}{2}\right)f_b\, e_s = 4\pi\left(\frac{N}{\gamma}\right)^2\frac{K}{\sqrt{2}}\lambda_u f_b\, e_s,$$
$$e_s = \frac{eE_s}{\sqrt{2}m_e c^2}. \tag{4.64}$$

A little college physics is necessary to understand the non-linear effects in FEL. In figure 4.17 we have reported the pendulum phase space plot (ζ, ν).

Figure 4.17 reports either orbits (contours in ζ, ν at constant H) and special points. We have an 'elliptic' fixed point ($\zeta = 0$, $\nu = 0$) around which nearby orbits wrap forming ellipses. The ζ coordinate runs from $-\pi$ to π and a second fixed point occurs at $(\pi, 0)$. In the case of the simple pendulum it corresponds to the pendulum pointing vertically up. This is called unstable 'hyperbolic' fixed point, because the trajectory departing from it takes hyperbolic forms The remaining orbits are periodic in time, and are called 'limit cycles'. A special pair of orbits leave the hyperbolic fixed point, and then eventually return to it. These are known as 'homoclinic' orbits or separatrix, because they separate the closed from the open curves.

Such a 'geometric' description should be embedded in a dynamic context, making a comparison with the simple pendulum (see figure 4.18)

$$H = \frac{1}{2}\dot{\vartheta}^2 + \frac{g}{l}(1 - \cos(\vartheta)), \tag{4.65}$$

the separatrix corresponds to $H = 2g/l$ and therefore

$$\dot{\vartheta} = \pm\sqrt{2\frac{g}{l}}\cos\left(\frac{\vartheta}{2}\right). \tag{4.66}$$

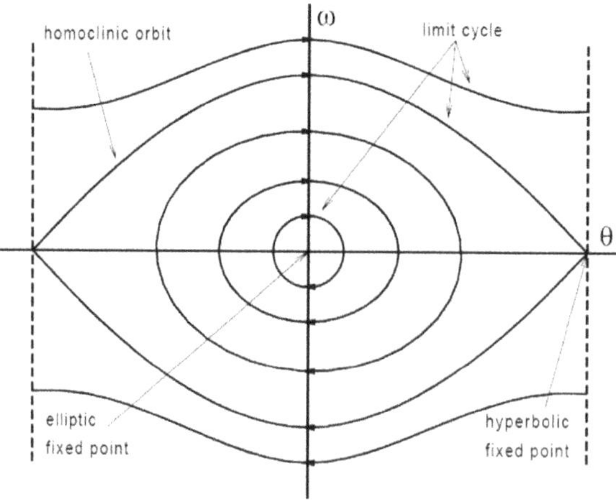

Figure 4.17. Trajectories and critical points in simple pendulum phase space.

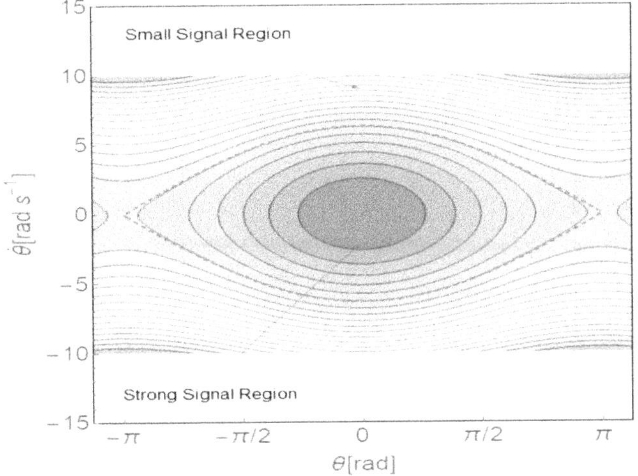

Figure 4.18. Phase space and dynamical regions, the separatrix in red-dashed line.

In the case of FEL, g/l is replaced by the field amplitude, which, while growing determines an increase of the separatrix height, which grows with the square root of the field intensity.

The phase space dynamics reported in figure 4.19 better clarifies what are the mechanisms determining saturation. Electrons, belonging to different buckets (sampled on the field wave length), are characterized by different colors.

At the undulator entrance the electrons exhibit a constant distribution in ζ, ν. When the electrons move inside the undulator the field intensity grows and the electrons are 'captured' inside the buckets, where they diffuse, by increasing, e.g., the associated relative energy spread or overlapping with those of the other buckets. The dynamics summarized in figure 4.19 reports all these features, including the energy loss of the beam (it is measured by the shift down of the electron 'cloud' inside the bucket) and the increase of the energy spread.

According to figure 4.19 the intensity growth (in the high gain regime) exhibits a kind of logistic curve, naively reproduced by the curve

$$P(z) = P_0 \frac{A(z)}{1 + \frac{P_0}{P_F}[A(z) - 1]},$$
$$A(z) = \frac{1}{9}\left[3 + 2\cosh\left(\frac{z}{L_g}\right) + 4\cos\left(\frac{\sqrt{3}}{2}\frac{z}{L_g}\right)\cosh\left(\frac{z}{2L_g}\right)\right]. \tag{4.67}$$

Where P_F is the 'saturated' power, defined as (see below)

$$P_F \cong \sqrt{2}\rho\ P_E \tag{4.68}$$

with P_E being the e-beam power.

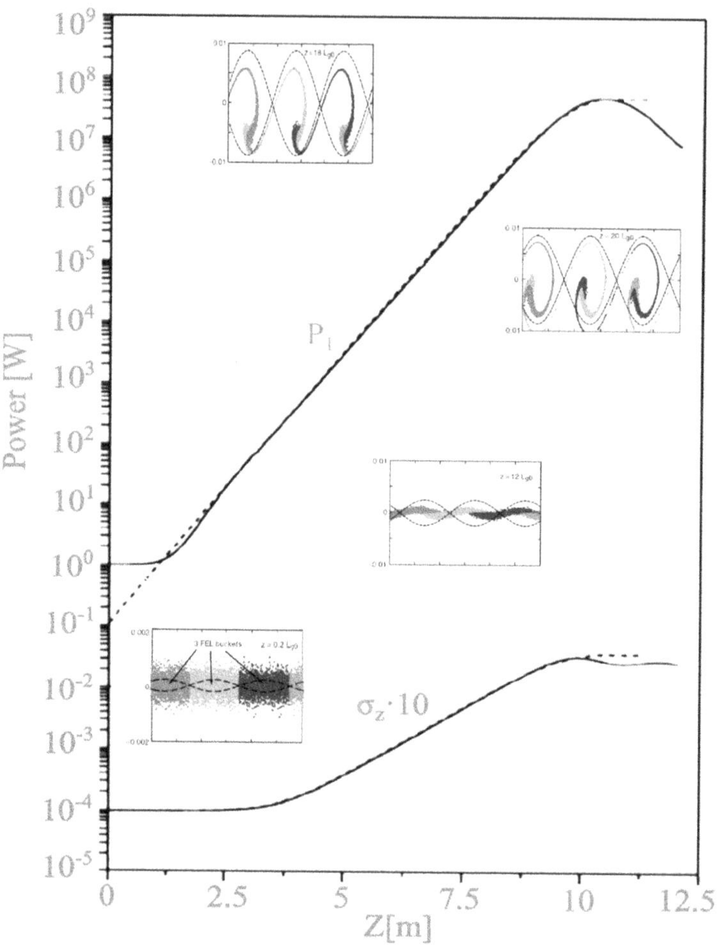

Figure 4.19. Bunching and phase-space evolution, while the e-beam is progressing through the undulator.

The induced energy spread is reproduced by an analogous expression, reported below

$$
\begin{aligned}
\sigma_i(z) &\cong 3C\sqrt{\frac{A(z)}{1 + 9B[A(z) - 1]}}, \\
C &= \frac{1}{2}\sqrt{\frac{\rho P_0}{P_E}}, \; B \cong \frac{1.24}{9}\frac{P_0}{P_F}, \\
\sigma_F &\cong \frac{C}{\sqrt{B}} \cong 1.6\rho.
\end{aligned} \tag{4.69}
$$

Regarding the final power we note that the saturated power is linked to the energy lost by the electrons through the relationship

$$P_F \cong I\ \Delta\mathcal{E} = P_E\left(\frac{\Delta\mathcal{E}}{\mathcal{E}}\right), \tag{4.70}$$

where $\Delta\mathcal{E}$ is the maximum energy loss induced by the FEL interaction. According to the previous discussion (regarding the high gain regime)

$$\frac{\Delta\mathcal{E}}{\mathcal{E}} \propto \Delta\tilde{\nu}_{\mathrm{MAX}} \cong \rho, \tag{4.71}$$

which justifies equation (4.68).

On the other hand, regarding the low gain regime

$$\frac{\Delta\mathcal{E}}{\mathcal{E}} \propto \Delta\nu_{\mathrm{Max}} \cong \frac{1}{4N}. \tag{4.72}$$

We have so far developed the phenomenological framework of FEL physics. The forthcoming sections are devoted to the relevant implementation.

4.7 Free electron laser oscillators

In the previous section we have left open a point regarding the definition of dimensionless field amplitude. To this aim we adopt a heuristic argument.

The square modulus of a is

$$\begin{gathered} |a|^2 = 8\pi^2\frac{I}{I_S}, \\ I_S\left[\frac{MW}{cm^2}\right] = 6.931\,2\cdot 10^2\frac{1}{2}\left(\frac{\gamma}{N}\right)^4\left(\lambda_u[cm]\frac{K}{2}f_b\right)^{-2} \end{gathered} \tag{4.73}$$

with I_S being the FEL saturation intensity, whose physical meaning will be discussed below. We first note that when $I/I_S = 1$

$$|a| = 2\sqrt{2}\pi, \tag{4.74}$$

and it corresponds to the separatrix height[3].

After these remarks we describe actual devices based on the principles we have just discussed.

We start with the FEL oscillators.

There was a sterile controversy during the emergence of the FEL, regarding the question whether it might be considered a laser or not [4]. Here we will not enter that dispute but simply note that an FEL oscillator possesses all the paradigmatic characteristics of any laser operating with a feedback cavity, see figure 4.20.

The laser field growth inside the cavity occurs through different steps. The first of these is the spontaneous emission, which triggers the stimulated part of the process.

[3] We recall that in the case of the simple pendulum we have $H = 2\,g/l = 8\pi^2/T^2$ and in dimensionless unit we find $(2\,g/l)T^2 = 8\,\pi^2$.

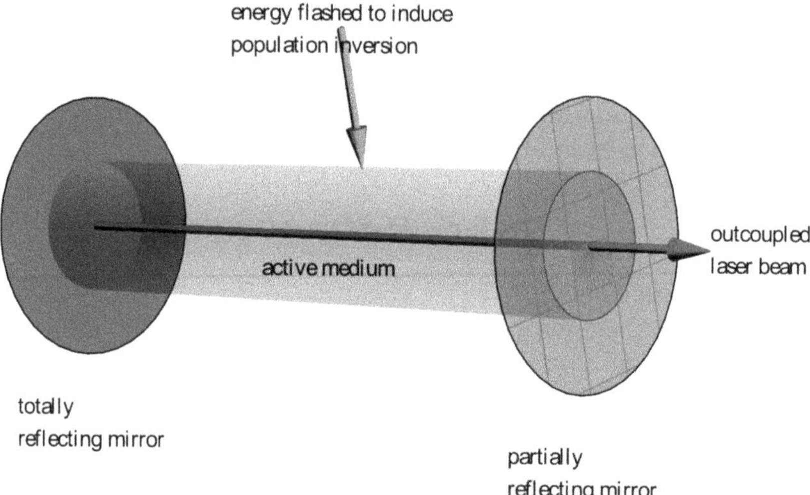

Figure 4.20. Laser oscillator and its components.

The spontaneous emission is isotropic, but confinement occurs only for those (transverse) modes[4] propagating along the cavity axis and contained inside the cavity itself, thus giving rise to the mechanism leading to the spatial coherence. Having assumed that some gain mechanism (due to the stimulated emission) exists, we expect that at any reflection back and forth (round trip r.t.) of the light in the optical cavity, the stored radiation intensity increases and only a fraction of it is transmitted outside. The evolution process should be ruled by the gain mechanism (including the losses) and by another effect, determining an ***intensity dependent feedback*** to the gain, eventually leading to the saturation.

The expression, quoted in bold characters, is a rather obscure statement pointing out that the laser gain is counteracted by the increase of the intensity itself, this effect determines the slow-down of the growth and eventually the stationary condition, when the gain equals the cavity losses η. The power density growth I_n inside the cavity at each round trip n can be accounted for by the following rate equation

$$I_{n+1} = (1 - \eta)[G(I_n) + 1]I_n, \tag{4.75}$$

where $G(I_n)$ is the gain including the field-intensity dependence, which will be specified later. The intra-cavity equilibrium condition is ensured by

$$G(I_E) = \frac{\eta}{1 - \eta}, \tag{4.76}$$

and the intra-cavity equilibrium intensity I_E is obtained if the explicit form of the gain function on the intra-cavity intensity.

[4] The selection of longitudinal modes will be discussed later in this chapter.

The most natural assumption is a rational type of function like

$$G(I) = \frac{g}{1 + \sum_{k=1}^{n} \alpha_k X^k}, \qquad X = \frac{I}{I_s}, \quad \sum_{k=1}^{n} \alpha_k = 1. \tag{4.77}$$

where I_s is the saturation intensity, which is a characteristic quantity of any laser device including FEL's, it corresponds to the laser power density halving the small signal gain g, namely

$$G(I_s) = \frac{1}{2} G_{\text{MAX}}. \tag{4.78}$$

Regarding the FEL oscillators the saturated gain exhibits the following intensity dependence

$$G(X) = \frac{G_M}{F(X)}, \qquad F(X) = 1 + a_1 X + a_2 X^2, \; a_1 = 2\left(\sqrt{2} - 1\right), \; a_2 = 3 - 2\sqrt{2}. \tag{4.79}$$

The equilibrium intensity calculated from equations (4.76) and (4.79) yields

$$I_E = \left(\sqrt{2} + 1\right)\left(\sqrt{\frac{1 - \eta}{\eta} G_M} - 1\right) I_S, \tag{4.80}$$

and that of equation (4.22) in the form of a discrete logistic curve

$$I_r = I_0 \frac{[(1 - \eta)(G_M + 1)]^r}{1 + \frac{I_0}{I_e}\{[(1 - \eta)(G_M + 1)]^r - 1\}}, \tag{4.81}$$

where I_0 is the initial seed intensity. An example of intra-cavity evolution with the number of round trips is shown in figure 4.21.

We have so far described the intra-cavity power growth, the extracted power depends on the active losses. To understand this point we proceed as follows and consider the function $F(X)$ in gain saturation at lowest order in X, the equilibrium intensity is accordingly given by

$$I_E = \left(\frac{1 - \eta}{\eta} G_M - 1\right) I_S. \tag{4.82}$$

Keeping $G_M \cong 0.85\, g_0$, neglecting term 1 in round brackets, we end up with the laser output intensity as

$$I_0 \cong \eta I_E \propto g_0 I_S. \tag{4.83}$$

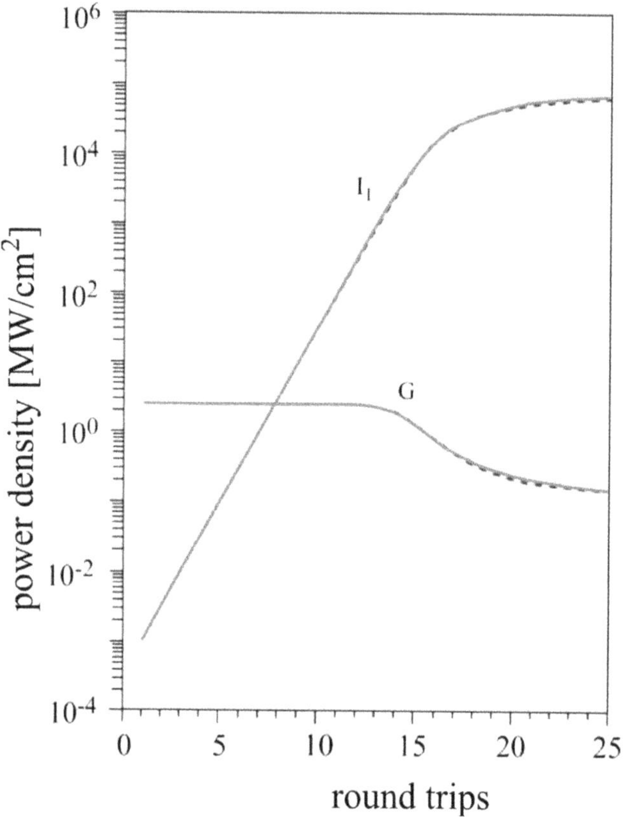

Figure 4.21. (a) Upper curve intra-cavity intensity evolution for a laser FEL oscillator operating with the following parameters $E = 108.24$ MeV, $\lambda_u = 2.8$ cm, $K = 2.10$, $N = 50$, $g_0 = 2.0$, $P_E = 2.012 \cdot 10^6$ MW cm^{-2}, $\sigma_{\varepsilon,0} = 10^{-4}$, $I_S = 8.08 \cdot 10^3$ MW cm^{-2}, $\lambda_0 = 1.008 \mu$m, $\eta = 0.10$. Continuous line: 1-D simulation, dot line equation (4.81). (b) Lower curve gain versus round trip (simulation (continuous) and analytical formula 4.79 (dot)).

We make the final observation and close the circle. We find from the definition of the small signal gain coefficient g_0 (equation (4.37)) and of the saturation intensity (4.73) that

$$g_0 I_S = \frac{1}{4N} \Pi_e, \tag{4.84}$$

with

$$\Pi_e = \frac{P_E}{2\,\pi\,\sigma_x \sigma_y} \equiv \text{e-beam power density}. \tag{4.85}$$

In other words, we have obtained that the laser power extracted from the optical cavity is a fraction $1/(4N)$ of the e-beam power.

4.8 High gain FELs and self-amplified-spontaneous emission devices: generalities

Between the end of the Twentieth and the beginning of the Twenty-first Century a kind of revolution took place. It was going to create a scientific environment capable of providing devices capable of providing coherent radiation sources in the x-ray region with a brightness exceeding those of conventional synchrotron radiation sources by many orders of magnitudes.

In figure 4.22 we have reported a figure, shown in the past ad nauseam. However, we did not shy away, because at the time it was displayed for the first time it provided a clear idea of the impact that sources of this type could have produced in the field of research.

High brightness means very intense short pulses of photons, flashing very small surfaces. In figure 4.23 we have reported the '***small and fast world***' which can be explored with such a high brightness probe.

The price to be paid, to build and operate such a device, is the use of a high energy accelerator providing a high energy beam, above the GeV region (hundreds of meters

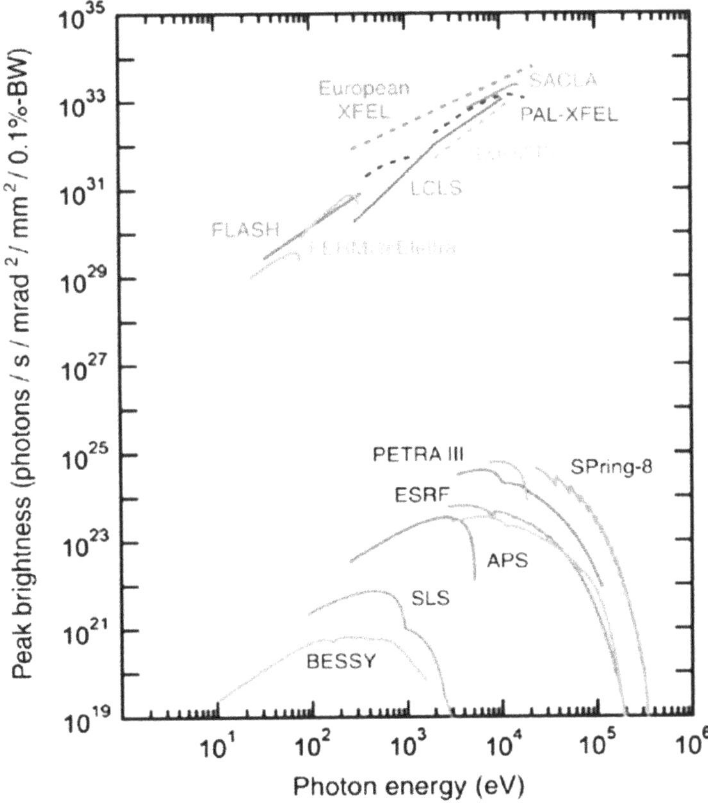

Figure 4.22. Comparison between synchrotron radiation sources and FEL (dated 2013). Reproduced from [15], copyright (2013) by JACoW. CC By 3.0.

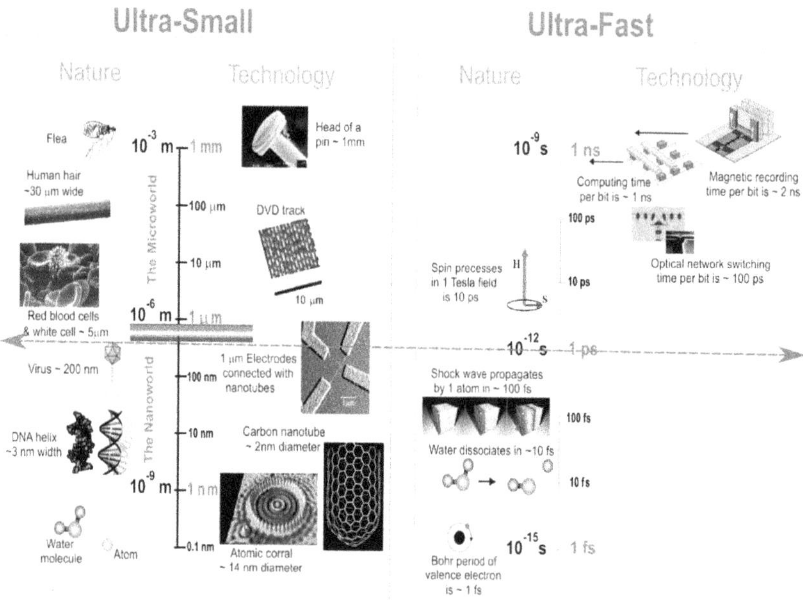

Figure 4.23. Small and fast matter regions which can be explored with laser probes, those below the red dashed-arrows can be studied with SASE FEL devices. Graphic courtesy of J Stöhr.

in length, for a normal gradient accelerator) high peak currents (above 1000 A), 'needle' transverse section e-beam...

We have seen in the previous sections that the pivotal quantity of high gain devices is the Pierce parameter. For hard x-ray operation (see figure 4.14) it ranges around 10^{-4}.

According to figure 4.23 and to equations (4.67) the field grows from an input seed and it is amplified to the value reported in equation (4.68); using the exponential growth part only, we find that the undulator length necessary to reach P_F is

$$
\begin{aligned}
Z_S &\cong \ln\left(\frac{9\,P_F}{P_0}\right) L_g, \\
P_F &= \frac{\gamma\, m_e c^2}{e} I,
\end{aligned}
\tag{4.86}
$$

where P_0 is the initial seed coherent power. A source providing such a quantity of power may not be available and therefore it is preferable to start the emission from the spontaneous emission noise, this is the reason why the process, we are interested in is called SASE (namely self-amplified spontaneous emission).

The term '***noise***' may not be appropriate. We note that at the beginning photons and electrons move in the same direction and after an undulator period the radiation will be ahead of the electrons by a quantity called the coherence length.

We can imagine the process as illustrated in figure 4.12, the electrons enter at random times inside the undulator, emit spontaneous radiation which is amplified up to the saturation.

A certain number of electrons may contribute (coherently) to the emission process and they are distributed on the coherence length scale which can be determined by the use of this argument. To appreciate such a characteristic length we go back to the definition of high gain detuning parameter and note that it can be written as

$$\tilde{\nu} = \frac{\lambda_c}{c}\delta\omega, \qquad (4.87)$$
$$\lambda_c = \frac{\lambda_0}{4\,\pi\,\sqrt{3}\rho}, \quad \delta\omega = \omega_0 - \omega,$$

where λ_c defines a specific length, which can be exploited to evaluate the number of electrons 'co-operating' to the emission process, such a number is given by

$$n_c = I_e\frac{\lambda_c}{c} = \frac{I_e}{2\,e\,\sqrt{3}\rho\omega}. \qquad (4.88)$$

Before going further it is worth stressing that, if the electrons come in bunches with length σ_z and if $\lambda_c < \sigma_z$, the ratio

$$N_s \propto \frac{\sigma_z}{\lambda_c}, \qquad (4.89)$$

represents the number of portion of bunches independently contributing to the SASE process (see below for a more accurate definition).

The power emitted by the cooperating electrons in one coherence length can be estimated as

$$P_n \approx \frac{1}{4\pi\sqrt{3}}\left(\frac{\delta\omega}{\omega}\right)\gamma mc^2\frac{c}{\lambda_c} \cong \rho^2\frac{\gamma mc^3}{\lambda_0}, \qquad (4.90)$$

and can be exploited as seed power to estimate the length of the undulator to reach the saturation as being

$$Z_S \cong \ln\left(\frac{9\,P_F}{P_n}\right)L_g \cong \ln\left(\frac{9\sqrt{2}\,I}{\rho e}\frac{\lambda_0}{c}\right)L_g = \ln\left(36\,\pi\sqrt{2}\,N_c\right)L_g. \qquad (4.91)$$

As a practical remark, we note that for most of the foreseeable cases the logarithm in the previous equation is $4\,\pi\,\sqrt{3} \cong 20$ and therefore

$$Z_S \cong 4\,\pi\,\sqrt{3}\,L_g = \frac{\lambda_u}{\rho}. \qquad (4.92)$$

It is evident that for FEL operating with $\lambda_u = 5$ cm and $\rho = 5 \cdot 10^{-4}$, undulators with length exceeding 100 m should be designed.

A more general discussion of the matter treated in this section can be found in appendix A. According to the discussion summarized here and detailed in appendix C, the field may grow as the amplification of an initial coherent seed or from a pre-bunched beam; in this case the associated logistic curve is written as

$$
\begin{aligned}
P(z) &= P_n \frac{B(z)}{1 + \frac{P_n}{P_F} B(z)}, \\
B(z) &= 2\left[\cosh(\tilde{z}) - \exp(-\tilde{z})\cos\left(\frac{\pi}{3} + \frac{\sqrt{3}}{2}\tilde{z}\right) - \exp(\tilde{z})\cos\left(\frac{\pi}{3} - \frac{\sqrt{3}}{2}\tilde{z}\right)\right].
\end{aligned} \tag{4.93}
$$

A comparison between evolution from seed or from bunching is provided in appendix C where a more detailed analysis of the condition giving rise to amplifier and SASE growth is given.

This chapter has been devoted to an introduction to the physics of the generators of coherent radiation from free electrons. Even though undulator based FELs do not find specific applications in the field of plasma magnetic fusion, the discussion has been preparatory to analogous devices, exploited for such purposes.

References

[1] Zhom H 2010 On the minimum size of demo *Fusion Sci. Technol.* **58** 613–24

[2] Sakamoto K, Kariya T, Oda Y, Minami R, Ikeda R, Kajiwara K, Kobayashi T, Takahashi K, Moriyama S and Imai T 2015 Study of sub-terahertz high power gyrotron for ECH&CD system of demo *IEEE Int. Conf. on Plasma Sciences (ICOPS) (Antalya)*

[3] Dattoli G, Doria A, Sabia E and Artioli M 2017 *Charged Beam Dynamics, Particle Accelerators and Free Electron Lasers* (IOP Plasma Physics Series) (Bristol: IOP Publishing)

[4] Dattoli G, Di Palma E, Pagnutti S and Sabia E 2018 Free electron coherent sources: from microwave to x-rays *Phys. Rep.* **739** 1–51

[5] Dattoli G, Del Franco M, Labat M, Ottaviani P L and Pagnutti S 2012 *Introduction to the Physics of Free Electron laser and Comparison with Conventional Laser Sources, Free Electron lasers* (Rijeka: IntechOpen)

[6] Elias L R, Madey J M J, Schwettman H A and Smith T I 1976 Observation of stimulated emission of radiation by relativistic electrons in a spatially periodic transverse magnetic field *Phys. Rev. Lett.* **36** 717

[7] Deacon D A G, Elias L R, Madey J M J, Ramian G J, Schwettman H A and Smith T I 1977 First operation of a free-electron laser *Phys. Rev. Lett.* **38** 892

[8] Galayda J N, Arthur J, Ratner D F and White W E 2010 x-ray free-electron lasers-present and future capabilities *J. Opt. Soc. Am.* **27** B106

[9] Colson W B 1990 Classical free electron laser theory *Laser Handbook* vol 6 (Amsterdam: North Holland)

[10] Ohnuki H and Ellaume P 2014 *Wigglers, Undulators and their Applications* (London: CRC Press)

[11] Clarke J A 2014 *The Science and Technology of Undulators and Wigglers* (Oxford: Oxford Science Publication)

[12] Smith S J and Purcell E M 1953 Visible light from localized surface charges moving across a grating *Phys. Rev.* **92**

[13] Schächter L and Ron A 1989 Smith-Purcell free-electron laser *Phys. Rev.* A **40**

[14] Marcuse D 1980 *Principle of Quantum Optics* (New York: Academic)

[15] Huang Z 2013 Brightness and coherence of synchrotron radiation and FELs *Proc. of the 4th Int. Particle Accelerator Conf. (Shanghai, China)*

IOP Publishing

G Dattoli, E Di Palma, S P Sabchevski and I P Spassovsky

Chapter 5

An overview of the gyrotron theory

5.1 Introduction

Gyrotrons are microwave vacuum tubes that belong to the family of fast-wave free-electron devices. The remarkable history of their discovery, the subsequent spectacular progress of their development, and the formulation of the underlying theory are well represented in the memorial paper dedicated to the 50th anniversary of their discovery [1]. According to the historical records and recollections presented there, the first gyrotron was invented, designed and tested in Nizhny Novgorod (formerly Gorky), Russia in 1964. Conceived and considered initially as millimeter- and submillimeter-wave tubes, in recent years gyrotrons have advanced towards the sub-terahertz and terahertz frequencies and crossed the symbolic threshold of 1 THz. Nowadays, gyrotrons are the most powerful sources of coherent radiation in this frequency range—a region of the electromagnetic spectrum which is still habitually referred to as the THz gap (although a more appropriate term would be 'the last frontier') due to the lack of powerful devices operating there. They have reached record high values of the product of the output power (P) and the squared frequency (f), which is a well-known figure of merit that characterizes the high-power sources of electromagnetic radiation. Figure 5.1 shows a comparison of gyrotrons with other microwave devices.

The current status of gyrotrons development is being presented annually in a very comprehensive and detailed state-of-the-art report issued by Professor Manfred Thumm [4]. The remarkable potential for filling the THz-gap demonstrated by gyrotrons has opened an avenue for many novel applications in the very broad and steadily expanding fields of high-power terahertz science and technology [5]. The advancements in this direction are presented in numerous papers and summarized in several very informative reviews [6–12]. Some of the applications are well-advanced and matured. The most notable among them is the usage of the gyrotrons for

doi:10.1088/978-0-7503-2464-9ch5

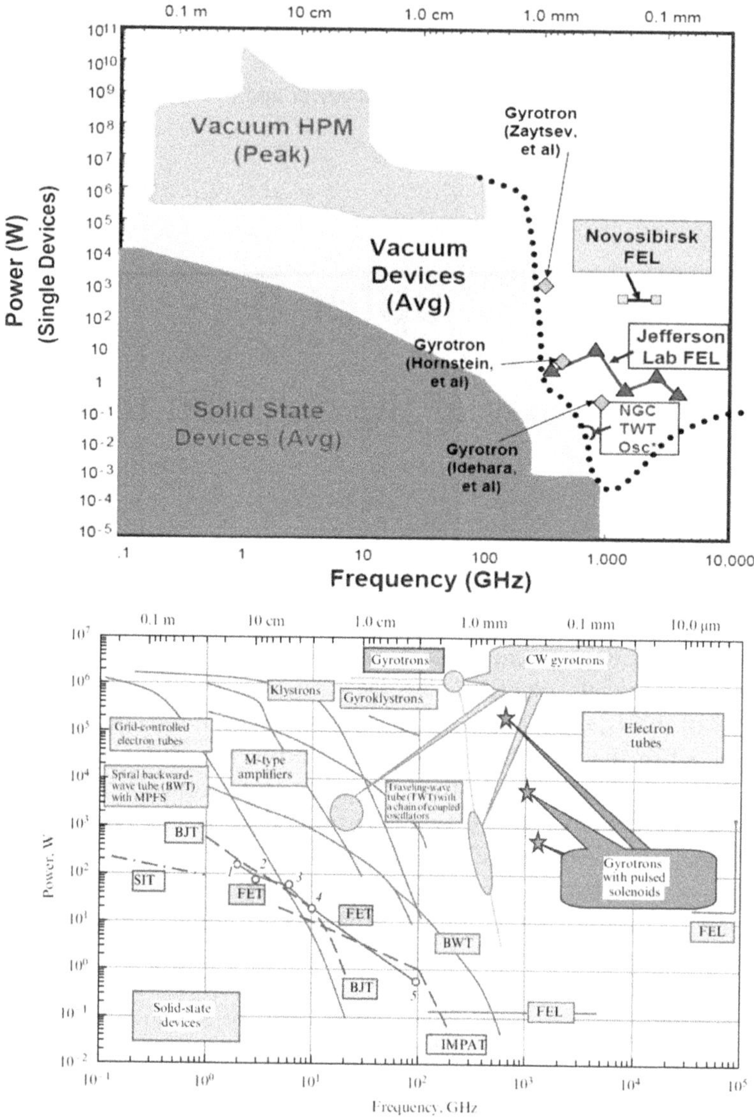

Figure 5.1. Top diagram: current status of the high-power microwave sources, output power versus frequency (reprinted from [2] with the permission of AIP Publishing). Bottom diagram: power output levels achieved in electron devices: IMPAT devices based on impact ionization avalanche transit-time diodes, BJT devices based on bipolar junction transistors, FET devices based on field effect transistors, SIT devices based on static induction transistors, MPFS magnetic periodic focusing system; 1 devices based on Schottky-gate field effect transistors (MESFET) in GaAs (manufactured by Fujitsu, Japan), 2 devices based on MESFETs in SiC (manufactured by Cree, USA), 3 Devices based on MESFETs in GaAs (manufactured by Toshiba, Japan), 4 devices based on pseudomorphic high electron mobility transistors (PHEMTs) in GaAs (manufactured by Raytheon, USA), 5 devices based on PHEMTs (manufactured by TRW, USA) (reproduced from [3]. Copyright Uspekhi Fizicheskikh Nauk 2016).

electron cyclotron resonance heating (ECRH) and electron cyclotron resonance current drive (ECRCD) of magnetically confined plasma in fusion devices (e.g., Tokamaks and stellarators) as well as for plasma control (for instance, suppression of the neoclassical tearing modes) and plasma diagnostic based on the collective Thomson scattering (CTS). Another industrial-grade technology, which benefits from gyrotrons as powerful sources of millimeter and submillimeter waves is the thermal treatment of advanced materials, most notably ceramic sintering.

Other groups of relatively new but very promising applications of gyrotrons includes spectroscopic studies, ion cyclotron ion sources, radars, gas discharge experiments, medical technologies, warfare military devices (e.g. active denial systems), etc. It is anticipated that the number of fields where gyrotrons are used will continue to expand and many prospective novel applications will emerge in the years to come.

Taking into account that the development, study, and applications of gyrotrons are very active fields of research, there is no surprise that a vast literature on these subjects exists. In this short book chapter we do not attempt to overview these numerous publications but rather try to present some essentials that the reader would need in order to understand the operational principles, theory, and technical implementation of gyrotrons. For a more detailed study, we refer to several comprehensive monographs on gyro devices [13–16] as well as to some excellent books on microwave tubes that also cover gyrotrons [17–24].

The general structure of the gyrotron and its subsystems is shown in figure 5.2. The electron-optical system (EOS) of the tube includes a magnetron-injection gun (MIG) and a system of magnetic coils. It generates a hollow helical electron beam which propagates in an adiabatically increasing magnetic field that transfers most of the beam's energy to the transverse rotational motion ('pumping of the beam'). In the resonant cavity, the kinetic energy associated with this motion is extracted and transferred to the generated wave as a result of a resonance interaction as will be explained later. The spent electron beam is dumped on the collector, where its residual energy is dissipated. The generated radiation which has a spatial structure determined by the operating cavity mode is reshaped by an internal mode converter (consisting of an waveguide antenna (launcher) and a system of quasi-optical mirrors) and transformed in a well-collimated Gaussian-like wave beam, which is radiated through an output vacuum window. Therefore, the main parts of the gyrotron are: (i) EOS with a magnetic system; (ii) resonant cavity; (iii) collector; (iv) internal quasi-optical converter; and (v) an output window. They will be explained in more detail after presenting the operational principles and the theory of the gyrotron.

5.2 Basic physical principles of gyrotron operation

5.2.1 Cyclotron resonance

Gyrotron operation is based on a physical phenomenon known as electron cyclotron resonance maser instability, which takes place when an ensemble of electrons gyrating in a magnetic field B with a cyclotron frequency Ω_c interacts with an electromagnetic wave with a close frequency $\omega \approx \Omega_c$. The cyclotron frequency

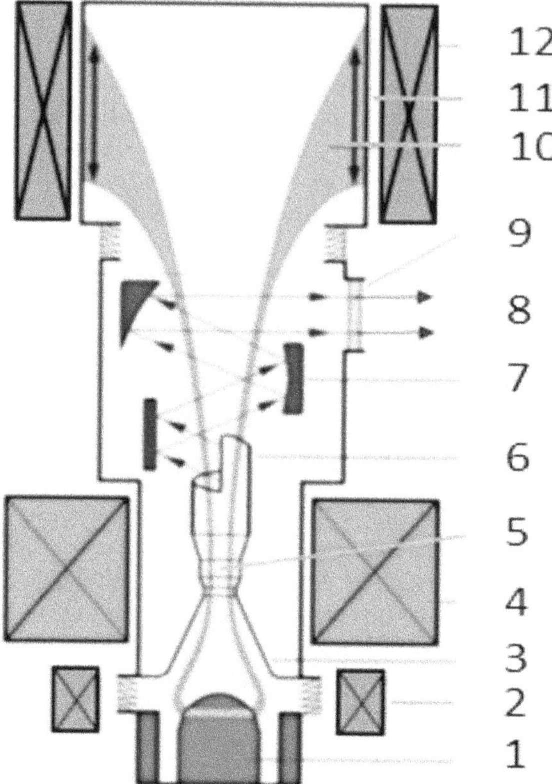

Figure 5.2. Schematic structure of a generic gyrotron and its components: 1—magnetron injection gun (MIG); 2—additional gun coils; 3—beam tunnel and compression region; 4—main coils of the superconducting magnet; 5—gyrotron cavity resonator; 6—launcher of the internal mode converter; 7—system of mirrors of the internal mode converter; 8—Gaussian wave beam; 9— output vacuum window; 10—helical electron beam; 11—water-cooled collector of the spent electron beam; 12—magnetic sweeping coils for smearing the spent electron beam. Reproduced from [25], copyright (2020) by the authors. CC BY 4.0.

$$\Omega_c = \frac{e}{\gamma m_0} B \tag{5.1}$$

is proportional to the intensity of the magnetic field and depends on the electron energy through the relativistic Lorentz factor γ, e and m_0 being the charge and the rest mass of an electron, respectively. Since generally beside the transverse velocity $v_\perp$ associated with their rotational motion the electrons also have an axial velocity v_z their orbits are helices (see figure 5.3) wound along the magnetic field lines and having a radius (Larmor radius)

$$r_L = \frac{v_\perp}{\Omega_c}. \tag{5.2}$$

Therefore, for one cyclotron period $T_c = 2\pi/\Omega_c$ the electron propagates in a longitudinal direction at a distance (cyclotron wavelength) $\lambda_c = T_c v_z = \frac{2\pi}{\Omega_c} v_z$.

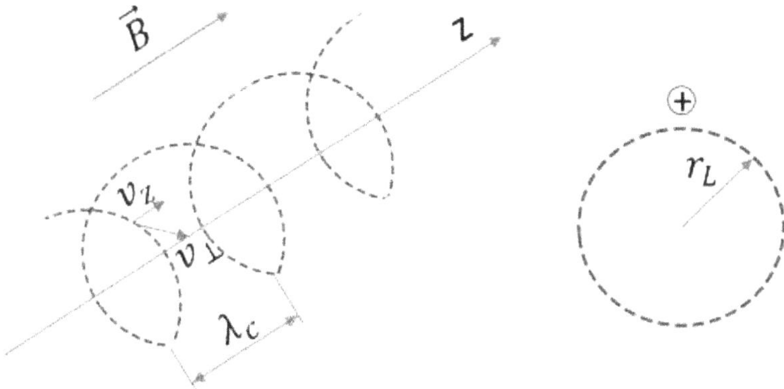

Figure 5.3. Helical orbit (trajectory) of an electron in a uniform magnetic field.

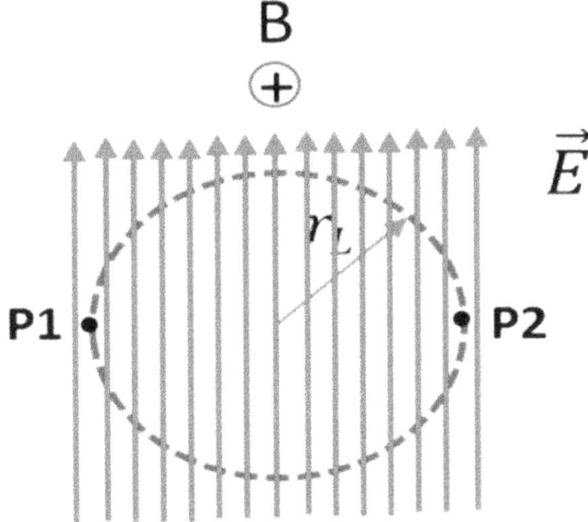

Figure 5.4. Analysis of the electron motion at cyclotron resonance.

Consider an electron gyrating in a constant magnetic field on the Larmor circle (figure 5.4) and interacting with a harmonic electric field $\vec{E} = \vec{A}\cos(\omega t)$ with amplitude $\vec{A}$ and an angular frequency ω equal to the cyclotron frequency Ω_c. Additionally, we assume that the field amplitude is small enough in order not to perturb significantly its motion during several cyclotron periods. If at the initial time t_1 the electron is located at point P1 and the electric field reaches its maximum value in a direction as shown in figure 5.4 then the Lorentz force $\vec{F} = -e\vec{E}$ is retarding at this position. At the time moment $t_2 = \frac{T_c}{2}$, i.e. after a half cyclotron period, this electron will reach point P2. Since for this time interval the direction of the electric field is reversed to the opposite, the electron again will be in a deaccelerating phase at that position. Then, after one more half period, at $t_3 = T_c$, the situation repeats and at point P1 the electron is deaccelerated again and so on. Analogously, an electron which is initially accelerated at $t_1 = 0$ will continue to face an accelerating phase of the field. Therefore, at $\omega = \Omega_c$ the circular motion of an electron is synchronized with the changing electric field in such a way that the interaction has a resonance nature and its result (acceleration or deceleration) accumulates with the time. Alongside the fundamental electron cyclotron resonance ($\omega = \Omega_c$) higher order resonances are also possible at the harmonics of the cyclotron frequency. Therefore, in the general case, the resonance condition is

$$\omega = s\Omega_c, \tag{5.3}$$

where the harmonic number n is an integer ($s = 1, 2, 3, \ldots$).

5.2.2 Azimuthal bunching of the electrons

The interaction of an electron with the electromagnetic field alters both its transverse velocity and the energy which, according to (5.1) and (5.2), changes also the cyclotron frequency and the Larmor radius. The former change stems from the relativistic dependence of the electron mass on the energy in accordance with the relation $m = \gamma m_0$. Thus, the cyclotron frequency of the accelerated electrons (for which γ has been increased) decreases while, in contrast, for the decelerated particles it increases. Such dependency of Ω_c on γ leads to an azimuthal bunching, as illustrated in figure 5.5. The electrons at the positions 1 and 5 do not undergo changes in their transverse velocities since $\vec{E} \cdot \vec{v}_\perp = 0$ and continue an unperturbed motion. In contrast, for the rest of the electrons in the accelerating (2, 3, and 4) and decelerating (6, 7, and 8) phases, where $\vec{E} \cdot \vec{v}_\perp < 0$ and $\vec{E} \cdot \vec{v}_\perp > 0$, respectively, both Ω_c and r_L will vary. More specifically, the electron at 2, 3, and 4 are in the accelerating phase of the electric field and their transverse (orbital) velocity increases as does the radius of the rotation while the cyclotron frequency decreases. As a result, such electrons shift towards the electron at position 1 and increase their orbital radii. For the electrons at 6, 7, and 8, $v_\perp$ decreases and Ω_c increases, which leads also to a shift towards the position 1 albeit at a smaller radius. The final effect

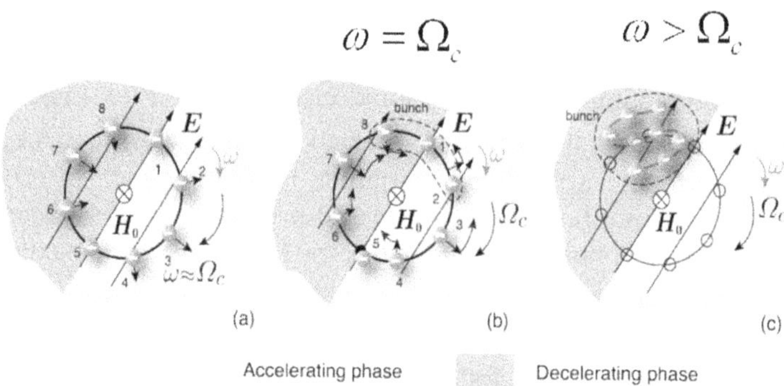

Figure 5.5. Schematic of the azimuthal electron bunching: (a) initial uniform distribution of the electrons on the Larmor circle; (b) electron bunching at cyclotron resonance; (c) electron bunching at gyrotron interaction ($\omega > \Omega_c$). Figure courtesy of V Bratman.

of such interaction is the formation of a bunch (group) of electrons condensed around point 1 (figure 5.5(b)). As will be explained in detail below, the gyrotrons operate at a frequency slightly higher than the cyclotron frequency ($\omega > \Omega_c$). In this case (see figure 5.5(c)) the bunch of electrons slips toward the decelerating phase where due to the bremsstrahlung effect their energy is transferred to the wave. Because the electrons are bunched in phase due to a relativistic effect and thus are synchronized with the wave, the radiation is coherent.

5.2.3 Beam-wave synchronism and Brillouin diagram

In gyrotrons, the interaction between the electron beam and the electromagnetic wave takes place inside a resonant structure, which in the simplest case is a part (section) of a regular waveguide delimited by down- (at the entrance) and up-taper (at the exit), respectively. The synchronism between the beam and the wave at the resonance condition is conveniently depicted by the Brillouin diagram (figure 5.6), which includes the dispersion curve of the operating transverse mode and the beam line defined as

$$\omega^2 = \omega_c^2 + c^2 k_z^2, \tag{5.4}$$

$$\omega = s\Omega_c + v_z k_z. \tag{5.5}$$

Here, ω is the circular frequency of the electromagnetic field, $\omega_c = c\frac{\chi_{m,n}}{R}$ is the cut-off frequency of the mode ($TE_{m,n}$ or $TM_{m,n}$) with an eigenvalue $\chi_{m,n}$ in a waveguide of radius R, k_z is the axial wavenumber, $\beta_z = v_z/c$ is the axial velocity of the beam electrons v_z, normalized to the speed of light in vacuum c. The possible resonances correspond to the intersections of the beam line (5.4) with the mode dispersion (5.5) at points with coordinates (see figure 5.6)

$$k_{z_{+,-}} = \frac{s\Omega_c\gamma_z^2}{c}\left(\beta_z \pm \sqrt{1 - \frac{\omega_c^2}{\gamma_z^2 s^2 \Omega_c^2}}\right), \tag{5.6}$$

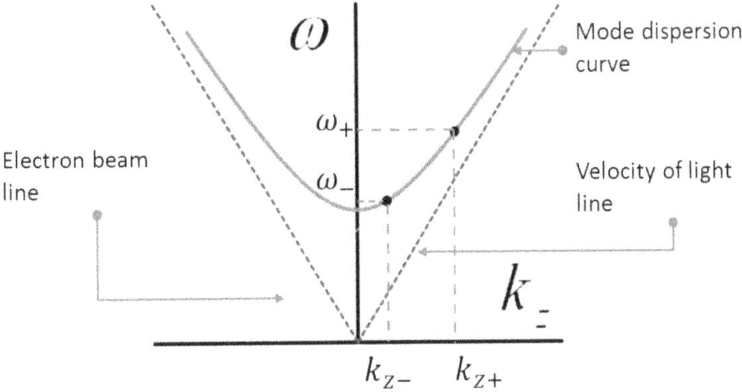

Figure 5.6. Brillouin diagram and beam–wave synchronism at resonance.

$$\omega_{+,-} = s\Omega_c \gamma_z^2 \left(1 \pm \beta_z \sqrt{1 - \frac{\omega_c^2}{\gamma_z^2 s^2 \Omega_c^2}} \right), \tag{5.7}$$

where $\gamma_z = 1/\sqrt{1 - \beta_z^2}$.

An important waveguide parameter in these equations is

$$\Psi = \frac{\omega_c^2}{\gamma_z^2 \Omega_c^2}. \tag{5.8}$$

When $\frac{1}{\gamma_z^2} < \Psi < 1$ there are two intersection points both in the right quadrant of the $\omega - k_z$ coordinate system, where $k_z > 0$ which corresponds to a forward propagating wave. A special case is the so-called *grazing condition* (coalescence between the field and the beam for which the phase velocity $v_{\text{ph}} = \omega/k_z$ equals the axial velocity of the electrons v_z). It is realized if $\Psi = 1$, i.e. when

$$\omega_c = \gamma_z \Omega_c. \tag{5.9}$$

In this particular case, illustrated in figure 5.7, the coordinates of the intersection point are

$$k_z = \frac{\Omega_c \gamma_z^2 \beta_z}{c} \text{ and } \omega = \Omega_c \gamma_z^2, \tag{5.10}$$

At $\Psi < 1/\gamma_z^2$ (and thus $\omega_c < \Omega_{\text{CH}}$), however, $k_{z-} < 0$ and the electron beam interacts with a backward propagating wave for which the group velocity $v_{\text{gr}} = \frac{d\omega}{dk_z} < 0$ while the phase velocity $v_{\text{ph}} > 0$ (figure 5.8).

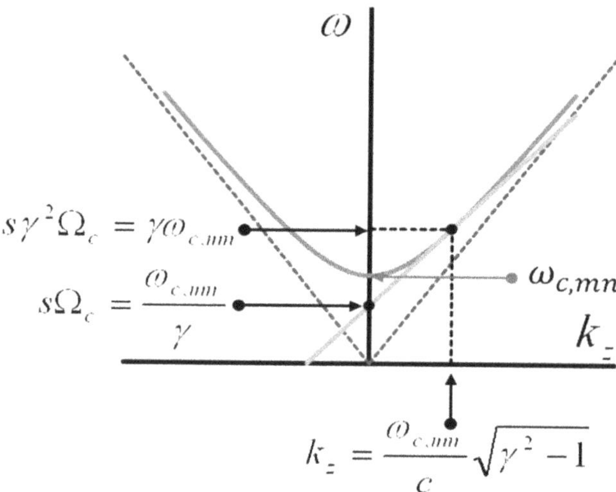

Figure 5.7. Brillouin diagram at grazing condition ($\omega_{c,mn} = c\chi_{mn}/R$ cut-off frequency of the mode, s—harmonic number).

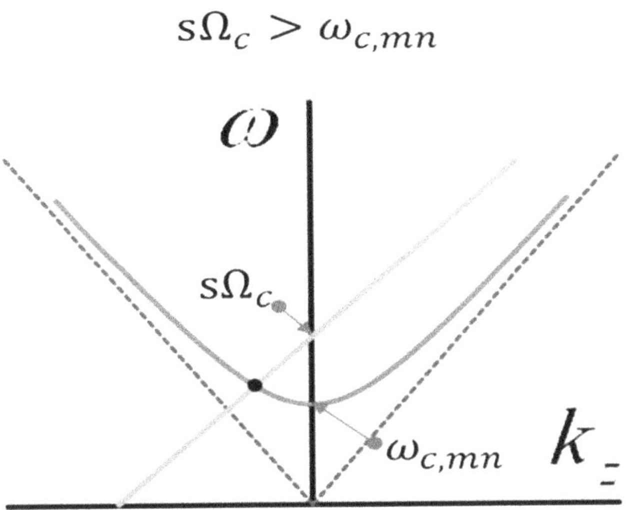

Figure 5.8. Brillouin diagram illustrating gyro-backward-wave interaction.

It is important to stress that the gyrotrons are *fast-wave devices* since the phase velocity of the wave in the cavity is superluminal, i.e. $v_{ph} > c$. This is a fundamental difference with the classical microwave tubes in which the electrons interact with the wave in slow-wave structures (helices, aperture loaded waveguides, etc) with characteristic dimensions of the order of the wavelength. At high frequencies, the elements of such structures become extremely tiny. Since the field intensity decreases rapidly with the distance from them, the electron beam should propagate close to their surfaces. This, however, puts limits on the beam current that could be used without damaging the delicate and fragile slow-wave structures. In contrast, the gyrotron interaction takes place in an overmoded (oversized) simple resonant cavity whose radius could be much greater than the wavelength. This allows high-current electron beams to be used and, eventually, high output powers to be achieved.

5.2.4 Types of gyro-devices

It should be mentioned that almost all slow-wave tubes have analogs in the family of the gyro-devices, as illustrated in figure 5.9 [26]. Among them, the gyrotron (aka gyro-monotron) is an oscillator while the gyro-TWT, gyro-BWT, etc, are amplifiers. A characteristic feature of the gyrotron is that it operates at frequencies that are close to the corresponding cut-off frequencies of the used cavities. Since for gyrotrons the Doppler shift term ($v_z k_z$) is small (see equation (5.5)) their output frequency is close to the cyclotron frequency or its harmonics.

Depending on the used electron beams and resonant structures, different types of gyrotrons exist. Some of them are shown in figure 5.10 and will be discussed in more detail in the following sections.

A special kind of gyro-device is the Cyclotron Auto-Resonance Maser (CARM) in which due to a specific mechanism the cyclotron resonance is maintained automatically. For the CARM which utilizes relativistic electron beams (and thus

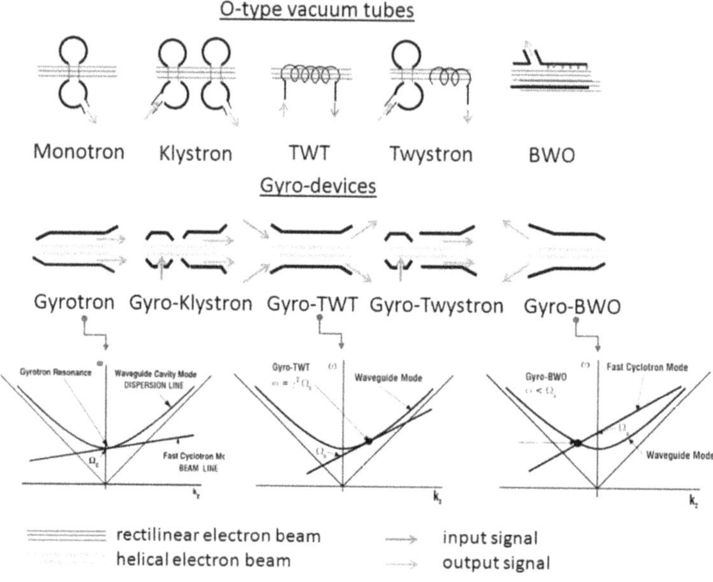

Figure 5.9. Schematics of gyro-devices and their slow-wave analogs.

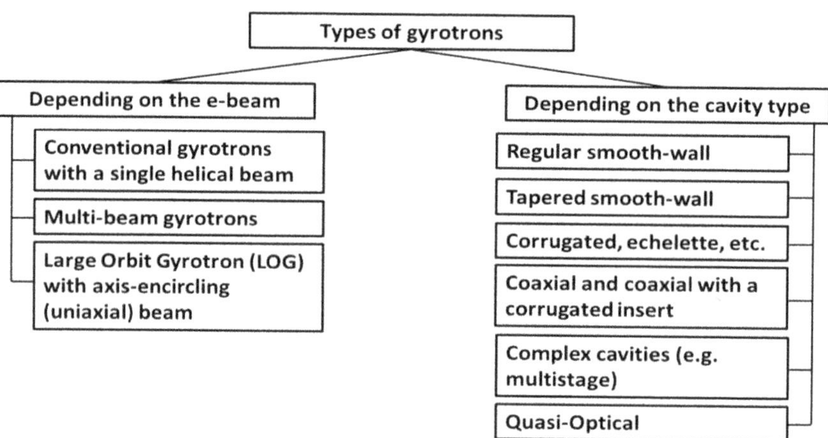

Figure 5.10. Varieties of gyrotrons.

large γ), the Doppler shift is significantly higher. This allows achieving generation at high frequencies using relatively low magnetic fields. Due to its significance, CARM will be discussed in chapter 6.

5.2.5 The efficiency of the interaction and output power

In gyrotrons, the energy is extracted only from the rotational motion of the beam electrons having a velocity ratio (pitch factor)

$$\alpha = \frac{v_{\perp 0}}{v_z}. \tag{5.11}$$

Therefore the maximum transverse energy $W_{\perp 0}$ of a gyrating electron, which is available for a transfer to the wave, is only a fraction of its total energy W_0 according to the relation

$$W_{\perp 0} = \frac{\alpha^2}{1 + \alpha^2} W_0. \tag{5.12}$$

Defining the total electronic efficiency η_e as a ratio of the energy $\Delta W = W_{\perp 0} - W_{\perp}$ extracted from the electron beam to the total initial kinetic energy W_0 gives

$$\eta_e = \frac{\Delta W}{W_0} = \eta_{e\perp} \frac{\alpha^2}{1 + \alpha^2}, \tag{5.13}$$

where

$$\eta_{e\perp} = \frac{W_{\perp 0} - W_{\perp}}{W_{\perp 0}}, \tag{5.14}$$

is the transverse (orbital) efficiency. It is obvious that in order to obtain high total efficiency, electron beams with large pitch factors should be used. In practice, values of $a \approx 1.5$ are attainable in the electron-optical systems (EOS) with magnetron injection guns (MIG) without appreciable reflections of the electrons. It should be mentioned that the above relations give the so-called single-particle efficiencies. For a real electron beam with a velocity spread distribution, the corresponding values are calculated by averaging (denoted below by <…>) over the whole ensemble of particles representing the beam

$$\eta_e = \frac{\gamma_0 - \langle \gamma \rangle}{\gamma_0 - 1}. \tag{5.15}$$

From energy conservation, the following power balance takes place

$$P_{tot} = P_{\Omega} + P_{out}, \tag{5.16}$$

where $P_{tot} = \eta_e P_{EB}$ is total power produced by the electron beam of power P_{EB}, P_{Ω} is the power dissipated at the cavity walls due to the ohmic losses, and P_D is the power of the output (diffracted) radiation. These quantities can be expressed using the corresponding loaded (total)-, ohmic- and diffractive-Q factors (Q_{tot}, Q_{Ω}, Q_D), respectively, and the stored energy in the cavity W, as

$$P_{tot} = \omega \frac{W}{Q_{tot}}, \; P_{\Omega} = \omega \frac{W}{Q_{\Omega}}, \; P_{out} = \omega \frac{W}{Q_D}. \tag{5.17}$$

Then, the total device efficiency can be written as

$$\eta = \frac{P_{\text{out}}}{P_{\text{EB}}} = \eta_e \frac{Q_{\text{tot}}}{Q_D} = \eta_e \left(1 - \frac{Q_{\text{tot}}}{Q_\Omega}\right) = \eta_e \frac{1}{1 + Q_D/Q_\Omega} = \eta_e \eta_r. \tag{5.18}$$

Here

$$\eta_r = \frac{Q_{\text{tot}}}{Q_D} = 1 - \frac{Q_{\text{tot}}}{Q_\Omega} = \frac{1}{1 + Q_D/Q_\Omega} \tag{5.19}$$

is the so-called circuit efficiency and $Q_{\text{tot}} = Q_D Q_\Omega/(Q_D + Q_\Omega)$. It can be seen that if Q_Ω exceeds significantly Q_D then $\eta_r \approx 1$. The diffractive Q-factor of the cavity can be estimated using the following simple formula

$$Q_D = \frac{4\pi}{(1 - R_i R_o)} \left(\frac{L}{\lambda}\right)^2, \tag{5.20}$$

where R_i and R_o are the reflection coefficients at the input and the output (exit) of the cavity. It is instructive to note the strong (quadratic) dependence of Q_D on the ratio of the cavity length L and the wavelength of the radiation λ. The ohmic Q-factor depends on the mode pattern (field distribution) through the eigenvalue $\chi_{m,n}$ and the azimuthal index of the mode $TE_{m,n}$, the cavity radius R, and the skin depth d according to the relations

$$Q_\Omega = \frac{R}{\delta}\left(1 - \frac{m^2}{\chi_{m,n}^2}\right), \tag{5.21}$$

$$\delta = \sqrt{\frac{\lambda}{\pi Z_0 \sigma}}, \tag{5.22}$$

Z_0 being the impedance of free space ($Z_0 \approx 377\ \Omega$), and σ the conductivity of the cavity wall. From the above formula, it is clear that the diffractive Q-factor is proportional to the square of the frequency ($Q_D \propto f^2$), while the ohmic one is proportional to the square root of the frequency ($Q_\Omega \propto \sqrt{f}$). Consequently, the ratio Q_D/Q_Ω increases rapidly with frequency and imposes a limit for the output power at high frequencies. One way to reduce the ohmic losses is by increasing the cavity size, which, however, makes the mode spectrum denser and therefore leads to a severe mode competition. The latter observation is a good example of the inevitable problem of the gyrotron design as a whole and that of the cavity in particular, namely the necessity of compromises ('trade-offs') between many contradicting requirements.

An approximate but very informative formula [27] gives an estimate of the total efficiency as a function of the main operational parameters, namely the pitch factor α, the eigenvalue of the working mode $\chi_{m,n}$, wavelength of the radiation λ in mm, the azimuthal mode index m, the specific power of the ohmic losses P_Ω in kWcm^{-2}, the output power P_{out} in kW, and the beam voltage U_α. It reads

Table 5.1. Factors limiting the efficiency and the output power of gyrotrons with and without CPD.

Gyrotrons without CPD	Efficiency
Maximal transverse efficiency	$\eta_{\perp}$ = 70%–85%
Pitch-factor limitation, $\alpha \leqslant 1.4$, reduction 35%	$\eta_e \leqslant$ 0.66, $\eta_{\perp}$ = 46%–56%
Electron beam velocity spread $\delta v_{\perp} > 0.3$, reduction 10%	$\eta_e \leqslant$ 0.9, $\eta_e(0)$ = 41%–50%
Restrictions of efficiency by $P_{\Omega} < 2$ kW cm^{-2}, reduction 10%,	$\eta_e \leqslant$ 0.9, $\eta_e(0)$ = 37%–45%
Ohmic cavity and transition losses, 5%	η = 0.95, η_e = 35%–43%
Losses due to RF space charge, 2%,	η = 34%–42%
Parasitic mode conversion in the cavity and transition, $\leqslant$ 0.3%	η = 34%–42%
Quasi-optical mode converter losses (old type), 10%	η = 31%–38%
Quasi-optical mode converter losses (new type), 2%	η = 35%–41%
Gyrotron output window losses, ~0.2%–3%	η = 30%–41%
Cyclotron reabsorption at the output transition, 5%–10%	η = 27%–39%
Gyrotrons with CPD	Efficiency
Efficiency of CPD-gyrotron without cyclotron re-absorption	η_1 = 1.6, η_0 = 48%–60%
Losses due to cyclotron re-absorption	η_1 = 38%–49%

$$\eta = \frac{0.26\alpha^{3/2}\chi_{m,n}{}^{5/4}(\lambda^{5/2}P_{\Omega})^{3/4}}{(1+\alpha^2)^{1/2}m^{1/4}U_a^{1/2}P_{\text{out}}^{1/2}}. \tag{5.23}$$

The analysis of the above formula shows that an increase of the working frequency (other parameters being the same) requires the utilization of high-order modes in overmoded resonance cavities.

The efficiency can be increased by partial recovery (recuperation) of the energy of the spent electron beam using a collector with potential depression (CPD). Table 5.1 taken from [27] shows the decrease of the efficiency due to various factors in both the gyrotrons with and without energy recovery.

5.2.6 Mode selection and coupling factor

As already mentioned above, due to the dense mode spectrum of the oversized gyrotron cavity, the mode competition becomes one of the most serious problems. In order to avoid excitation of the unwanted (parasitic) competing modes, a careful selection of the operating (design) mode is mandatory. There are several methods for mode selection that can be classified as electrodynamical, electronic as well as hybrid.

The first one (*electrodynamical mode selection*) is based on the fact that in the gyrotron cavity (which usually comprises a slightly irregular waveguide section and, possibly, adjacent tapers) the diffractive losses of the lowest-order axial modes ($TE_{m,n,l}$, $l = 1$), that have only one variation of the field along the axis (and also the lowest group velocity) will be much smaller than that of the higher-order axial modes with $l > 1$. This is favorable for the modes with $l = 1$ and at moderate beam currents, only those modes will be excited.

The *electronic mode selection* is based on the following considerations. First of all, it is clear that only a small fraction of the possible modes for which the resonance conditions are satisfied by proper tuning of the magnetic field can be excited. It is instructive to recall also that in the start-up stage when the accelerating voltage increases, the relativistic cyclotron frequency decreases so that the 'excitation window' [1] shifts down from higher to lower frequencies.

The second basis of the electronic selection can be explained considering the coupling between the electron beam and the corresponding cavity mode, which is characterized by the function

$$G = \frac{J_{m\pm s}^{2}\left(\frac{\chi_{m,n} R_g}{R}\right)}{\left(\chi_{m,n}^{2} - m^{2}\right) J_{m,n}^{2}(\chi_{m,n})}, \tag{5.24}$$

where R_g is the radius of the guiding centers of the electron beam orbits. The negative and the positive signs of the Bessel function $J_{m\pm s}$ correspond to electron beams that are co- and counter-rotating with respect to the wave, respectively. The strongest beam–wave interaction takes place at the maxima of G that correspond to the maxima of the cavity field. Since the latter have different positions (albeit very often too close) for different modes, by a proper selection of the beam injection radius the desired mode could be excited. In high-power gyrotrons operating at the fundamental resonance, the optimum beam radius coincides with the first maximum of G and is given by

$$R_{g,\mathrm{opt}} = \frac{\chi_{m\pm 1,\,1}}{\chi_{m,n}} R. \tag{5.25}$$

The mode selection at the initial design stage practically always begins with the analysis of the beam–wave coupling G for various modes paying attention to their separation and to the location of their maxima. Such analysis is complemented with the calculation of the corresponding starting current I_{st} for each mode. The latter is the threshold value of the beam current, which is necessary for excitation of the mode and is given by [28]

$$I_{\mathrm{st}} = \frac{\gamma \beta_z \beta_{\perp}^{2(2-s)} \left(\chi_{m,n}^{2} - m^{2}\right) J_{m}^{2}(\chi_{m,n}) \int_0^{\xi_{\mathrm{out}}} |f(\xi)|^2 \, d\xi}{0.47 \times 10^{-3} Q_D J_{m\pm s}^{2}\left(\frac{\chi_{m,n} R_g}{R}\right) \Gamma}, \tag{5.26}$$

$$\Gamma = -\left(s + \frac{\partial}{\partial \Delta}\right) \left| \int_0^{\xi_{\mathrm{out}}} f(\xi) e^{is\Delta\xi} \right|^2,$$

where

$$\Delta = \frac{2}{\beta_{\perp}^{2}}\left(\frac{\omega - s\Omega_c}{\omega}\right), \tag{5.27}$$

is the frequency mismatch, $f(\xi)$ is the field profile, and ξ is the dimensionless longitudinal coordinate.

Naturally, the starting current for the operating mode should be the smallest in relation to the starting current of the competitors.

Additional means for mode selection are offered by some advanced gyrotron concepts such as large orbit gyrotron (LOG), coaxial gyrotron, double-beam gyrotron, quasi-optical gyrotron, and gyrotrons with a complex (e.g. slotted, split, vane loaded) cavity. In the LOG, for example, the beam couples only with the co-rotating modes having azimuthal index equal to the harmonic number. This significantly rarifies the mode spectrum and allows operation at very high-harmonics $s > 2$. In double-beam gyrotrons (with two generating or one generating and one absorbing beam) the beams are injected at properly selected maxima of the field in order to excite the desired mode while, at the same time, preventing the oscillations of the competing modes. In the coaxial gyrotron a special kind of *radial mode selection* is provided by an insert tapered towards the collector end of the system. Its presence lowers the diffractive Q-factor of the modes with large radial index and thus suppresses their excitation.

The *azimuthal mode selection* can be realized in cavities with perturbed azimuthal symmetry. For instance, in the split cavity gyrotrons, the two halves can be considered as two mirrors similarly to that in the quasi-optical gyrotron. In such configuration, the modes with one azimuthal variation at the reflector surface have highest diffraction Q.

5.2.7 Different approaches and physical models describing the operation of the gyrotron

The operation of gyrotrons can be described in the framework of both the quantum theory and the relativistic electrodynamics. Although the former is not very practical for the design of these devices, it provides deep physical insight. Moreover, historically the quantum mechanical treatment made in the pioneering works of Twiss, Schneider, and Gaponov [26, 29–31] precedes all other approaches. That is why we will describe it first.

5.2.7.1 Quantum description of gyrotron operation

From the quantum-mechanical point of view, an electron in a magnetic field has a discrete spectrum of energy levels (Landau levels). The latter have values that are quantized according to

$$E_n = \left(n + \frac{1}{2}\right)\hbar\Omega_c, \tag{5.28}$$

where $\hbar$ is the Plank constant and n is an integer. Such equidistant levels are identical to that of a simple harmonic quantum oscillator and cannot yield a stimulated emission. Taking into account the relativistic dependence of the cyclotron frequency on the energy leads to non-equidistant energy levels

$$E_n = mc^2\left[1 + (2n + 1)\left(\frac{\hbar\Omega_c}{mc^2}\right)\right]^{1/2} - mc^2, \tag{5.29}$$

as illustrated in figure 5.11(a). Thus, the frequencies corresponding to induced (stimulated) absorption ($n \rightarrow n + 1$) and induced radiation ($n \rightarrow n - 1$) become slightly different. Provided the frequency is tuned as described above (see equation (5.5)) the induced emission will prevail over the induced absorption. This is illustrated in figure 5.11(b), which shows that at an appropriate detuning $X = (\omega_{n,n+1} - \omega)t$ the absorption of a relativistic gyrating electron may become negative, corresponding to a net stimulated emission (gain) instead of absorption. Making this observation in his short but seminal paper [30], Schneider has formulated the following prophetic anticipation: 'It does not appear unlikely that this effect could be used for a new type of maser, which would require no microwave 'pump' and no low-temperature operation.'

5.2.8 Physical models in the framework of the relativistic electrodynamics

There exist a vast number of physical models that describe the operation of gyrotrons (figure 5.12). Generally, each of them includes a system of equations of motion that govern the relativistic dynamics of the beam electrons and a system of equations that describe the electromagnetic fields during the beam–wave interaction. The first one (the dynamical part of the physical model) originates from the Lorentz force equation while the second one (the field part of the model) stems from the Maxwell equations. The most adequate are the self-consistent physical models that involve a direct solution of the Maxwell equations with a simultaneous integration of the relativistic equations without any approximation besides the discretization and interpolation of the physical quantities in the used computational domain

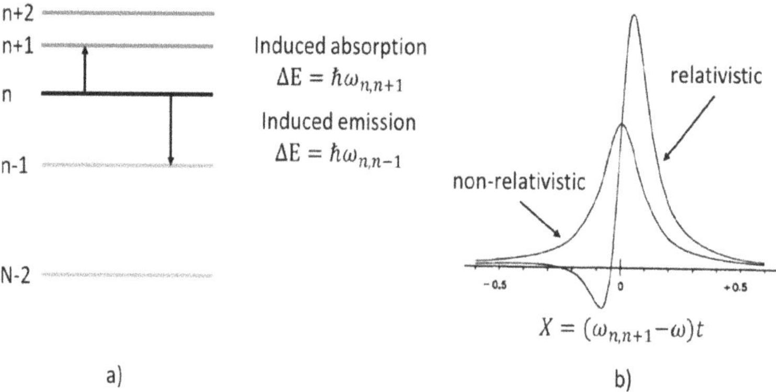

Figure 5.11. Non-equidistant energy levels of a relativistic gyrating electron (a); and the cyclotron resonance energy absorption of a non-relativistic and relativistic electron as a function of the detuning from resonance $X = (\omega_{n,n+1} - \omega)t$; t being the time of interaction [30].

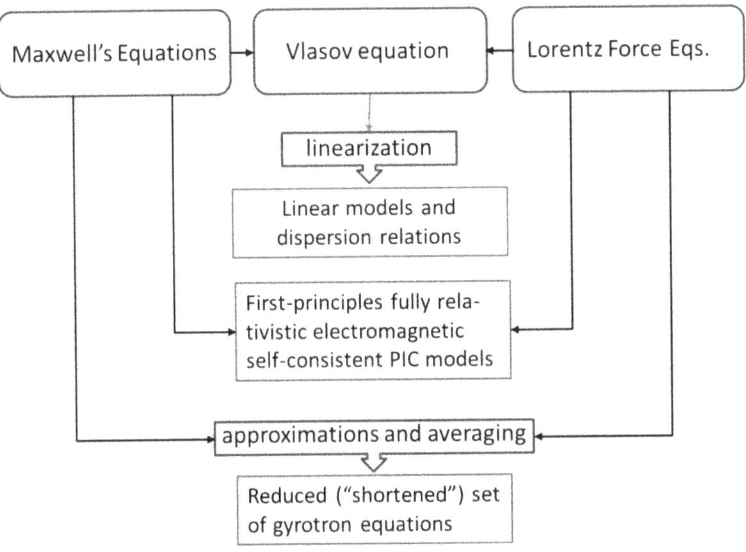

Figure 5.12. Genealogy of the physical models describing the operation of gyrotrons.

(in 2D or 3D). Such 'first principles' models are being widely used in numerous particle-in-cell (PIC) codes as modeling and simulation tools for computer-aided-design (CAD), optimization and numerical studies of gyrotrons. Their main strength is that they take into account directly all important physical factors (and underlying phenomena) and provide exact solutions. At the same time, such codes require significant computational resources in order to scan a wide parameter space of the possible initial data. From the other side, the more simple modes derived using appropriate approximations (e.g. small-signal) and averaging are easier to use and their results have more straightforward interpretation. In practice, however, the whole hierarchy of models implemented in various numerical codes is used in a sequence starting from the known analytical relations and elementary linear formulation and culminating with the most sophisticated self-consistent models. Below we present some of the most widely used physical models very briefly and refer the reader to an excellent companion of books on the subject and the references to the original papers in them [13, 24].

In a cylindrical coordinate system (with unit vectors $\vec{r}$, $\vec{\theta}$, and $\vec{z}$) the components of the electric field $\vec{E} = \vec{E}_r\vec{r} + \vec{E}_\theta\vec{\theta}$ of the mode $TE_{m,n}$ are given by

$$E_r(r, \theta, z, t) = (\vec{z} \times \nabla\psi_{mn})_r F_{mn}(z)\exp(\mathrm{i}\omega t), \tag{5.30}$$

$$E_\theta(r, \theta, z, t) = (\vec{z} \times \nabla\psi_{mn})_\theta F_{mn}(z)\exp(\mathrm{i}\omega t), \tag{5.31}$$

where the so-called membrane function

$$\psi_{mn} = C_{mn}J_m(k_{mn}r)\exp(\mathrm{i}m\theta),$$

is a solution of the corresponding wave equation for an uniform cylindrical waveguide. Here $C_{mn} = 1/\sqrt{\pi(\chi_{mn}^2 - m^2)}J_m(\chi_{mn})$ is a normalization coefficient, and $k_{mn} = \chi_{mn}/R$ is the transverse wave number. The axial dependence of the field intensity (aka longitudinal field profile) is described, in the most general case, by the complex function $F_{mn}(z) = A_{mn}(z)\exp(-i\Phi_{mn}(z))$ with amplitude $F_{mn}(z)$ and phase $\Phi_{mn}(z)$. Below we will omit the indices m, n that label the modes assuming that they are implicated by default. In the simplest models, $F(z)$ is postulated (e.g. as sinusoidal or Gaussian function) while in the more adequate formulations it is determined by the solution of the Helmholtz equation for the axial distribution of the complex field amplitude $F = \mathrm{Re}(F) + \mathrm{iIm}(F)$

$$\frac{d^2F}{dz^2} + k_z^2(\omega, z)F = S, \tag{5.32}$$

where $k_z^2(\omega, z) = k^2 - k_\perp^2 = \left(\frac{\omega}{c}\right)^2 - k_\perp^2$, k_z and $k_\perp$ are, respectively, the axial and the transverse components of the complex wave propagation vector $k = \frac{\omega}{c}$. The complex eigen-frequency $\omega = Re(\omega) + iIm(\omega) = \omega_r + i\omega_i$ is represented by the relation

$$\omega = \omega_r\left(1 + \frac{1}{2Q_D}\mathrm{i}\right).$$

The transverse wave-number $k_\perp$ depends on the cavity radius $R(z)$ which changes slowly along the axis (the so-called slightly irregular string equation approximation). In the cold-cavity models, the source term S of the above inhomogeneous Helmholtz equation is neglected and therefore the field profile does not depend on the electron beam and its interaction with the wave in the cavity. In the self-consistent physical models, however, this term depends on the efficiency of the interaction, as will be illustrated below considering several of the most prominent models. In an open gyrotron cavity, the Helmholtz equation is supplemented by the following, so-called radiation boundary conditions (a.k.a. Sommerfeld's boundary conditions)

$$\left[\frac{dF}{dz} - \mathrm{i}k_zF\right]_{z=z_{\mathrm{in}}} = 0, \left[\frac{dF}{dz} + \mathrm{i}k_zF\right]_{z=z_{\mathrm{out}}} = 0, \tag{5.33}$$

that correspond to an *evanescent wave* at the entrance of the cavity ($z = z_{\mathrm{in}}$) and to an *outgoing wave* BC at its exit ($z = z_{\mathrm{out}}$).

As already mentioned, any model includes also a dynamical part which describes the motion of the beam electrons. Their orbits are obtained by integration of the relativistic equation of motion

$$m_0\frac{d\vec{p}}{dt} = -e\left(\vec{E} + \frac{\vec{p}}{\gamma} \times \vec{B}\right), \tag{5.34}$$

where $\vec{p} = \gamma\vec{v}$ is the normalized momentum of an electron.

The above generic formulation has numerous implementations which differ in the mathematical derivations, used approximations, and normalizations of the variables of the model.

As a whole, however, they are equivalent. Here, as examples, we will describe briefly some of the most widely used physical models that have proved their adequacy throughout a long history of successful application since their formulation.

The first model of this series to be described here includes the gyro-averaged Yulpatov's shortened equation of electron motion and equations describing the excitation of resonance modes in the cavity [32]. The former describes the change of the transverse electron momentum $p_\perp$ and the phase $\theta(p_x + ip_y = p_\perp e^{\theta})$ under the action of the high-frequency field. It is written for the complex energetic variable

$$a = \sqrt{1 - \frac{2}{\beta_{\perp 0}^2}\left(1 - \frac{p_\perp^2}{p_{\perp 0}^2}\right)} \exp[\mathrm{i}(\theta - \theta_0 - \omega_0 t)], \tag{5.35}$$

and reads

$$\frac{da}{d\zeta} - i\left(\Delta + |a|^2 - 1\right)a = i\left\{\sum_{j=1}^{j} a^{s_j-1}F_j f_j(\zeta)\exp\left[\mathrm{i}\left(\Psi_j - s_j\theta_0\right)\right]\right\}^*, \tag{5.36}$$

with an initial condition $a(0) = 1$. The summation is over all considered modes labeled by the index j. In this equation, $\zeta = \pi(\beta_{\perp 0}^2/\beta_{z0})(\omega_0 z/c)$ is the normalized axial coordinate, and $\Delta = \frac{2}{\beta_{\perp 0}^2}\left(1 - \frac{\Omega_{c0}}{\omega_0}\right)$ is the normalized frequency detuning between the initial cyclotron frequency and the frequency over which the gyro-averaging is performed.

The equations of the cavity excitation are given in the following form

$$\begin{aligned} \frac{dF_j}{d\tau} &= \left(\Phi'_j - \frac{s_j}{2Q_j}\right)F_j, \\ \frac{d\Psi_j}{d\tau} &= \Phi''_j + \frac{\omega_j}{\omega_0} - s_j, \end{aligned} \tag{5.37}$$

where $\tau = \omega_0 t$ and Q_j is the Q-factor of the jth mode. The complex quantity $\Phi_j = \Phi'_j + \mathrm{i}\Phi''_j$, which characterizes the power of the beam–wave interaction of the jth mode is

$$\Phi_j = -\mathrm{i}\int_{S_\perp}\left\{\int \frac{I_j}{F_j} W\left(\vec{R}_{\perp 0}, \vec{\beta}_0\right)d\vec{\beta}_0 \frac{1}{2\pi}\int_0^{2\pi}\left[\int_0^{\zeta_k} a^{*s_j} f_j^* \exp\left[-\mathrm{i}\left(\Psi_j - s_j\theta_0\right)\right]d\zeta\right]d\theta_0\right\}dS_\perp. \tag{5.38}$$

The dimensionless function $W(\vec{R}_{\perp 0}, \vec{\beta}_0)$ describes the distribution of the guiding center radii $\vec{R}_{\perp 0}$ of the electrons and their initial velocities $\vec{\beta}_0$. Here, for a cylindrical cavity, the current parameters I_j, the field amplitudes F_j, and phases Ψ_j of the modes are given by the following expressions

$$I_j = 0.47 \times 10^{-3} I \left(\frac{\beta_{\perp 0}^{2(s_j-2)}}{\beta_{z0}} \right) \left(\frac{s_j^{s_j+1}}{2^{s_j} s_j!} \right)^2 J^2_{m_j \pm s_j}\left(2\pi \frac{R_0}{\lambda} \right)$$
$$\times \left[J^2_{m_j}(\chi_j)\left(\chi_j^2 - m_j^2\right) \int_0^{\xi_k} \left| f_j \right|^2 d\zeta \right]^{-1}, \tag{5.39}$$

$$F_j = 4\frac{A_j}{H_0} \beta_{\perp 0}^{s_j-4} \frac{s_j^{s_j}}{2^{s_j} s_j!} J_{m_j \pm s_j}\left(2\pi \frac{R_0}{\lambda} \right), \; \Psi_j = -(m_j \pm s_j)\Psi + \alpha_j, \tag{5.40}$$

where I is the beam current in amperes, A_j and α_j are time-dependent amplitudes and phases of the modes.

In Fliflet's model [33], equation (5.34) is reduced to the following two equations for the normalized transverse momentum $p_\perp$ and the slow time scale variable Λ

$$\frac{dp_\perp}{dt} = -A_{mn}\left(s, R_g\right) B^{(-)}_{mn}(s, R_L(p_\perp)) F(z) \cos(s\Lambda - \Psi), \tag{5.41}$$

$$\frac{d\Lambda}{dt} = \frac{1}{p_\perp} A_{mn}\left(s, R_g\right) B^{(+)}_{mn}(s, R_L(p_\perp)) F(z) \sin(s\Lambda - \Psi) + \frac{\omega}{s} - \frac{\Omega_{c0}}{\gamma}, \tag{5.42}$$

where

$$A_{mn}\left(s, R_g\right) = k_{mn} C_{mn} J_{m-s}\left(k_{mn} R_g\right), \tag{5.43}$$

$$B^{(\pm)}_{mn}(s, R_L) = \frac{e}{2m_0}[J_{s-1}(k_{mn} R_L(p_\perp)) \pm J_{s+1}(k_{mn} R_L(p_\perp))]. \tag{5.44}$$

The axial dependence of the field amplitude (in the framework of the theory of weakly irregular waveguide) is given by the following inhomogeneous Helmholtz equation

$$\frac{d^2 F_{mn}}{dz^2} + k_z^2 F_{mn} = -\mathrm{i}\mu\omega I_b A_{mn}\left(s, R_g\right) \times \frac{1}{2\pi} \int_0^{2\pi} \alpha J_{s-1}\left(\frac{k_{mn} p_\perp}{\Omega_{c0}} \right) \exp(-\mathrm{i}s\Lambda)\, d\Lambda_0. \tag{5.45}$$

with radiation boundary conditions (see equation (5.33)).

Integrating the above system of equations, for an ensemble of electrons representing the beam the efficiency of the interaction is calculated by averaging

$$\eta_\perp = \left\langle \frac{\gamma_0 - \gamma}{\gamma_0 - 1} \right\rangle. \tag{5.46}$$

The generalized nonlinear harmonic gyrotron theory of Danly and Temkin [34] is widely used for calculation of the efficiency for gyrotron oscillators at harmonics of the cyclotron frequency. It is applicable to a wide range of operating conditions and is formulated using a minimal set of convenient generalized parameters. Their physical model is reduced to the following equations.

$$\frac{du}{d\xi} = 2\left(\frac{2^s s!}{s^s \beta_{\perp 0}^{s-1}}\right) Ff(\xi) \frac{p_\perp'}{\beta_{\perp 0}} J_s'(sp_\perp') \sin\theta, \tag{5.47}$$

$$\frac{d\theta}{d\xi} = \Delta - u - s\left(\frac{2^s s!}{s^s \beta_{\perp 0}^{s-1}}\right) Ff(\xi) \frac{\beta_{\perp 0}\left(1 - \frac{\beta_{\perp 0}^2 u}{2}\right)}{p_\perp'^2} J_s(sp_\perp') \cos\theta. \tag{5.48}$$

Here the normalized energy variable u and the normalized coordinate ξ are

$$u = \frac{2}{\beta_{\perp 0}^2}\left(1 - \frac{\gamma}{\gamma_0}\right), \tag{5.49}$$

$$\xi = \pi \frac{\beta_{\perp 0}^2}{\beta_{z0}} \frac{z}{\lambda}. \tag{5.50}$$

The normalized interaction length is, therefore, $\mu = \pi\beta \frac{\beta_{\perp 0}^2}{\beta_{z0}} \frac{L}{\lambda}$. The slow-time scale phase variable is

$$\theta = \omega t - s\phi, \tag{5.51}$$

ϕ being the electron phase. The normalized field amplitude F is defined by

$$F = \frac{E_0}{B_0} \beta_{\perp 0}^{s-4} \left(\frac{s^{s-1}}{s! 2^{s-1}}\right) J_{m\pm s}(k_\perp R_e), \tag{5.52}$$

where the plus and the minus signs correspond to the two possible rotations of the mode and E_0, B_0 are the amplitudes of the high-frequency electric field and the static magnetic field, respectively. The normalized frequency detuning is given by $\Delta = \frac{2}{\beta_{\perp 0}^2}\left(1 - \frac{\Omega_{c0}}{\omega}\right)$. The argument in the Bessel function in (5.48), and its derivative in (5.47) is $sp_\perp' = s\frac{\gamma\beta_\perp}{\gamma_0}$. The initial conditions for the integration of the above differential equations are $\theta = \theta_0 \in [0, 2\pi]$ and $u = 0$. The efficiency is given by

$$\eta = \frac{\gamma_0 - \gamma}{\gamma_0 - 1} = \left[\beta_{\perp 0}^2 / 2(1 - \gamma_0^{-1}\right]\eta_\perp, \tag{5.53}$$

where the transverse efficiency $\eta_\perp$ is calculated by averaging over the initial phases of the electrons $\eta_\perp = \langle u(\xi_{out}\rangle_{\theta_0}$. As is evident from the equations, the transverse efficiency depends only upon four parameters, namely (F, μ, Δ, and $\beta_{\perp 0}$). Further, Danly and Temkin reduce the variables to three (F, μ, Δ) using the approximation $p_\perp' \approx \beta_{\perp 0}(1 - u)^{1/2}$, which is valid for a weakly relativistic electron beam when the condition $s\beta_{\perp 0}^2/2 \ll 1$ is satisfied. This approximation allows simplifying the governing equations to

$$\frac{du}{d\xi} = 2Ff(\xi)(1 - u)^{s/2}\sin\theta, \tag{5.54}$$

$$\frac{d\theta}{d\xi} = \Delta - u - sFf(\xi)(1 - u)^{s/2-1}\cos\theta. \tag{5.55}$$

In this self-consistent model, the field amplitude is inferred from the energy balance equation. The produced power is

$$P = \eta I_A U = (mc^2/e)\left(\frac{\gamma_0\beta_{\perp 0}^2}{2}\right)\eta_\perp I_A, \tag{5.56}$$

where I_A and U are the beam current and voltage. Assuming a Gaussian field profile, the energy balance equation takes the form

$$F^2 = \eta_\perp I, \tag{5.57}$$

where the normalized current parameter is defined by

$$I = 0.238 \times 10^{-3}\left(\frac{QI_A}{\gamma_0}\right)\beta_{\perp 0}^{2(s-3)}\left(\frac{\gamma}{L}\right)\left(\frac{s^s}{2^s s!}\right)^2\frac{J_{m\pm s}^2(k_\perp R_e)}{(\xi_{mn}^{-2} - m^2)J_m^2(\xi_{mn})}, \tag{5.58}$$

and I_A is in amperes.

The results of the calculation can be presented conveniently as contours of constant transverse efficiency in either $(F - \mu)$ or $(I - \mu)$ space ($\eta_\perp(F, \mu) = \text{const}$, and $\eta_\perp(I, \mu) = \text{const}$).

We will end up this brief description of Danly&Temkin model by an apologetic warning. Although we preserve most of the original notations for the harmonic number we use s instead of n in order to be consistent with the notations adopted in this chapter.

As already mentioned, in the vast literature on gyrotrons there are numerous and practically countless but equivalent physical models. The briefly described three models above have been selected because of their both historical and methodological values. The derivations of their equations, which have not been presented here, are indeed very insightful and the reader is referred to the cited original papers for self-study. Over the years, these models have been implemented in many computer codes used for numerical studies and computer-aided design (CAD) of gyrotrons for various applications. Therefore, they are well-validated and their adequacy is proved in practice.

5.3 Electron-optical systems of gyrotrons

5.3.1 Conventional EOS

Efficient operation of gyrotrons requires the formation of high-quality helical electron beams with appropriate parameters (current, voltage, pitch factor, injection radius) and low velocity spread. Such beams are generated by EOS with MIG and a system of solenoids that provide an adiabatically increasing magnetic field and possibly additional coils for fine tuning of the magnetic field profile. First, we will describe briefly the EOS with MIG used in the conventional gyrotrons whose beams consist of off-axis helical electron orbits (figure 5.11(a)) and then continue with EOS generating axis-encircling (aka uniaxial) electron beams (figure 5.11(b)) that are used in the LOGs.

In the most common types of MIGs, the beam current is extracted from a conical emitting ring of a thermionic cathode in a temperature-limited regime. A schematic of MIG is presented in figure 5.13. Other emitters, for example operating in a space-charge limited regime or those with field emission arrays, ferroelectric cathodes, etc, are rare. The formation of a hollow helical electron beam takes place in crossed accelerating electric field and a static magnetic field in either diode or triode axially-symmetric configuration. The magnetic field increases adiabatically from the cathode towards the cavity, where it attains the maximum value in a flat-top region. Due to the conservation of the first adiabatic invariant ($p_{\perp}^2/B$ = const) the transverse velocity of the gyrating electron increases at the expense of the axial velocity. This process is known as adiabatic pumping of the beam since it leads to an increase of the velocity ratio (pitch factor) and therefore of the rotational energy which takes place in the beam–wave interaction (energy exchange).

A comprehensible starting point for the electron-optical theory of MIG [17, 35] is the analysis of the conservation of the angular momentum of an electron in a static axially-symmetric electric $E(r, z)$ and magnetic $B(r, z)$ (assuming a paraxial approximation $B_z = B_z(z)$)

$$p_\theta = \gamma m_0 r^2 \dot{\theta} - e r A_\theta = \text{constant}. \tag{5.59}$$

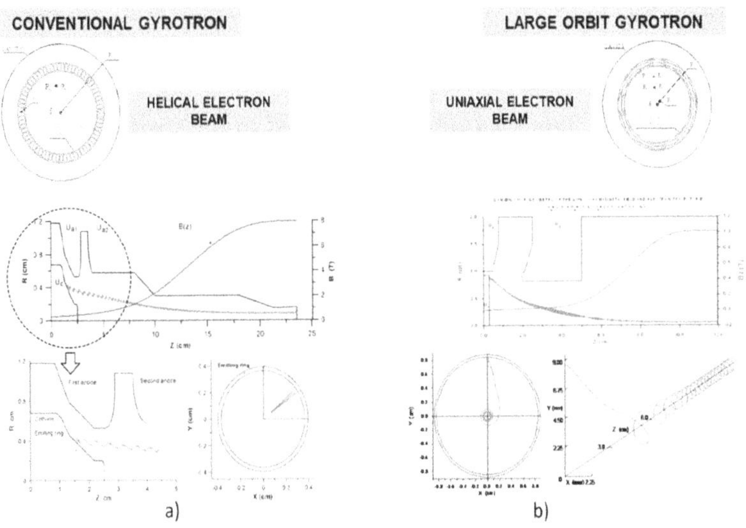

Figure 5.13. Electron orbits in conventional (a) and LOG (b).

Here r is the instantaneous radial coordinate of the electron gyrating with an azimuthal angular velocity $\dot{\theta}$ with respect to the axis of symmetry, and A_θ is the azimuthal component of the magnetic vector potential, which can be obtained from the following relation for the magnetic flux density within a circle of radius r and axial coordinate z

$$\Phi_B(r, z) = 2\pi \int_0^r RB_z(R, z)dR = 2\pi r A_\theta(r, z). \tag{5.60}$$

In a paraxial approximation $\Phi_B(r, z) \approx \pi r^2 B(z)$. Considering L_θ at two arbitrary axial positions z_1, z_2 directly leads to the famous Busch's theorem, which links the change of the magnetic field to the corresponding change of the angular momentum in the following form

$$\Delta p_\theta = p_\theta(z_2) - p_\theta(z_1) = \frac{e}{2\pi}[\Phi_B(z_2) - \Phi_B(z_1)] \approx \frac{e}{2}[r^2(z_2)B_z(z_2) - r^2(z_1)B_z(z_1)]. \tag{5.61}$$

At the radius $r = \sqrt{R_g^2 - R_L^2}$ (R_g, R_L being the radius of the guiding center and Larmor radius) the angular velocity $\dot{\theta}$ of the electron is zero and thus the angular momentum is also zero as on the cathode surface. This leads to the following relation between the starting radius r_c of the electron on the emitter where the magnetic field intensity is denoted by B_c and the radius of its guiding center at any other position, where the magnetic field is B,

$$B_c r_c^2 = BR_g^2\left(1 - \frac{R_L^2}{R_g^2}\right). \tag{5.62}$$

Taking into account that in the conventional (small orbit) gyrotrons $R_L \ll R_g$ the latter relation reduces to $B_c r_c^2 \approx BR_g^2$ or, introducing the magnetic compression ratio $b = B/B_c$, to

$$R_g \approx r_c/\sqrt{b}. \tag{5.63}$$

The drift velocity of the guiding center (normalized to the speed of light) in crossed electric and magnetic fields is given by

$$\beta_{g,d} = \frac{|\vec{E} \times \vec{B}|}{c\,|\vec{B}|^2}. \tag{5.64}$$

Assuming zero total initial velocity of the electron at the emitter (which is a sum of the drift β_{dc} and the rotational $\beta_{\perp c}$ velocities) leads to the conclusion that these velocities have equal magnitude and opposite directions. Therefore,

$$|\beta_{dc}| = |-\beta_{\perp c}| = \frac{E_c}{cB_c}\cos\varphi_c, \tag{5.65}$$

where E_c is the electric field normal to the emitting surface, B_c is the axial magnetic field at the cathode, and φ_c is the slant angle of the cathode surface with respect to the axis of symmetry. From the adiabatic invariant, it follows that the transverse velocity of the gyrating electron transforms along the axis z according to the relation

$$\beta_{\perp}(z_2) = \beta_{\perp}(z_1)\frac{\gamma(z_1)}{\gamma(z_2)}\sqrt{\frac{B(z_2)}{B(z_1)}}. \tag{5.66}$$

This equation shows that in an increasing magnetic field the transverse velocity also increases. Taking into account the energy conservation it follows that the axial velocity will, respectively, decrease. When an electron loses entirely its axial velocity ($\beta_z = 0$) it is reflected and reverses the direction of propagation as illustrated in figure 5.14 bouncing back and forth in a magnetic mirror trap. From the above relations, it follows that in the adiabatic approximation the transverse velocity of the electron in the resonant cavity is given by

$$\beta_{\perp} \approx \frac{1}{\gamma c} b^{1/2} \frac{E_c \cos\varphi_c}{B_c} = \frac{1}{\gamma c}\frac{B^{1/2}}{B_c^{3/2}} E_c \cos\varphi_c. \tag{5.67}$$

The adiabatic theory outlined briefly above is used at the initial design stage in order to find the basic parameters of the MIG. In practice, many and often contradictory requirements need to be satisfied. In this respect, a very helpful method is that based on the so-called 'trade-off equations' and proposed by Baird and Lawson [36]. The reader is referred to their original paper for more detail. This approach, however, neglects many important physical factors and phenomena, most notably the velocity spread, space-charge effects, emitting surface roughness, departure from adiabaticity, etc. A more advanced and adequate method, which allows taking them into account is trajectory analysis (aka ray tracing), which is based on a self-consistent model.

The physical model used for trajectory analysis of the EOS includes a system of relativistic equations of electron motion and Poisson equation for the electrostatic potential with appropriate boundary conditions. They are solved self-consistently in

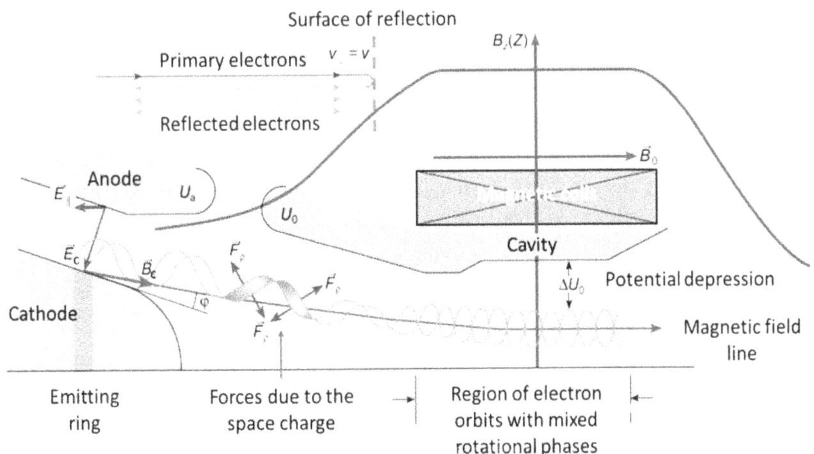

Figure 5.14. Schematic of a typical triode MIG (figure courtesy V Manuilov, adopted from [37]).

an iterative procedure until a convergence of the solution is reached. There are different but equivalent forms in which the relativistic equation of motion can be written the main distinction being in the convenience of their implementation (programming) in the numerical codes. As an example here we will present the model realized in many computer programs, including the famous EGUN (computer code for trajectory analysis of electron guns). Taking into account the axial symmetry of the EOS, it is formulated in two dimensions for the electrostatic potential $\phi(Z, r)$ and three components of the electron velocities (i.e. in a 5D phase space and thus 2.5D physical model) in a polar coordinate system (R, θ, Z). The relativistic equations of electron motions are given by

$$\ddot{R} = \frac{1}{m_0}(1 - \beta^2)^{1/2}\left[f_R - \frac{e}{c^2}\dot{R}(\dot{R}E_R + R\dot{\theta}E_\theta + \dot{Z}E_z)\right] + R\dot{\theta}^2, \tag{5.68}$$

$$\ddot{\theta} = \frac{1}{m_0 R}(1 - \beta^2)^{1/2}\left[f_\theta - \frac{e}{c^2}R\dot{\theta}(\dot{R}E_R + R\dot{\theta}E_\theta + \dot{Z}E_z)\right] - \frac{2\dot{R}\dot{\theta}}{R}, \tag{5.69}$$

$$\ddot{Z} = \frac{1}{m_0}(1 - \beta^2)^{1/2}\left[f_Z - \frac{e}{c^2}\dot{Z}(\dot{R}E_R + R\dot{\theta}E_\theta + \dot{Z}E_z)\right], \tag{5.70}$$

where

$$f_R = eE_R + e(R\dot{\theta}B_z - \dot{Z}B_\theta), f_\theta = eE_\theta + e(\dot{Z}B_R - \dot{R}B_z),$$
$$f_Z = eE_{RZ} + e(\dot{R}B_\theta - R\dot{\theta}B_R),$$

and $E_Z = -\frac{\partial\phi}{\partial Z}$, $E_R = -\frac{\partial\phi}{\partial R}$, B_Z, B_R, and B_θ are the components of the electrostatic and magnetic fields, respectively. The electrostatic potential distribution obeys the Poisson equation

$$\frac{1}{R}\frac{\partial}{\partial R}\left(R\frac{\partial\phi}{\partial R}\right) + \frac{\partial^2\phi}{\partial Z^2} = -\frac{\rho}{\varepsilon_0}, \tag{5.71}$$

where $\rho = \rho(Z, R)$ is the space-charge density and ε_0 is the dielectric constant of vacuum. Taking into account the axial symmetry, the computational domain for the potential is half of the meridional cross-section of the EOS. Along its contour, Dirichlet boundary conditions are specified on the electrodes and Neumann type along the axis. The Poisson equation can be solved by various methods but most frequently in the practice the finite-differences method (FDM) with a successive over-relaxation (SOR) or the finite-element method (FEM) are used. The iterative procedure starts with the solution of the Laplace equation $(\rho = 0)$ for the electrostatic potential. Then the equations of motion are integrated and the space charge is allocated to the nodes of the used grid (computational mesh). After these first steps, the Poisson equation is solved for the obtained space-charge distribution and the electron trajectories are integrated for the new electrostatic field. The latter two steps are repeated iteratively until the solution converges to the self-consistent one (when in two successive iterations the changes in the solution become negligible.

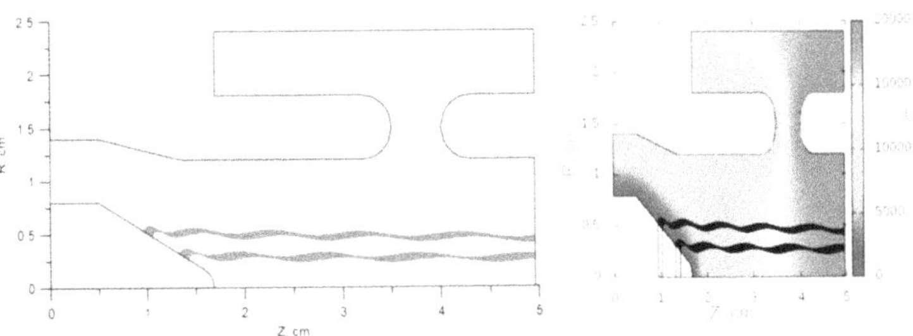

Figure 5.15. Ray tracing of the EOS for 0.8 THz double-beam gyrotron: configuration of the MIG and electron trajectories (left) and potential distribution in the gun (right) performed using GUN-MIG/CUSP code of the GYROSIM problem-oriented software package.

Some of the most prominent 2-1/2D codes for electron gun and beam tunnel simulations are DAFNE [38] (Centre de Recherches en Physique des Plasmas, École Politechnique Fédérale de Lausanne) and ESRAY [39] (Institute for Pulsed Power and Microwave Technology, Karlsruhe Institute of technology), as well as EPOSR [40], EPOS-V [41] (Institute of Applied Physics, Russian Academy of Sciences, Nizhny Novgorod), MAGY [42] (Institute for Plasma Research, University of Maryland, College Park) and GUN-MIG/CUSP [43]. As an illustration, figure 5.15 shows results from the trajectory analysis of the EOS for 0.8 THz double beam gyrotron carried out using the GUN-MIG/CUSP code.

5.3.2 EOS of LOG

Axis-encircling electron beams used in the LOG can be produced by electron guns with magnetic field reversal (cusp guns) or by EOS with kickers.

The formation of a uniaxial electron beam in a cusp gun can be explained by the conservation of the angular momentum. From Busch's theorem it is evident that since the magnetic flux densities before $\Phi_B(r, z_1)$ and after $\Phi_B(r, z_2)$ the field reversal have different signs (for example $\Phi_B(r, z_1) < 0$ and $\Phi_B(r, z_2) > 0$) the electrons passing through the cusp obtain an additional twist because, in this case, we have (compare with equation (5.61))

$$\Delta p_\theta = \frac{e}{2\pi}[\Phi_B(z_2) + |\Phi_B(z_1)|] \approx \frac{e}{2}[r^2(z_2)B_z(z_2) + r^2(z_1)|B_z(z_1)|]. \tag{5.72}$$

The analysis [44] shows that an off-axis electron orbit before the cusp transforms to an axis-encircling one after the field reversal. For an ideal cusp (abrupt change of the magnetic field), for example, the off-axis orbit of the gyrating electron with a Larmor radius R_{L1} and guiding center radius R_{g1} (for which $R_{L1}<R_{g1}$) transforms to an axis-encircling orbit with $R_{L2} = R_{g1}$ and $R_{g2} = R_{L1}$. The direction of the off-centering R_{g2}, however, is determined by the instantaneous position of the electron at the cusp transition and in any real beam can be quite random. Therefore, the actual beam width after the cusp is twice the maximum Larmor radius before it. In any real (nonideal) cusp with a finite width of the field reversal, additional off-centering takes place, which introduces a ripple (pulsation) of the beam envelope.

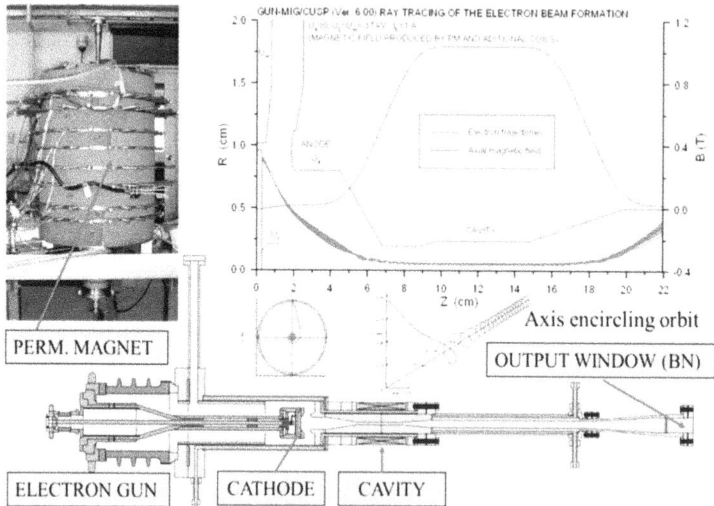

Figure 5.16. Electron-optical system for LOG with a permanent magnet.

In the practical design of a cusp electron gun, the latter is a severe problem and the minimization of the beam ripple among the main goals.

As an example, figure 5.16 shows the EOS of LOG with a permanent magnet [45]. In it, the point of the field reversal is located close to the cathode surface and gradual instead of abrupt transition is used. In such a configuration the magnetic field profile (and most importantly the position of the reversal, the slope and the length of field distribution) has been optimized for generation of a high-quality electron beam with appropriate parameters (current, voltage, pitch-factor, injection radius) with small velocity spread and ripple.

Another possibility for the generation of axis-encircling beams is offered by the EOS with kickers. Initially, a quasi-Pierce type gun creates a thin electron beam with rectilinear trajectories. The magnetic system of the kicker form two local magnetic fields directed perpendicularly to the axis of symmetry and separated by a distance equal to half a cyclotron wave. The first magnetic field abruptly 'kicks' the beam apart from the axis and the electrons start to follow circular orbits with centers shifted with respect to the symmetry axis. The second transverse magnetic field imparts an azimuthal momentum to the electrons and forming in such a way a hollow axis-encircling electron beam rotating around the symmetry axis with cyclotron frequency. A schematic of an EOS with kickers is shown in figure 5.17.

5.3.3 Cathodes of MIG for gyrotrons

The materials used for fabrication of emitters for the thermionic cathodes of MIG are of paramount importance for the overall operational performance and the lifetime of the entire EOS. One of the most prominent cathodes is the tungsten dispenser cathode, which is produced by compressing and sintering a tungsten powder into cylindrical billets from which a conical emitting section is

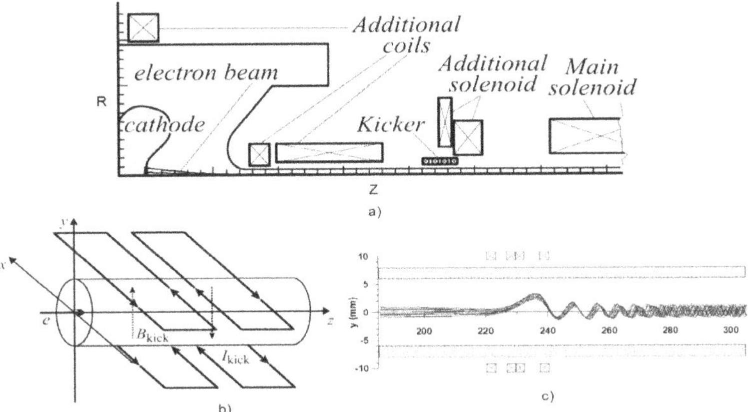

Figure 5.17. Schematic of an EOS for generation of axis-encircling beams (a), arrangement of the kicker coils (b), and trajectories of the beam electrons in the kicker (c).

manufactured. In them, the porous tungsten matrix is impregnated with a mixture consisting of barium oxide (BaO), calcium oxide (CaO), and aluminum oxide (Al_2O_3). The composition of these components in a proportion 5: 3: 2 is known as B-type and that in a proportion 4: 1: 1 as S-type. In the so-called M-type dispersion cathodes an additional coating of Os, Ir or Re is used to further lower the work function. The heating of the cathode during its operation causes migration of the barium towards the emitting surface where it lowers the work function of the emitter. Depletion of barium from the surface is due to the evaporation and ion bombardment. Both phenomena limit the lifetime of the cathode. It is well known, however, that the main factor that affects the lifetime of the emitter is the cathode loading. Therefore, there is always a trade-off between the extracted current and the lifetime. Although M-type cathodes coated with Os/Ru or Ir have superior emission characteristics compared to B-type and S-type, they are more susceptible to gas poisoning at poor vacuum conditions. For such high-current emitters, emission cooling effects can occur, which lower the temperature of the emitter. One of the latest advanced cathodes is the so-called Control Porosity Dispenser (CPD) cathode in which a thin (20–30 μm) tungsten film is used in which tiny holes are drilled by a laser beam. While the work function of a standard CPD cathode is around 2.0 eV, when covered by Os/Ru it is reduced to about 1.8 eV. Another very promising (though not yet widely available) novel type of cathode is the Scandate cathode, which offers saturated current density as high as several hundred amperes per square centimeter. The barium Scandate cathode employs scandium oxide (Sc_2O_3) to lower the work function of the emitter and allows operation at lower temperatures than the standard dispenser cathode for the same current density. Moreover, since the Ba–Sc mix is homogeneous throughout the tungsten matrix, it is free from many of the problems associated with the thin-film coated emitters (e.g. M-cathode). A distinct advantage is the ability of the Ba–Sc cathode to withstand much stronger ion bombardment. Additionally, such cathodes have excellent resistance to possible surface abrasions encountered during inspection, packaging, jiggling and general handling, as well as an increased resistance to exposure to moisture. Important requirements that must be met by the cathodes are: (i) a smooth emitting surface with minimum roughness; (ii) a uniform work function; (iii) uniform heating (with temperature variations of less than few degrees) over the entire surface); (iv) thermal stability that prevents a temperature drift of the configuration (distortions, buckling and creasing) and parameters of the emitter with time; and (v) uniform distribution of the electric field over the emitter.

Besides these advanced thermionic cathodes, very often lanthanum hexaboride (LaB_6) emitters are used in the MIG of gyrotrons because they provide stable electron emission at relatively low temperatures and have a long lifetime.

Completely different from the thermionic cathodes are ferroelectric cathodes (FECs) in which the charge separation and emission are achieved by rapid switching of the spontaneous, ferroelectric polarization. The polarization switching can be induced by applying an electric field, mechanical pressure or thermal heating by laser radiation. The emitted current densities are of the order of 100 A cm^{-2} and the pulse lengths are shorter than the excitation pulses. There are many advantages of using a ferroelectric cathode, such as room temperature operation, control of the emission by an external trigger pulse and, most importantly, high electron current emission. Several applications of such emitters in MIGs for gyrotrons have been reported recently.

An indispensable part of any gyrotron EOS is the collector where the energy of the spent electron beam is dissipated. In the simplest case, they are water-cooled beam tunnels at ground potential and covered by tick shielding to stop the generated X-ray radiation. Sometimes, in order to decrease the thermal loading of the collector a system of sweeping coils are used for smearing the beam and making the distribution of the dissipated energy more uniform. In the more advanced collectors with depressed potential part of the kinetic energy of the spent electron beam is recuperated, which leads to a significant enhancement of the total efficiency. In a single-stage depressed collector, the overall efficiency of the gyrotron is given by the relation [46]

$$\eta_{\text{out}} = \frac{P_{\text{out}}}{U_A I_B - U_{\text{DC}} I_{\text{DC}}} = \frac{\varepsilon_{\text{out}} \eta_e}{1 - \eta_{\text{DC}}(1 - \eta_e)}, \tag{5.73}$$

where U_A, I_B are the accelerating voltage and current of the electron beam while U_{DC}, I_{DC} are collector depression voltage and the current at the collector, respectively, η_e is the electronic efficiency, $\varepsilon_{\text{out}} = \frac{P_{\text{out}}}{P_{\text{RF}}}$, P_{RF} is the generated RF-power, and the collector efficiency is $\eta_{\text{DC}} = \frac{U_{\text{DC}} I_{\text{DC}}}{U_A I_B - P_{\text{RF}}}$. The efficiency of the single-stage collector is limited by the fact that there are electrons with different residual energies in the spent beam. This problem can be solved using multi-stage collectors in which different energy groups are separated.

5.4 Quasi-optical systems of the gyrotrons

In the typical regimes of operation, high-order modes are excited in the gyrotron cavity. They are characterized by high azimuthal and radial indices, circular polarization and radiation pattern in the form of a hollow cone. Such radiation is inconvenient for many applications that require a well-collimated wave beam. Moreover, such radiation is difficult to transmit because of the significant losses in the transmission lines. That is why the advanced gyrotrons are equipped with internal (built-in) mode converters (situated between the cavity and the output window) that transform the operating cavity mode into a linearly polarized Gaussian beam. Such a beam is appropriate to be used directly as a free space TEM_{00} mode, or to be transmitted as a low-loss hybrid HE_{11} mode in highly over-moded corrugated waveguides. The efficiency of this conversion is crucial for high-power gyrotrons, recalling that minimizing conversion and diffraction losses means not simply increasing the output power, but far more, solving severe cooling problems and avoiding the device damage.

The external mode converters used with the earlier gyrotrons have been in the form of rippled-wall or corrugated circular waveguides and serpentine structures. They are characterized by high efficiency but have large size and narrow bandwidth, which makes them inappropriate for implementation in the high-power gyrotrons. The internal mode converters used nowadays are mainly of two varieties known as Vlasov and Denisov type quasi-optical converter, respectively (see figure 5.18) [47–49].

The former of them represents a smooth surface circular waveguide, with a cut (stepped, slant or helical one) acting as a launcher (aka Vlasov antenna) and combined with a parabolic reflector that forms the Gaussian wave beam. Since the electromagnetic field in both the azimuthal and the axial directions is approximately uniform, the losses due to the diffraction from the edges of the aperture (cut) are high in a Vlasov antenna and a beam with side lobes is formed. A modification of such a system, which employs two reflectors (one elliptic and one parabolic mirror), is used to focus the wave beam in two perpendicular planes onto a spot. The Vlasov type converters have moderate efficiency (typically 80%) and require large mirrors

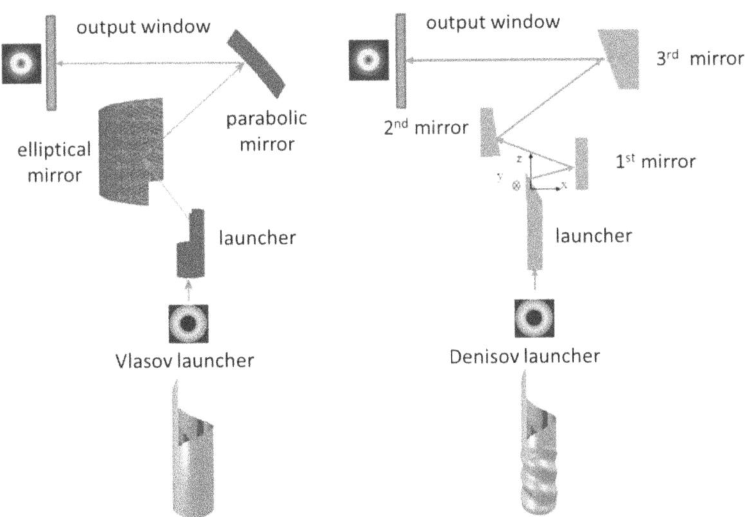

Figure 5.18. Schematics of QO mode converters with Vlasov (left) and Denisov type launchers (right).

for higher order modes. The Denisov converter consists of a pre-bunching section (instead of the smooth surface circular waveguide in Vlasov converters) with a helical-cut aperture and up to four mirrors (cylindrical, quasi-parabolic and phase correcting ones). The pre-bunching (or field pre-shaping) section represents a rippled-wall (dimpled) waveguide of relatively short length. Its role is to transform the gyrotron cavity mode into a bundle (mixture) of modes. Their interference leads to bunching of the wave field in both the axial and the azimuthal directions forming a Gaussian beam distribution. A longitudinal bunching could take place provided the modes have equal caustic radii and interference length close to the length of the launcher cut, L_c. At the same time, the azimuthal bunching requires modes that have equal caustic radii and similar Bessel zeros. These requirements impose the following mode selection rules for the differences $\Delta\beta$ and Δm between the propagation constants of the modes and their azimuthal indices, respectively

$$\Delta\beta = \pm\frac{2\pi}{L_c} \text{ and } \Delta m = \pm\frac{\pi}{\theta}, \tag{5.74}$$

where the transverse section angle between two reflections of the ray, representing the mode, $\theta = \arccos(m/\chi_{mn})$ is defined by the azimuthal index m and the eigenvalue χ_{mn} of the mode that forms a Gaussian intensity profile at the aperture, prior to the reflectors (in the Vlasov type converters such profile is formed after the uniformly profiled beam is reflected from the parabolic mirror). The Denisov converters have higher conversion efficiency (about 95% and higher) and smaller mirror size that makes them suitable to be embedded in the gyrotron tube. One advanced modification of the Denisov converter has two sections—one dimpled, the other smooth. The first section creates the bundle of modes with appropriate amplitudes, while the other one provides the necessary phase shifts between the modes.

The ray-optics description and the basic geometrical relations of a Vlasov type launcher are presented in figure 5.19. On the unrolled surface of the launcher there are Brillouin zones shown in figure 5.20, where each ray is reflected once. These zones suggest how to cut the launcher in such a way as to ensure that the entire wave beam emerges from the radiating aperture. As an illustration of this procedure, also shown in figure 5.20 are the emerging rays from a launcher simulated by GO&ART (Geometrical Optics and Analytical Ray Tracing) code [50].

For analysis of the Denisov type converter, usually the coupled modes theory is used [51–54]. The coupling between L modes as they propagate along the longitudinal axis z can be described by a set of simultaneously coupled differential equations of the following general form (also known as generalized telegraphist's equation)

$$\frac{dA_k}{dz} = jk_{z_k}(z)A_k + j\sum_{\substack{l=1\\ l\neq k}}^{L} C_{k,l}(z)A_l,\ k = 1, 2, \dots, L; \tag{5.75}$$

where A_k is the complex amplitude of the kth mode (TE_{mp} or TM_{mp}) and C_{kl} is the coupling coefficient between it and the mode denoted by the subscript l (TEnq or TMnq). Here k_{z_k} is the longitudinal wave number (propagation constant) of the kth

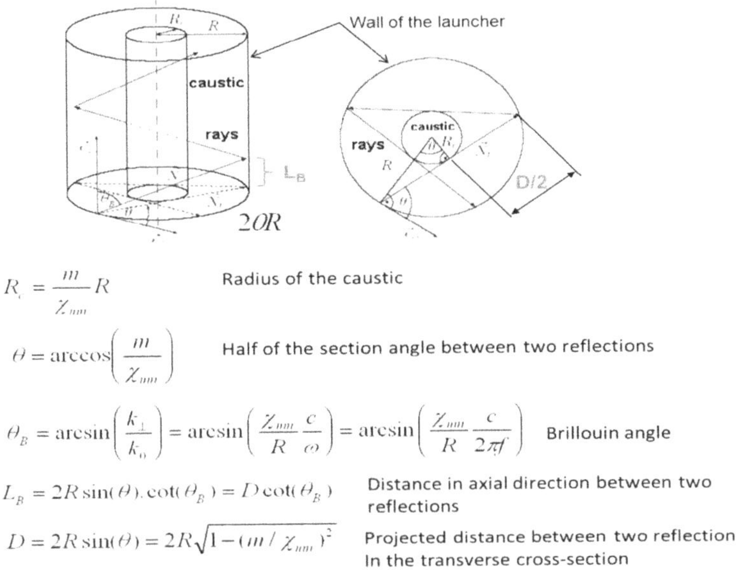

Figure 5.19. Basic geometrical relations of the ray optics representation.

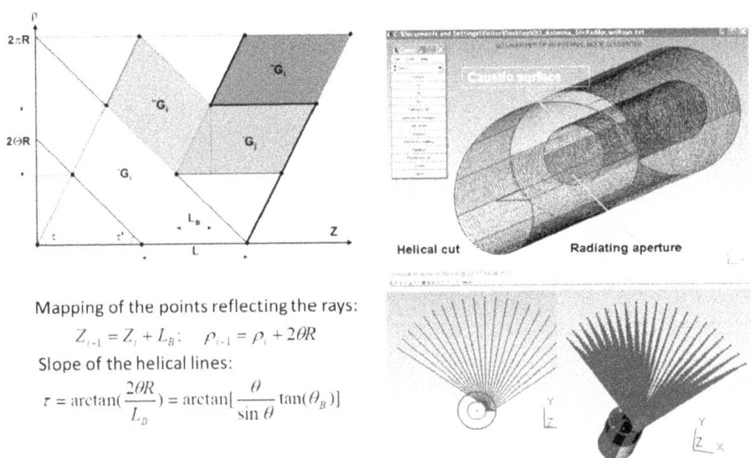

Figure 5.20. Brillouin's zones on the unrolled surface of the launcher (left panel) and a launcher with a helical cut (right panel) together with the rays emerging from the radiating aperture simulated by GO&ART package.

mode, which is a function of the waveguide dimensions. A waveguide with wall perturbations (deformations) $d(\phi, z)$ of its radius $R(z)$, that depend on both the azimuthal coordinate ϕ and the axial coordinate z can be described by the equation

$$r(\phi, z) = R(z) + d(\phi, z). \tag{5.76}$$

In the most general case of I perturbations with amplitudes δ_i and parameters l_i (l_i being the order of the azimuthal wall perturbation, and $l_i\phi$ being the phase angle of the ith perturbation), and $\Delta\beta_i$ (defined by the geometrical period $\lambda_i = 2\pi/\Delta\beta_i$ of the ith perturbation) we have

$$r(\phi, z) = R(z) + \sum_{i=1}^{I} d_i(\phi, z) = R_0 + \alpha z + \sum_{i}^{I} \delta_i \cos(l_i\phi + \Delta\beta_i z), \tag{5.77}$$

where R_0 is the initial radius of the launcher and α is the slope of its taper. More specifically, in the helical converter of the Denisov type

$$r(\phi, z) = R_0 + \alpha z + \delta_1 \cos(\Delta\beta_1 z - l_1\phi) + \delta_2 \cos(\Delta\beta_2 - l_2 z), \tag{5.78}$$

where

$$\Delta\beta_1 = k_{z_{mn}} - k_{z_{m\pm1,n}},\ l_1 = \pm 1, \tag{5.79}$$

$$\Delta\beta_2 = k_{z_{m,n}} - k_{z_{m\pm\Delta m, n\mp\Delta n}},\ l_2 = \pm\Delta m, \tag{5.80}$$

and m, n being the azimuthal and the radial indices of the mode TE_{mn}, respectively. In the above equations, the term with l_1 describes the longitudinal bunching, while that with $l_2 = \pm\Delta m$ the azimuthal bunching. The positive signs for l_1 and l_2 correspond to helical deformations with right-hand rotation, otherwise with left-hand rotation. Since both $\Delta\beta_1$ and $\Delta\beta_2$ have two slightly different values corresponding to plus and minus signs, an average for each of them is used in the calculations. For such perturbations, the coupling coefficients can be obtained from the equations

$$C_{kl}(z) = K_{k,l}\frac{\alpha_{k,l}(z)}{R(z)}, \tag{5.81}$$

$$\alpha_{k,l}(z) = \frac{1}{2\pi}\int_0^{2\pi} d_i(\phi, z)\exp[-j(m-n)\phi]\,d\phi. \tag{5.82}$$

The coupling factors $K_{k,l}$ for a periodically corrugated launcher are

$$\begin{aligned}
K_{k,l} &= \frac{mn(k_0^2 - k_{z_{m,p}}k_{z_{n,q}}) - \left(\dfrac{\chi_{mp}\chi_{nq}}{R(z)}\right)^2}{\sqrt{k_{z_{m,p}}k_{z_{n,q}}(\chi_{mp}^2 - m^2)(\chi_{nq}^2 - n^2)}} && \text{for } TE_{mp} \text{ to } TE_{nq}\\
K_{k,l} &= \frac{jmk_0(k_{z_{m,p}} - k_{z_{n,q}})}{\sqrt{k_{z_{m,p}}k_{z_{n,q}}(\chi_{mp}^2 - m^2)}} && \text{for } TE_{mp} \text{ to } TM_{nq}\\
K_{k,l} &= \frac{jnk_0(k_{z_{m,p}} - k_{z_{n,q}})}{\sqrt{k_{z_{m,p}}k_{z_{n,q}}(\chi_{nq}^2 - n^2)}} && \text{for } TM_{nq} \text{ to } TE_{mp}\\
K_{k,l} &= \frac{k_0^2 - k_{z_{m,p}}k_{z_{n,q}}}{\sqrt{k_{z_{m,p}}k_{z_{n,q}}}} && \text{for } TM_{mp} \text{ to } TM_{nq}
\end{aligned} \tag{5.83}$$

Here χ_{ij} are the corresponding eigenvalues, defined by $J_i'(\chi_{ij}) = 0$ for the TE_{ij} modes and by $J_i(\chi_{ij}) = 0$ for the TM_{ij} modes, where J_i is a Bessel function of ith order and J_i' is its derivative, i.e. χ_{ij} is the jth zero of the corresponding Bessel function or its derivative, respectively. The coefficients $\alpha_{k,l}$ are given by

$$\alpha_{k,l}(z) = \frac{dR(z)}{dz} = \alpha, \text{ for the case when } l_i = m - n = 0 \tag{5.84}$$

$$\alpha_{k,l}(z) = \frac{\delta_n}{2}\exp(\pm j\Delta\beta_i z), \text{ for the case when } l_i = \pm(m - n) \tag{5.85}$$

$$\alpha_{k,l}(z) = 0, \text{ for the case when } |l_i| \neq |m - n|. \tag{5.86}$$

The above system is integrated for the amplitudes of a mixture of modes that obey the Bragg's conditions and forms a Gaussian beam.

The selection of modes amplitudes is based on the observation that a Gaussian distribution can be approximated by a raised cosine, which can be represented as a superposition of three plane waves with appropriately matched amplitudes and relative phases according to the relations

$$\frac{\sqrt{2}}{\sqrt{\pi}w}e^{-2\frac{z^2}{w^2}} \approx \frac{b}{3\pi}(1 + \cos(bz))^2 = \frac{b}{3\pi}\left(1 + \frac{e^{+jbz}}{2} + \frac{e^{-jbz}}{2}\right)^2. \tag{5.87}$$

This implies that the distribution of the amplitudes should be

$$f(\theta, z) = \left(1 + \frac{e^{+j\frac{\pi}{\theta}z}}{2} + \frac{e^{-j\frac{\pi}{\theta}z}}{2}\right)\left(1 + \frac{e^{+j\frac{\pi}{L}z}}{2} + \frac{e^{-j\frac{\pi}{L}z}}{2}\right). \tag{5.88}$$

The solution can be conveniently represented as a density map of the H_z field component on the unrolled (unwound) surface of the launcher. This map shows clearly the Brillouin zones and allows designing an appropriate cut. This is illustrated in figure 5.21 with an example for the operating mode $TE_{28,\,8}$ of a coaxial gyrotron for ITER developed at KIT-IHM.

5.5 Output windows of the gyrotrons

The output window is one of the most critical parts of the tube, especially in high-power gyrotrons. It must transmit with minimal losses the generated wave beam and withstand severe heating. This imposes very stringent requirements on the dielectric, mechanical, and thermos-physical properties of the materials used. The most important parameters that determine both power transmission and absorption are the loss factor $\tan\delta$ and the relative dielectric permittivity ε_r. In order to minimize the power reflections, the thickness of the window must obey the relation

$$d = N\lambda/(2\varepsilon_r^{1/2}), \tag{5.89}$$

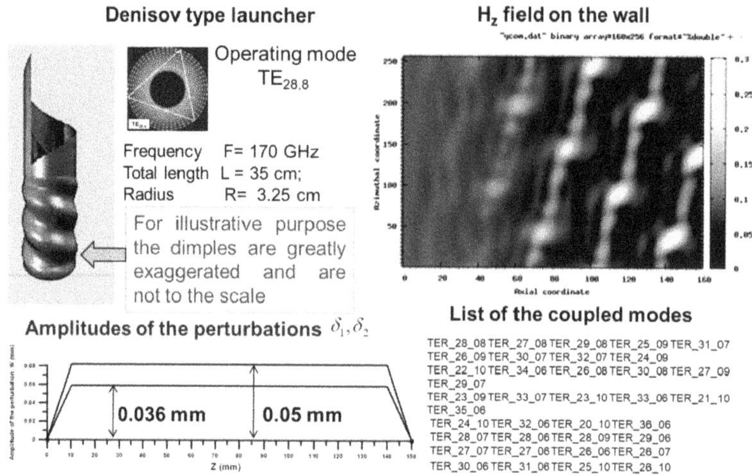

Figure 5.21. Results of the numerical analysis of a Denisov type launcher in the 2 MW/170 GHz coaxial gyrotron obtained by the coupled modes theory.

where N is an integer and λ is the free-space wavelength. For a given thickness, there is a series of frequencies (multipassband window) for which $N_i\lambda_i = 2d\varepsilon_r^{1/2}$. The total absorbed power in the dielectric disk is proportional to $d(\varepsilon_r + 1)\tan\delta$.

The most advanced output windows that are used in the high-power gyrotrons are the ultra-broadband Brewster's windows. The Brewster angle for reflection-free transmission of linearly polarized waves through the dielectric disk depends only on ε_r and is given by

$$\theta_{\text{Brewster}} = \arctan(\varepsilon_r^{1/2}). \tag{5.90}$$

The most frequently used dielectric materials are BN, Si_3N_4, sapphire, and PACVD (plasma-assisted chemical vacuum deposition) diamond. The latter has the most superior properties but is expensive. CVD diamond windows have enabled the power level to reach 2 MW while such traditionally used material as sapphire is limited to only 150 kW. Worldwide, commercially available CVD-diamond disks are offered by only two suppliers: Element 6 (De Beers) and Diamond Materials (visit: https://www.iaf.fraunhofer.de/en/researchers/diamond-devices.html). The current level of the technology has demonstrated the possibility to fabricate high-quality diamond disks with the following parameters: (i) diameter $\leqslant$ 140 mm and thickness $\leqslant$ 2.5 mm (up to 572 carat); (ii) $\tan\delta \approx 2 \cdot 10^{-5}$; (iii) $\varepsilon_r = 5.67$ (constant in the temperature range 300—400 K); (iv) very small thermal expansion; (v) thermal conductivity $\approx$1900 W m^{-1} · K at $T = 300$ K; (vi) mechanical strength $\sigma_B \approx 400$ MPa; (vii) available brazing technology for joining. Besides the choice of an appropriate material, the design of the output window must stipulate an effective cooling and mechanical strength. Some of the experimentally studied solutions to the cooling problems are: (i) liquid-edge-cooled window; (ii) gas-surface-cooled single disk, and (iii) liquid-surface-cooled double disk window.

Figure 5.22. Design of an output window at Brewster angle made of synthetic diamond for KIT 1 MW coaxial gyrotron for ITER (Picture courtesy M Thumm).

As an example, figure 5.22 shows the design of a diamond elliptically brazed output window at Brewster angle for KIT 1 MW Frequency Step-Tunable (112–166 GHz) Gyrotron for ITER and the results of its numerical analysis. It has been proven that the current designs of the single-disk diamond windows are capable of handling the thermal and mechanical loading of the megawatt class gyrotrons. Despite the high quality of the material, however, in some experiments, failures due to arching that destroy the window have been reported.

Depending on the specific application, after the output window, the gyrotron radiation has to be transmitted using one or another transmission line. In the case of gyrotrons for fusion (as well as for some other high-power applications), this is a corrugated waveguide system operating in HE_{11} mode. In order to couple efficiently the TEM_{00} mode Gaussian distribution to the HE11 mode of the corrugated waveguide a Match Optics Unit (MOU) that includes mirrors for beam shaping is being used.

References

[1] Nusinovich G S, Thumm M K A and Petelin M 2014 The gyrotron at 50: historical overview *J. Infrared Millim. Terahertz Waves* **35** 325–81

[2] Booske J H 2008 Plasma physics and related challenges of millimeter-wave-to-terahertz and high power microwave generation *Phys. Plasmas* **15** 055502

[3] Glyavin M Y, Denisov G G, Zapevalov V E, Koshelev M A, Tretyakov M Y and Tsvetkov A I 2016 High-power terahertz sources for spectroscopy and material diagnostics *Phys.-Usp.* **59** 595–604

[4] Thumm M 2020 State-of-the-art of high power gyro-devices and free electron masers *J. Infrared Millim. Terahertz Waves* **41** 1–140

[5] Idehara T, Saito T, Ogawa I, Mitsudo S, Tatematsu Y and Sabchevski S 2008 The potential of the gyrotrons for development of the sub-terahertz and the terahertz frequency range–a review of novel and prospective applications *Thin Solid Films* **517** 1503–6

[6] Bratman V, Glyavin M, Idehara T and Kalynov Y 2009 Review of subterahertz and terahertz gyrodevices at IAP RAS and FIR FU *IEEE Trans. Plasma Sci.* **37** 36–43
[7] Kumar N, Singh U and Singh T P 2011 A review on the applications of high power, high frequency microwave source: gyrotron *J. Fusion Energy* **257**
[8] Bratman V L, Bogdashov A A, Denisov G G and Glyavin M Y 2012 Gyrotron development for high power THz technologies at IAP RAS *J. Infrared Millim. Terahertz Waves* **33** 715–23
[9] Glyavin M Y, Denisov G G, Zapevalov V E and Kuftin A N 2014 Terahertz gyrotrons: state of the art and prospects *J. Commun. Technol. Electron.* **59** 792–7
[10] Glyavin M Y, Idehara T and Sabchevski S P 2015 Development of THz gyrotrons at IAP RAS and FIR UF and their applications in physical research and high-power THz technologies *IEEE Trans. Terahertz Sci. Technol.* **5** 788–97
[11] Idehara T and Sabchevski S P 2017 Gyrotrons for high-power terahertz science and technology at FIR UF *J. Infrared Millim. Terahertz Waves* **38** 62–86
[12] Idehara T and Sabchevski S P 2018 Development and application of gyrotrons at FIR UF *IEEE Trans. Plasma Sci.* **46** 2452–9
[13] Kartikeyan M V, Borie E and Thumm M K A 2003 *Gyrotrons High Power Microwave and Millimeter Wave Technology* (Berlin: Springer)
[14] Nusinovich G 2004 *Introduction to the Physics of Gyrotrons* (Baltimore, MD: Johns Hopkins University Press)
[15] Chu K R 2004 The Electron Cyclotron Maser *Rev. Mod. Phys.* **6** 489
[16] Du C-H and Liu P-K 2014 *Millimeter Wave Gyrotron Traveling Wave Tube Amplifiers* (Berlin: Springer)
[17] Tsimring S E 2007 *Electron Beams and Microwave Vacuum Electronics* (New York: Wiley-Interscience)
[18] Gilmour A S 2011 *Klystrons, Traveling Wave Tubes, Magnetrons, Crossed-Field Amplifiers, and Gyrotrons* (Boston, MA: Artech House)
[19] Barker R J, Luhmann N C, Booske J H and Nusinovich G S 2005 *Modern Microwave and Millimeter-Wave Power Electronics* (New York: Wiley-VCH)
[20] Grigoriev A D, Ivanov V A and Molokovsky S I 2018 *Microwave Electronics* (Cham: Springer) 489 p
[21] Benford J, Swegle J A and Schamiloglu E 2015 *High Power Microwaves* 2nd edn (New York, London: CRC Press)
[22] Barker R J and Schamiloglu E (ed) 2001 *High-power Microwave Sources and Technologies* (New York: Wiley-IEEE Press)
[23] Faillon G, Kornfeld G, Bosch E and Thumm M K 2008 *Microwave Tubes. In Vacuum Electronics* (Berlin, Heidelberg: Springer)
[24] Carter R G 2018 *Microwave and RF Vacuum Electronic Power Sources* (Cambridge: Cambridge University Press)
[25] Idehara T, Sabchevski S, Glyavin M and Mitsudo S 2020 The gyrotrons as promising radiation sources for THz sensing and imaging *Appl. Sci.* **10** 980
[26] Flyagin V A, Gaponov A V, Petelin M I and Yulpatov V K 1977 The gyrotron *IEEE Trans. Microw. Theory Tech.* **25** 514–21
[27] Zapevalov V E 2006 The gyrotron: constraints on output-power and efficiency increase *Quantum Electron.* **49** 779–85

[28] Dumbrajs O, Idehara T, Saito T and Tatematsu Y 2012 Calculations of starting currents and frequencies in frequency-tunable gyrotrons *Jpn. J. Appl. Phys.* **51** 126601
[29] Twiss R Q 1958 Radiation transfer and the possibility of negative absorption in radio astronomy *Aust. J. Phys.* **14** 564–79
[30] Schneider J 1959 Stimulated emission of radiation by relativistic electrons in a magnetic field *Phys. Rev. Lett.* **2** 504–5
[31] Gaponov A V 1959 Interaction between electron fluxes and electromagnetic waves in waveguides *Izv. Vyssh. Uchebn. Zaved. Radiofiz* **2** 450–62
[32] Yulpatov V K 1981 Shortened equations for the self-oscillations of the gyrotron *Gyrotron: Collection of Scientific Papers* vol 25; A V Gaponov-Grehov (Gorki: Institute of Applied Physics of RAN) pp 26–40
[33] Fliflet A W, Read M E, Chu K R and Seeley R 1982 A self-consistent field theory for gyrotron oscillators: application to a low Q gyromonotron *Int. J. Electron.* **53** 505–21
[34] Danly B G and Temkin R J 1986 Generalized nonlinear harmonic gyrotron theory *Phys. Fluids* **29** 561–7
[35] Piosczyk B 1993 *Electron Guns for Gyrotron Applications Gyrotron Oscillators: Their Principles and Practice* ed C J Edgcombe 1st edn (London: Taylor & Francis)
[36] Baird J M and Lawson W 1986 Magnetron injection gun (MIG) design for gyrotron applications *Int. J. Electron.* **62** 81–7
[37] Manuilov V N 2011 Electron beams for masers on cyclotron resonance and free-electron lasers *Soros Educ. J.* **17** 81–7 in Russian
[38] Tran T M, Whaley D R, Merazzi S and Gruber R 1991 DAPHNE, a 2D axisymmetric electron gun simulation code *Proc. of the 16th Int. Conf. on Infrared and Millimeter Waves (Lausanne, Switzerland)* pp 122–4
[39] Illy S and Borie E 1999 Investigation of beam instabilities in gyrotron oscillators using kinetic theory and particle-in-cell simulation *J. Plasma Phys.* **62** 95–115
[40] Lygin V K and Manuilov V N 1987 Elektronnaya tekhnika Ser. 1 *Elektron. SVCh* **7** 36–8
[41] Tsimring S E and Zapevalov V E 1996 Experimental study of intense helical electron beams with trapped electrons *Int. J. Electron.* 199–205
[42] Botton M and Antonsen T M Jr 1998 MAGY: A time-dependent code for simulation of slow and fast wave microwave sources *IEEE Trans. Plasma Sci.* **26** 882–92
[43] Sabchevski S, Idehara T, Glyavin M, Ogawa I and Mitsudo S 2005 Modelling and simulation of gyrotrons *Vacuum* **77** 519–25
[44] Rhee M J and Destler W W 1974 Relativistic electron dynamics in a cusped magnetic field *Phys. Fluids* **17** 519–25
[45] Zapevalov V 2003 Design of a large orbit gyrotron with a permanent magnet system *Int. J. Infrared Millim. Waves* **24** 253–60
[46] Ling G, Piosczyk B and Thumm M K 2000 A new approach for a multistage depressed collector for gyrotrons *IEEE Trans. Plasma Sci.* **28** 606–13
[47] Thumm M and Kasparek W 2002 Passive high-power microwave components *IEEE Trans. Plasma Sci.* **30** 755–86
[48] Blank M, Kreischer K and Temkin R J 1996 Theoretical and experimental investigation of a quasi-optical mode converter for a 110-GHz gyrotron *IEEE Trans. Plasma Sci.* **24** 1058–66
[49] Jin J 2007 Quasi-optical mode converter for a coaxial cavity gyrotron *Wiss. Ber.* **FZKA-7264** 1058–66

[50] Sabchevski S, Idehara T, Damyanova M, Zhelyazkov I, Balabanova E and Vasileva E 2007 Modelling, simulation and computer-aided design (CAD) of gyrotrons for novel applications in the high-power terahertz science and technologies *J. Phys.: Conf. Ser.* **992** 012001–66
[51] Pretterebner J, Mübius A and Thumm M 1992 Improvement of quasioptical mode converters by launching an appropriate mixture of modes *Proc. of the 17th Int. Conf. on Infrared and Millimeter Waves* vol 1929 pp 40–1
[52] Pretterebner J 2003 Compact quasi-optical antennas in oversized waveguide *PhD Thesis (in German)* Institut für Plasmaforschung, Universität Stuttgart
[53] Drum O 2002 Numerische Optimierung eines quasi-optischen Wellentypwandlers fur ein frequenzdurchstimmbares Gyrotron *Wiss. Ber.* 1058–66
[54] Prinz H O 2007 Dreidimensionale feldberechnung für elektrisch große geometrien und ihre anwendung für multifrequente hochleistungsgyrotrons *Wiss. Ber.* **FZKA-7351** 1058–66

IOP Publishing

High Frequency Sources of Coherent Radiation for Fusion Plasmas

G Dattoli, E Di Palma, S P Sabchevski and I P Spassovsky

Chapter 6

CARM theory and relevant phenomenology

6.1 Introduction

The Cyclotron Auto-Resonance Maser (CARM), like the gyrotron, belongs to the CRM gyro-devices wave generators family (see figure 5.9) with phase velocity close to the speed of light.

The CARM, as with all the other free electron devices, is based on the instability mechanism, we described either for UFEL and gyrotrons. The core of the system is the e-beam emitted from the electron gun and moving through the interaction region surrounded by a static magnetic field, coupling and amplifying a suitable electromagnetic wave. The beam is finally deposited in the collector, where the charge is recovered [1, 2]. The difference with the gyrotron lies on the frequency operation, well above the cut-off of the cavity mode (see figure 5.6). The operating wavelength shifts towards lower value, thus allowing reduction of the external magnetic field surrounding the interaction region.

The auto-resonance mechanism, which characterizes the CARM, derives from the fact that an electron which is injected in resonance with the wave, satisfying the condition $\beta_{\text{ph}} = v_{\text{ph}}/c \approx 1$, will remain in resonance though it loses energy to the wave.

Starting from equation (5.5), with the previous β_{ph} approximation, the following resonance condition holds

$$\omega \approx \frac{s\Omega_c}{\gamma(1 - \beta_z\beta_{\text{ph}})}. \tag{6.1}$$

The quantity at the denominator is a constant of motion, which, as shown below, characterizes the evolution of an electron coupling to a constant amplitude TE wave.

doi:10.1088/978-0-7503-2464-9ch6

The electron dynamics in a CARM device is reduced to the coupled energy equation

$$\frac{d\gamma}{dt} = -\frac{e}{m_0 c^2}\vec{v} \cdot \vec{E}, \tag{6.2}$$

and z-component of the Lorentz equation

$$\frac{dp_z}{dt} = -e(\vec{v} \times \vec{B}) \cdot \hat{z} = e\vec{v} \cdot (\hat{z} \times \vec{B}), \tag{6.3}$$

with $\vec{E}$ and $\vec{B}$ being the electric and magnetic field of the TE wave and $p_z = m_0 c\gamma\beta_z$ the electron axial momentum. The condition

$$\frac{d}{dt}\left(\gamma - \frac{p_z\beta_{\text{ph}}}{m_0 c}\right) = 0, \tag{6.4}$$

follows from the combination of equations (6.2) and (6.3), in which the electric and magnetic field vectors are linked by

$$\vec{E} = c\beta_\phi(\hat{z} \times \vec{B}). \tag{6.5}$$

The quantity $\gamma(1 - \beta_z\beta_{\text{ph}})$ remains, therefore, constant during the interaction. The particle remains in resonance with the wave, thus increasing the device efficiency as discussed in the next sections.

The auto-resonance condition in a more general case (for $k_z \neq \omega/c$), as discussed in [3], can be analyzed starting from the electron-wave phase slippage.

In accordance with the resonance condition equation (5.5) we introduce the transit angle of electron gyration with respect to the Doppler shift wave frequency satisfying the condition

$$\Theta = (\omega - k_z v_z - s\Omega)T < 2\pi, \tag{6.6}$$

where $T = L/v_z$ is the electron transit time, L is the length of the interaction region and $\Omega = \Omega_c/\gamma$.

Two different bunching mechanisms can be inferred from equation (6.6):

(a) Longitudinal bunching induced by the dependence of v_z on the energy $\mathcal{E}$ ($v_z(\mathcal{E})$).

(b) Azimuthal bunching induced by the relativistic dependence of the gyro frequency on the electron energy ($\Omega(\mathcal{E})$).

Starting from the phase slippage modulation, induced by the energy change, a bunching parameter can be defined in the following.

From equation (6.6) we derive

$$\delta\Theta_{\text{dyn}} = \mu\frac{\omega L}{v_z}\frac{\delta\gamma}{\gamma}, \tag{6.7}$$

where

$$\mu = \frac{\gamma}{\omega}\left(k_z\frac{\partial v_z}{\partial \gamma} + \frac{\partial \Omega}{\partial \gamma}\right)_{\gamma=\gamma_0}, \tag{6.8}$$

is the bunching parameter with

$$\delta\mathcal{E} = mc^2\delta\gamma. \tag{6.9}$$

Since, the energy change $\Delta\mathcal{E} = \hbar\omega$ induces a variation in the axial momentum $\Delta p_z = \hbar k_z$ we have

$$\frac{\Delta\mathcal{E}}{\Delta p_z} = \frac{\omega}{k_z} = v_{\text{ph}}, \tag{6.10}$$

with simple calculations, we find that the first derivative, with respect to gamma in equation (6.8), becomes

$$\frac{\partial v_z}{\partial \gamma} = \frac{1}{\gamma}\left(\frac{c^2}{v_{\text{ph}}} - v_z\right). \tag{6.11}$$

Now, using equation (6.11) with the condition (6.6) the bunching parameter assumes the following expression

$$\mu = \frac{c^2}{v_{\text{ph}}^2} - 1. \tag{6.12}$$

This term (vanishing when the phase velocity is c) regulates the auto-resonance mechanism.

The CARM operation occurs close to the exact auto-resonance ($c/v_{\text{ph}} \ll 1$) which allows one at the same time to realize a substantial extraction of the electron energy and a subsequent shift of the electron bunch into a proper decelerating phase as stated by Petelin in [1]. The calculation performed in [1] is limited to the ultra-relativistic electrons, a more general case involving arbitrary initial energy has been discussed in [4] with an optimization of the v_{ph} parameter with respect to the initial energy in order to extract all the power from the particles.

Now, assuming that for the dynamical phase slippage Θ_{dyn} holds the condition of the equation (6.6), the power extraction efficiency η can be expressed in terms of the bunching parameter as follows

$$\eta = \frac{\mathcal{E}_0 - \bar{\mathcal{E}}}{\mathcal{E}_0 - mc^2} = \frac{\Delta\gamma}{\gamma}\frac{1}{1 - \gamma_0^{-1}} = \frac{2\pi v_z}{\mu\omega L}\frac{1}{1 - \gamma_0^{-1}} = \frac{1}{1 - \gamma_0^{-1}}\frac{1}{\mu\Gamma N}, \tag{6.13}$$

where

$$\begin{aligned} \Gamma &= \frac{\omega}{\Omega}, \text{ frequency conversion factor,} \\ N &= \frac{\Omega_c L}{2\pi v_z}, \text{ number of cyclotron rotation.} \end{aligned} \tag{6.14}$$

We have now all the elements to compare gyrotron and CARM operation:

(a) For gyrotron we have

$$\begin{aligned} &\omega = \Omega \rightarrow \Gamma = 1, \\ &v_{\text{ph}} \gg c \rightarrow \mu = 1, \end{aligned} \tag{6.15}$$

and the efficiency is written as

$$\eta \sim \left(\frac{1}{1 - \gamma_0^{-1}} \frac{1}{N}\right). \tag{6.16}$$

We can accordingly distinguish between non-relativistic and relativistic regime of operation. In the first case $\gamma \approx 1 + \beta^2/2$ the efficiency may be extremely high, even of the order of unit, if $N \approx 1/\beta^2 \gg 1$. In the other case it can be drastically reduced, indeed $\eta \sim 1/N \ll 1$.

(b) For CARM we have

$$\begin{aligned} &v_{\text{ph}} \rightarrow c \rightarrow \mu \ll \gamma_0^{-2}, \\ &\Gamma = \frac{\omega}{\Omega} \approx \gamma_0^2, \end{aligned} \tag{6.17}$$

and large efficiency ($\eta \sim 1$) is assured for $N = 1/\mu$.

We have just grasped the surface of a more complicated story, which will be carefully discussed below. The real efficiency of any free electron source should be evaluated by calculating the efficiency of one electron and averaging it over all initial distributions of electrons in the coordinate and momentum configurations, as detailed in the following sections.

Since the original suggestion of CARM by Petelin [1] in 1974, several experiments were conducted globally, to test the relevant capabilities in terms of efficiency, in the 1980s. In table 6.1 we have summarized the various experiments, along with their operating parameters and performance in terms of efficiency.

Table 6.1. CARM experiments.

Institution	Frequency	Voltage	Current	Mode	V phase	Efficiency
IAP	32–35 GHz	400 keV	50–100 A	$TE_{1,1}$	1.2–1.4 c	~15%
			~5 kA (total)			(0.2%)
IAP	38 GHz	500 keV	50 A	$TE_{1,1} + TE_{2,1}$	1.25 c	26%
			~4 kA (total)	CARM-Gyrotron		(0.65%)
LLNL	220 GHz	2 MeV	1 kA	$TE_{1,1}$	1.05 c	2.5%
MIT	~30 GHz	~450 keV	80 A	$TE_{1,1}$	1.05 c	5%
MELBA	25 GHz	400 keV	1.2 kA	$TE_{1,1}$	1.2 c	1.5%

The first CARM experiment was carried out at the Institute of Applied Physics (IAP/RAS) in Gorky (USSR) [2, 3, 5, 6] and that was then followed by several experiments in the US. Initially at MIT a 35 GHz CARM amplifier of 10 MW with a low level efficiency of 3% was realized [7] using a 1.5 MeV, 260 A e-beam with an energy spread less than 2%. Later, several CARM experiments in the oscillator configurations, were conducted in several labs in the US, all characterized by an efficiency of the order of few %. These seminal activities were completed at the Accelerator Research Center (ARC) facility of Lawrence Livermore National Laboratory (LLNL) [8], at Plasma fusion Center of the MIT [9, 10], and at Michigan Electron Long Beam Accelerator (MELBA) with a pulse length of a few microseconds and employing a cavity resonator with high quality factor. These efforts brought important results from the scientific and technological point of view and the last CARM oscillator experiment performed by Bratman *et al* [11, 12] in the second half of the 1990s reached a considerable value of efficiency 26% at 7.9 mm wavelength.

The output efficiency was unsatisfactory with respect to the theoretical provisions and therefore unsuitable for magnetic fusion heating. The motivations of such a failure can be traced back to the qualities of the e-beam used in the experiment. We will address this point in the following and discuss how present technology allows us to provide beams with characteristics suitable for CARM operation, thus allowing their use in plasma fusion.

Regarding, e.g., the future DEMO reactor, a different design stage has been undergone with consequent re-adjustment of the corresponding ECRH&CD operating frequency, from the early 30 GHz proposal to the present day options, foreseen above 200 GHz [13–15], and those frequency values mixed with the efficiency and power requirements create some problems to be reached by a gyrotron device.

As already mentioned, a serious disadvantages of both the CARM and the FEL, as compared to the gyrotron, is their sensitivity to the spread in electron velocity, which affected all the quoted experiments. The presence of the Doppler-shift term in the resonance condition, equation (5.5), if one side enhanced the efficiency on the other, made the system vulnerable to the electron velocity spread. It is therefore evident that all the components of the system (modulator, gun, cavity, magnetic field ramp) require a new design conception, as discussed in the forthcoming sections of this chapter.

6.2 U-FEL, gyrotron, CARM interaction: a common point view

In this section U-FEL, CARM and gyrotron devices will be described by putting in evidence the common physical mechanisms underlying the operation of different types of free electron coherent generators. We proceed by identifying a set of key parameters characterizing these devices (gain, saturation intensity...) and model the relevant dynamics in terms of these macroscopic quantities.

U-FEL, gyrotron and CARM utilize free electrons in a vacuum to convert energy from a DC power source to an RF signal. The basic processes governing the electromagnetic (EM) radiation is the bremsstrahlung effect when charged particles move with a variable velocity.

In particular the magnetic bremsstrahlung occurs for electrons moving along curvilinear trajectories in an external magnetic field. Such radiation is produced when electrons, that are initially uncorrelated and producing spontaneous radiation with random phase, are gathered into microbunches which subsequently radiate in phase. This field can be either constant (in the case of CARM and gyrotron) or periodic (in the case of U-FEL). A constant magnetic field can be produced by a permanent magnet or solenoid, while a periodic field can be produced by a periodic array of magnet elements.

CARM and gyrotron are sources of coherent magnetic bremsstrahlung based on electron oscillations in a constant external applied magnetic field.

The resonance condition of the beam–wave interaction for CARM devices appears when the Doppler-shifted wave frequency is close to the frequency of electron oscillations or one of its harmonics [1, 16]

$$\omega = s\frac{\Omega}{\gamma} + k_z v_z, \tag{6.18}$$

where s is the harmonic number, v_z the electron axial velocity and k_z the axial component of the wave vector.

This resonance condition can be fulfilled for any wave phase velocity ($v_{\mathrm{ph}} = \omega/k_z$). The radiated waves can be either fast ($v_{\mathrm{ph}} > c$) or slow ($v_{\mathrm{ph}} < c$) wave.

Fast and slow wave devices (reported in table 5.9) are characterized by different bunching mechanisms [17]. In the case of fast wave it is manly induced by the $v_\perp \cdot E_\perp$ product, where $v_\perp$ is the electron transverse velocity, $E_\perp$ is the wave transverse electric component. The electric force effect changes the relativistic electron mass by modulating the particles cyclotron rotation in equation (6.18) inducing, therefore, a bunching. In the slow-wave devices the dominant effect is the axial bunching caused by the axial $v_\perp \times B_\perp$ Lorentz force, where $B_\perp$ is the transverse magnetic component of the wave. The Lorentz force modulating the particle axial velocity will induce a bunch acting on the last term in equation (6.18) [18–21].

To make a comparison of the gyro-device (like gyrotron and CARM) with U-FEL we should recall some formulae derived in chapter 4.

U-FELs, as discussed previously, are the most common devices based on radiation from electrons oscillating in periodic external fields. In figure 6.1 we have illustrated the typical configuration employed in the FEL interaction.

While in CRMs the electron oscillation frequency is just the electron cyclotron frequency, in U-FELs the electron oscillation frequency is $\Omega_{\mathrm{FEL}} = k_u v_z$, where $k_u = 2\pi/\lambda_u$ is the undulator period.

According to the discussion in chapter 4, the wavelength characterizing the emission process inside the undulator is written as

$$\omega_{\mathrm{FEL}} \cong 2\frac{\gamma^2}{1+\alpha^2}\omega_u, \tag{6.19}$$

where $\alpha = K_{\mathrm{FEL}}/\sqrt{2}$ with $K_{\mathrm{FEL}} \propto eB_0\lambda_u$ being the undulator strength which takes into account the effect of the transverse motion on the longitudinal velocity [22]

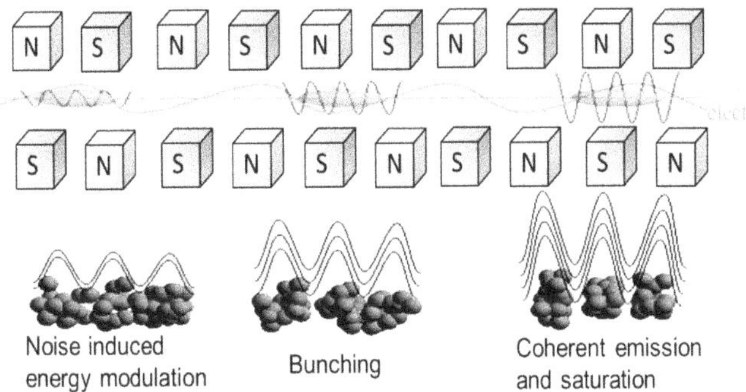

Figure 6.1. The FEL lasing process: energy modulation, bunching and coherent emission.

$$\beta_\perp \approx \frac{K_{\text{FEL}}}{\sqrt{2}\gamma},$$
$$\beta_z \approx 1 - \frac{1}{\gamma^{*2}}, \tag{6.20}$$

where $\gamma^* = \gamma\Big/\sqrt{1 + K_{\text{FEL}}^2/2}$.

The previous formula can be extended to CARM by noting that the relevant '***resonance***' condition, can be determined by using the same argument as before about constructive interference, which occurs whenever the accumulated slippage between radiation and electrons, in a helix period, equals the wavelength λ.

We recall that in a CARM a moderately relativistic e-beam, moves, inside a wave guide under the influence of an axial magnetic field, executing a helical path with a cyclotron frequency $\Omega_c = \frac{e\,B}{m_e}$.

The kinematical variables of the e-beam are specified by the longitudinal (v_z) and transverse ($v_\perp$) velocity components, linked to the relativistic factor γ by

$$\beta_z^2 + \beta_\perp^2 = 1 - \frac{1}{\gamma^2},$$
$$\beta_{z,\perp} = \frac{v_{z,\perp}}{c}, \tag{6.21}$$
$$\alpha = \frac{v_\perp}{v_z},$$

where α is the already defined pitch factor.

The electrons, with longitudinal velocity v_z, interact with a co-propagating electromagnetic field characterized by a wave-vector k_z, linked to the wave phase velocity v_{ph} by

$$k_z = \frac{\omega}{v_{\text{ph}}}, \tag{6.22}$$

In this case the link between helix period and guiding magnetic field is provided by

$$\Lambda = \frac{c}{\Omega_c}, \tag{6.23}$$

we impose the resonance condition as

$$(v_{\text{ph}} - v_z)\frac{\Lambda}{c} = \lambda, \tag{6.24}$$

where we have used the phase velocity v_{ph} to determine the radiation electron slippage. This is nothing other than a different form of the condition (6.18).

The above equation derived by using a kinematical argument and the analogy with U-FEL, has been the pivotal element of the discussion. The physical origin of the previous identity can, however, be understood on the basis of different arguments, involving, e.g., momentum (electron and fields) conservation.

We can further elaborate the previous identities, denoting by ω_{CARM} the resonant frequency, and we obtain, from equation (6.18)

$$\omega_{\text{CARM}} = \frac{\Omega}{1 - \frac{v_z}{v_{\text{ph}}}}. \tag{6.25}$$

It is worth stressing that, the phase velocity being dependent on the field frequency, equation (6.25) is not an explicit solution for ω, but only an approximation.

Before further pushing the analogy between U-FEL and CARM, we dwell on the physical meaning of the previous equations.

The CARM resonance condition can also be derived by requiring the matching between equation (6.18) and the waveguide dispersion relation

$$\omega^2 = c^2\left(k_\perp^2 + \beta_{\text{ph}}^2 k_z^2\right), \tag{6.26}$$

where $k_\perp$ is the transverse mode wave number, associated with the cutoff frequency $\omega_{\text{cutoff}} = ck_\perp$. It is easily checked that, from (6.18) and (6.25), one gets

$$\omega_\pm \cong \frac{\Omega}{1 \mp \frac{\beta_z}{\beta_{\text{ph}}}}. \tag{6.27}$$

The down-shifted intersection (see figure 6.2), yielding the gyrotron mode [23], is treated in chapter 5.

The upper shifted counterpart ω_+ is the resonant (CARM) frequency and, to better understand its role, it will be rewritten as

$$\begin{aligned} \omega_{\text{CARM}} &\cong \frac{\Omega}{1 - \frac{\beta_z}{\beta_{\text{ph}}}} = \frac{\Omega}{1 - \frac{1}{\beta_{\text{ph}}}\sqrt{1 - \frac{1}{\gamma_z^2}}} \cong 2\beta_{\text{ph}}\,\gamma_z^2\Omega, \\ \gamma_z &= \frac{\gamma}{\sqrt{1 + (\gamma\beta_\perp)^2}}, \end{aligned} \tag{6.28}$$

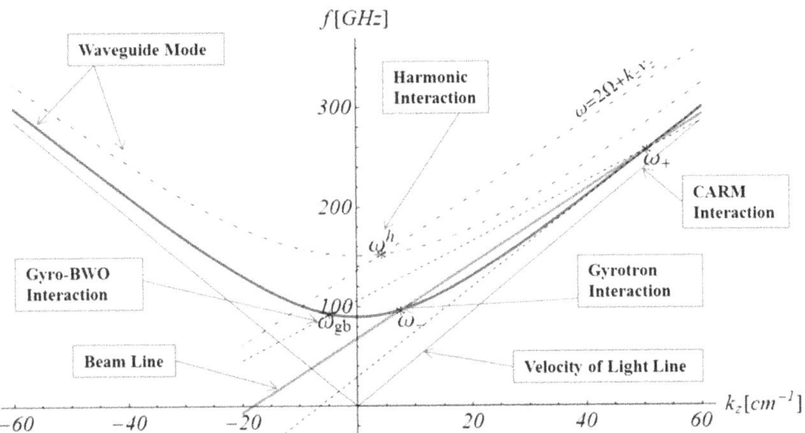

Figure 6.2. Brillouin diagram for the different conditions of electron cyclotron resonance selections provided by the intersections of the different beam lines (straight lines) with the dispersion curves of the operating cavity mode.

according to the assumption that γ_z be sufficiently large that $\sqrt{1 - \frac{1}{\gamma_z^2}} \cong 1 - \frac{1}{2\gamma_z^2}$. Equation (6.28) accounts for the frequency Doppler up-shift mechanism, characterizing most of the free electron devices.

It is important to emphasize that, at least formally, we have established so far an important analogy between CARM and U-FEL, namely $\Lambda \leftrightarrow \lambda_u$, which justifies the remark that the two devices are 'topologically' equivalent.

The role of the transverse velocity needs a more accurate comment. In the case of the U-FEL the transverse component, induced by the Lorenz force, is the tool allowing the coupling with the co-propagating electromagnetic (transverse) field (see equation (6.20)).

In the case of the CARM, the role of the transverse velocity component is the same as that of the undulator strength in the undulator FEL, as results comparing equations (6.20) and (6.28)

$$K_{\mathrm{CARM}} = \gamma\beta_{\perp}. \tag{6.29}$$

This velocity component should be induced during the e-beam preparatory phase, before the injection into the cavity.

What has been described so far are the physical conditions underlying the 'spontaneous' emission, which is the prerequisite for the onset of the coherent emission process. As is well known, it occurs via the bunching mechanism. The interaction of the electrons, with the cavity electric field mode, determines their energy modulation, which transforms into a density modulation, followed by a coherent RF emission when the electrons are bunched on a scale comparable to the RF electric field wavelength.

This description encompasses all devices of FEL type, CARM is, however, made peculiar by the fact that the auto-resonance is guaranteed even near saturation because any increase of Ω is balanced by a corresponding decrease of the longitudinal velocity[1].

A further important quantity, characterizing U-FEL, is the number of undulator periods, which is associated with the oscillations executed by an electron, while traveling inside the undulator. In the case of CARM it can be linked to the number of helical turns of the electrons inside the magnet. Accordingly, we get

$$\Omega \frac{L}{\beta_z c} = 2\pi N, \qquad N \cong \frac{L}{\Lambda}. \tag{6.30}$$

We have fixed the main element of our strategy and in the following section we will see how the correspondences, we have established, may provide an effective tool to evaluate the CARM evolution.

6.2.1 Small signal theory: FEL versus CARM analytical solution

In this section we will push further the analogy with U-FEL, by showing that the equation describing the CARM field evolution in the linear regime, can be written by taking advantage of the simplified expression valid in the former case.

In the analysis of the previous section we did not include any consideration regarding the interaction of the wave with the e-beam. The dispersion relation in equations (6.20)–(6.26) is appropriate for the 'cold' wave guide condition, which merely applies to the kinematic of the mode propagation.

The CARM dynamics, associated with the radiation intensity growth in the wave guide, undergoes different phases characterized by the amount of the field power density.

The weak coupling regime is characterized by a power level well below the threshold of the saturation intensity (namely the power density halving the small signal gain) and the relevant theory can be treated using perturbative methods and, to some extent, useful information can be drawn using analytical means.

We have stressed that the mechanisms leading to the CARM process are closely similar to those leading to U-FEL, we can therefore suspect that a closely analogous set of equations can be exploited to describe both devices.

The CARM beam-wave interaction can be described with a self-consistent physical model, as discussed by different authors [5, 23–25], in terms of three dimensionless parameters which take into account the appropriate kinematic conditions matching the electron longitudinal and wave group velocity and the cyclotron frequency. Within such a context a pivotal parameter is the frequency detuning δ defined as

[1] The efficiency enhancement is induced in U-FELs by tapering the undulator, by reducing e.g. the undulator period, in order to maintain the resonance condition in equation (4.6) fixed when β_z decreases, thus realizing the effect naturally entangled with the CARM operating mechanism.

$$\delta = 1 - \frac{\beta_z}{\beta_{\text{ph}}} - \frac{\Omega}{\omega_r}, \tag{6.31}$$

where ω_r is the resonance angular frequency, $\beta_{\text{ph}} = v_{\text{ph}}/c$, $\beta_z = v_z/c$, are the phase velocity and longitudinal velocities of the electrons, respectively, normalized to the speed of light, $\Omega = eB/(m\gamma)$ is the relativistic cyclotron frequency with B the external magnetic field. CARM being a laser-like device, a further quantity of crucial importance is the small signal gain coefficient which, within this framework, can be written as

$$I_g = \frac{2\mu_0|e|}{mc\gamma}\frac{\beta_{\text{ph}}}{\beta_\perp^4}\frac{\left(1 - \frac{\beta_z}{\beta_{\text{ph}}}\right)^3}{\left(1 - \beta_{\text{ph}}^{-2}\right)} I_0[CJ]^2, \tag{6.32}$$

where $\beta_\perp = v_\perp/c$ is the transverse velocities of the electrons normalized to the speed of light, I_0 is the beam current and $[CJ]$ is the beam–wave overlapping coefficient. Finally the peculiar nature of the CARM offers the possibility of achieving large efficiency, as a consequence of the auto-resonance mechanism, and the quantity controlling such an effect is the recoil parameter reported below

$$b = \frac{\beta_\perp^2}{2\beta_z\beta_{\text{ph}}\left(1 - \frac{\beta_z}{\beta_{\text{ph}}}\right)}, \tag{6.33}$$

Furthermore, the CARM beam–wave interaction is described by the following set of differential equation [5, 24] (see Ceccuzzi *et al* [26] for the relevant approximations)

$$\begin{aligned}
\frac{dF}{d\zeta} &\approx I_g\langle e^{i\theta}\rangle_{\theta_0} + \left[\bar{I}_g\langle ue^{i\theta}\rangle_{\theta_0}\right],\ \bar{I}_g = \left(b - \frac{1}{2}\right)I_g,\\
\frac{du}{d\zeta} &\approx Re[Fe^{-i\theta}],\\
\frac{d^2\theta}{d\zeta^2} &\approx (b\Delta - 1)|F|cos(\theta + \phi) - \left\{b\frac{d}{d\zeta}Q(\zeta) - \frac{1}{2}\frac{d}{d\zeta}Re(iFe^{-i\theta})\right\},\\
\frac{dQ(\zeta)}{d\zeta} &\approx -Re\left(i\frac{dF_s}{d\zeta}e^{-i\theta}\right),
\end{aligned} \tag{6.34}$$

where the normalized u, θ variables are associated with the electron energy and the electron–wave phase, respectively, Q is the axial momentum correction, ζ is the normalized space variable and $<\ldots>_{\theta_0}$ accounting the average on the initial phase distribution. The terms within square and curly brackets and the second equation, accounting for the energy variation, can be neglected in the small or weakly saturated regime. We are, therefore, left with a pendulum like equation and the first, accounting for the field amplitude evolution, in full analogy with the U-FEL case.

A significant result from such a treatment is the derivation of a modified dispersion relation including the interaction of the electrons with the wave guide modes linearizing the Maxwell–Vlasov equation [25]. According to [23, 25] we find

$$\frac{\omega^2}{c^2} - \left(k_\perp^2 + k_z^2\right) + \tilde{\varepsilon}_{\rm mnl}\frac{2k_\perp^2(\omega - k_z v_z)}{\left(\omega - \frac{\Omega_c}{\gamma} - k_z v_z\right)} - \tilde{\varepsilon}_{\rm mnl}\frac{k_\perp^2\beta_\perp^2\left(\omega^2 - c^2k_z^2\right)}{\left(\omega - \frac{\Omega_c}{\gamma} - k_z v_z\right)^2} = 0, \quad (6.35)$$

where $\tilde{\varepsilon}_{\rm mnl}$ plays the role of coupling parameter, with (m, n) the interacting waveguide mode number and l the harmonics of the electron cyclotron frequency due to the axial magnetic field. It depends on the beam current and on the geometrical parameters of the waveguide itself and will be specified later in this section. In equation (6.35) the terms containing the coupling $\varepsilon_{\rm mnl}$ are those ruling the field electron evolution, we simplify the analysis by neglecting the first because the second is dominating near the resonance. We are therefore left with

$$\frac{\omega^2}{c^2} = \left(k_\perp^2 + k_z^2\right) - \varepsilon_{\rm mnl}\frac{k_\perp^2\left(\omega^2 - c^2k_z^2\right)}{\left(\omega - \frac{\Omega_c}{\gamma} - k_z v_z\right)^2}. \quad (6.36)$$

The previous identity is the crucial element of the forthcoming discussion and, for later convenience, we set

$$\begin{aligned} \tilde{k}_z &= k_z + \delta_{k_z}, \\ k_z &= \left(\frac{\omega^2}{c^2} - k_\perp^2\right)^{\frac{1}{2}} = \frac{\gamma\omega - \Omega_c}{\gamma v_z}, \end{aligned} \quad (6.37)$$

with δ_{k_z} representing the deviation of the field longitudinal wave vector, induced by the coupling with the electrons. Inserting equation (6.37) into (6.36) we find that δ_{k_z} is specified by the following fourth order degree algebraic equation

$$\beta_z^2\delta_{k_z}^4 + 2\beta_z^2 k_z\delta_{k_z}^3 + \varepsilon_{\rm mnl}k_\perp^2\delta_{k_z}^2 + 2\varepsilon_{\rm mnl}k_\perp^2 k_z\delta_{k_z} - \varepsilon_{\rm mnl}k_\perp^4 = 0, \quad (6.38)$$

whose roots specifies the evolution of the CARM field amplitude along the coordinate z, according to

$$E(z) \propto \sum_{j=1}^{4} e_j e^{i(\delta_{k_z})_j z}, \quad (6.39)$$

where j refers to the roots of equation (6.38) and e_j are integration constants, fixed by the conditions

$$\begin{aligned} &E(0) = 1, \\ &\left(\left(\frac{d}{dz}\right)^k E(z)\right)_{z=0} = 0, \text{ with } k = 1, 2, 3. \end{aligned} \quad (6.40)$$

Neglecting the opposite propagation wave $\omega - kc$, the dispersion relation (6.35) will be reduced of one order and taking into account a small value for δ_{kz} results [23]

$$\beta_z^2 x^4 + 2\beta_z^2(t^2 - 1)^{1/2}x^3 + \epsilon_{mnnl}^{TE}x^2 + 2\epsilon_{mnnl}^{TE}(t^2 - 1)^{1/2}x - \epsilon_{mnnl}^{TE} = 0, \qquad (6.41)$$

whose imaginary solution is given by the following expression

$$\Gamma_{mnl} = K\left(\frac{\epsilon_{mnnl}^{TE}}{\beta_z^2(t^2 - 1)^{(1/2)}}\right)^{1/3} \text{ with } K = \frac{3^{1/2}}{2^{4/3}}. \qquad (6.42)$$

In figure 6.3 are reported the imaginary solution of the fourth order equation (6.38) and third order equation (6.42) polynomial versus the normalized resonance frequency $(\Omega/(ck_{11}))$ for an e-beam energy of 1 MeV with 500 A of current interacting with the TE_{11} mode of a waveguide having a radius $r_w = 1.4$ cm [25]. At the resonance condition $(\Omega/(ck_{11}) = 2.87)$, achieved with an external axial magnetic field $B_0 = 4.01$ kG, the approximation, we used, appears reasonable, because the solutions do not exhibit significant differences.

The linearized field growth along the longitudinal coordinate can accordingly be obtained by plotting $|E(z)|^2$, as shown in figures 6.4 and 6.5.

The evolution curve has the well-known shape, characterizing also the self-amplified spontaneous emission (SASE) FEL operating mode, namely a lethargy region where the system (electrons and radiation) organizes the coherence and a linear (in logarithmic scale) growth with a characteristic gain length L_g. In the case of CARM such a quantity is specified by [25]

$$L_g^{-1} = 2\Gamma_{mnl} = \sqrt{3}\left[\frac{\epsilon k_\perp^4}{16k_z\beta_z^2}\right]^{1/3}. \qquad (6.43)$$

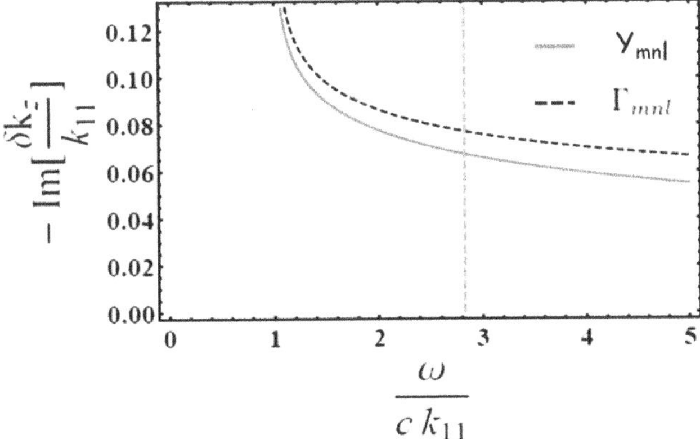

Figure 6.3. The imaginary part of the complex conjugate solution of the fourth degree polynomial (black-dashed line) and of the approximate cubic polynomial (red-continuous line) versus the normalized frequency. The vertical green line is the normalized resonance frequency.

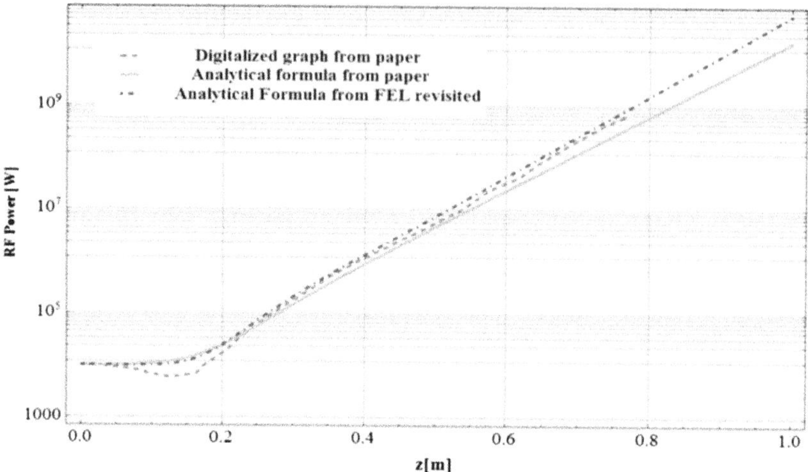

Figure 6.4. Comparison between U-FEL (blue) and CARM (green) SSR (small signal growth) intensity growth curves with the result reported in the Wurtele paper [25].

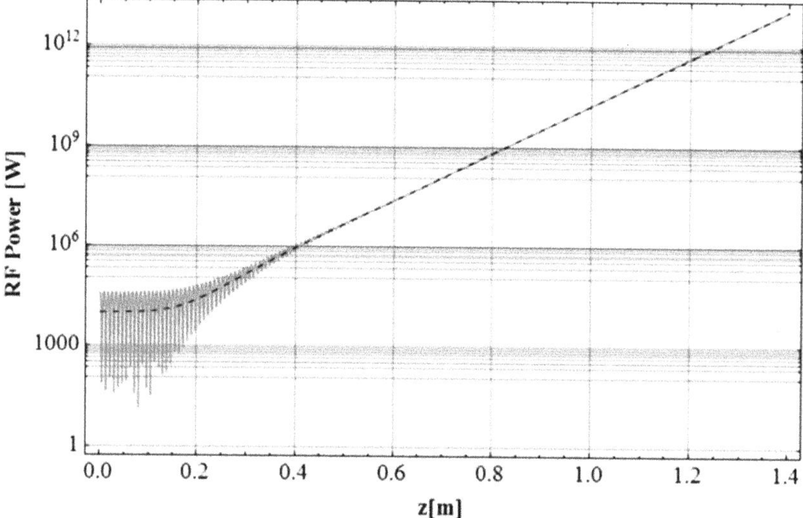

Figure 6.5. The growth intensity of the small signal CARM by solving the forth order equation (6.38) (continuously red line) and neglecting that of the real root (dashed black line).

Let us now invoke the analogy with the U-FEL, whose gain length is defined as [27]

$$L_g = \frac{\lambda_u}{4\,\pi\,\sqrt{3}\,\rho}, \tag{6.44}$$

with ρ being the Pierce parameter linked to the small signal gain coefficient g_0 by the identity [27]

$$\rho = \frac{(\pi\, g_0)^{1/3}}{4\,\pi\, N}. \tag{6.45}$$

The use of the correspondences established in the previous section and the comparison between equations (6.43) and (6.44) allows the following identification

$$\rho = \frac{\Lambda\,\Gamma}{4\,\pi\,\sqrt{3}}. \tag{6.46}$$

The dependence of U-FEL field amplitude on the longitudinal coordinate has been shown to be provided by [22, 27] (see also appendix C)

$$\begin{aligned}
a(\tau) = {} & \frac{a_0}{3(\nu + p + q)} e^{-\frac{2}{3} i \nu \tau} \Big\{ (-\nu + p + q) e^{-\frac{i}{3}(p+q)\tau} + \\
& + 2(2\nu + p + q) e^{\frac{i}{6}(p+q)\tau} \\
& \left[\cosh\left(\frac{\sqrt{3}}{6}(p - q)\tau \right) + i \frac{\sqrt{3}\,\nu}{p - q} \sinh\left(\frac{\sqrt{3}}{6}(p - q)\tau \right) \right] \Big\},
\end{aligned} \tag{6.47}$$

$$p = \left[\frac{1}{2}(r + \sqrt{d}) \right]^{1/3}, \qquad q = \left[\frac{1}{2}(r - \sqrt{d}) \right]^{1/3},$$

$$r = 27\,\pi\, g_0 - 2\,\nu^3, \qquad d = 27\,\pi\, g_0[27\,\pi\, g_0 - 4\,\nu^3],$$

and the intensity evolution is just given by $|a(\tau)|^2$.

The various parameters entering the above expression are recognized as

$\nu \equiv$ Detuning parameter.
$z \equiv$ Longitudinal coordinate.
$L \equiv N\,\lambda_u \equiv$ Interaction length.
$\tau \equiv$ Dimensionless time.

We can now get the correspondence with the CARM variables by defining the normalized detuning $\bar{\nu}$ parameter as

$$\bar{\nu} = \frac{\nu}{(27\pi\, g_0)^{1/3}}, \tag{6.48}$$

and then, using the relations (6.43)–(6.45), writing the dimensionless time in the following

$$\tau = \frac{z}{L} = \frac{z}{N l_g 4\, p\, \sqrt{3}\rho} = \frac{2\,\Gamma z}{\sqrt{3}(p g_0)^{1/3}}, \tag{6.49}$$

ending up with

$$\nu\tau = 2\sqrt{3}\Gamma\bar{\nu}z. \tag{6.50}$$

The complex amplitude (6.47), assumed now to be a function of the normalized detuning parameter $\bar{\nu}$ and of the inverse gain length Γ, will be exploited to describe the small signal growth of the radiation field amplitude in the following

$$a(z, \Gamma, \bar{\nu}) = \frac{a_0}{3} e^{-i\frac{4\Gamma\bar{\nu}z}{\sqrt{3}\beta_z}} \cdot \left\{ A_1 e^{-i\frac{2\Gamma z A_3^{(+)}}{\sqrt{3}}} + 2A_2 e^{i\frac{\Gamma z A_3^{(+)}}{\sqrt{3}}} \cdot \left[\cosh(\Gamma z A_3^{(-)}) + i\frac{\sqrt{3}\bar{\nu}}{A_3^{(+)}} \sinh(\Gamma z A_3^{(-)}) \right] \right\}, \tag{6.51}$$

with

$$\begin{aligned} A_1 &:= A_1(\bar{\nu}) = \frac{(-\nu + p + q)}{\nu + p + q} = \\ &= \frac{\left(\sqrt[3]{1 - 2\bar{\nu}^3 + \sqrt{1 - 4\bar{\nu}^3}} + \sqrt[3]{1 - 2\bar{\nu}^3 - \sqrt{1 - 4\bar{\nu}^3}} - \sqrt[3]{2}\bar{\nu}\right)}{\left(\sqrt[3]{1 - 2\bar{\nu}^3 + \sqrt{1 - 4\bar{\nu}^3}} + \sqrt[3]{1 - 2\bar{\nu}^3 - \sqrt{1 - 4\bar{\nu}^3}} + \sqrt[3]{2}\bar{\nu}\right)}, \\ A_2 &:= A_2(\bar{\nu}) = \frac{(2\nu + p + q)}{\nu + p + q} = \\ &= \frac{\left(\sqrt[3]{1 - 2\bar{\nu}^3 + \sqrt{1 - 4\bar{\nu}^3}} + \sqrt[3]{1 - 2\bar{\nu}^3 - \sqrt{1 - 4\bar{\nu}^3}} + 2\sqrt[3]{2}\bar{\nu}\right)}{\left(\sqrt[3]{1 - 2\bar{\nu}^3 + \sqrt{1 - 4\bar{\nu}^3}} + \sqrt[3]{1 - 2\bar{\nu}^3 - \sqrt{1 - 4\bar{\nu}^3}} + \sqrt[3]{2}\bar{\nu}\right)}, \\ A_3^{(\pm)} &:= A_3^{(\pm)}(\bar{\nu}) = \frac{1}{\sqrt[3]{2}}\left(\sqrt[3]{1 - 2\bar{\nu}^3 + \sqrt{1 - 4\bar{\nu}^3}} \pm \sqrt[3]{1 - 2\bar{\nu}^3 - \sqrt{1 - 4\bar{\nu}^3}}\right), \\ (p \pm q)\tau &= \frac{\sqrt{3}}{\sqrt[3]{2}} 2\Gamma z\left(\sqrt[3]{1 - 2\bar{\nu}^3 + \sqrt{1 - 4\bar{\nu}^3}} \pm \sqrt[3]{1 - 2\bar{\nu}^3 - \sqrt{1 - 4\bar{\nu}^3}}\right) = \\ &= 2\sqrt{3}\Gamma z A_3^{(\pm)}. \end{aligned} \tag{6.52}$$

In figure 6.4 we have provided a comparison between the prediction of the CARM theory and of the U-FEL scaling equations, given in equation (6.47). The agreement is satisfactory and further comments will be given below.

We should put in evidence that the linear solution obtained solving the dispersion relation (6.36) has been regularized neglecting the oscillating root of the equation (6.38) as reported in figure 6.5 the comparison of the amplitude signal with and without the oscillating solution.

The following two remarks are in order to complete the previous discussion:

(a) The dispersion relations for CARM and U-FEL lead to fourth and third degree algebraic equations, respectively. This is a consequence of the fact that the CARM field equations have been derived without the assumption of paraxial approximation, while, in the case of U-FEL the small signal problem is solved by the approximation of slowly varying envelope (SVE). This assumption leads to a treatment involving algebraic equations of one degree lower. In adapting U-FEL to CARM theory, according to the previous prescriptions and to [24], we did not find particular differences, except for the lethargic parts, where the SVE approximation is not fully justified and smoothens the field oscillations.

(b) Equations (6.36)–(6.43) have been written without fixing the waveguide mode structure, we can, however, factorize the ε coupling parameter as the product of two terms, namely

$$\varepsilon = \Xi \, \mathrm{f}, \\ \mathrm{f} = \frac{4\,\pi\,\beta_\perp^2}{\gamma\,\beta_z}\left(\frac{I_b}{I_A}\right), \tag{6.53}$$

where $I_{b,A}$ denote the beam and Alfvén current, the parameter Ξ summarizes the details of the cavity mode and the effect due to the geometrical overlapping between electrons and wave-guide modes and will be more carefully discussed in the following.

6.3 Non-linear regime and saturated power

In the previous section we have developed quite a straightforward formalism to prove that most of the scaling formulae developed within the framework of FEL theory can also be adapted to the study and design of CARM devices, at least for the case of small signal regime. In this section we include the non-linear contributions and show that the logistic curve model [27, 28] is an effective tool to study the evolution of the system up to the saturation. The logistic growth curve belongs to the family of S-shaped curves, the model has been shown to be very effective in reproducing the evolution of any system undergoing a dynamical behavior ruled by an equation of the type

$$\frac{dP}{dz} = \frac{P}{l_g}\left[1 - \frac{P}{P_S}\right], \tag{6.54}$$

even though both CARM and U-FEL satisfy more complicated non-linear equations as to the growth of the power density. Equation (6.54) captures the essential physics of the problem, namely a linear growth followed by a quadratic non-linearity when the power approaches P_S which denotes the saturated power. The solution of equation (6.54) can be written as

$$P(z) = P_0 \frac{e^{z/l_g}}{1 + \frac{P_0}{P_S}[e^{z/l_g} - 1]}. \tag{6.55}$$

The definition of the CARM saturated power P_S is easily given by just following the prescription of [27], we therefore set

$$P_S \cong \eta \, P_E, \tag{6.56}$$

where P_E is the e-beam power and η the efficiency of the device in turn provided by

$$\begin{aligned} &\eta = \eta_{sp}\eta_C, \\ &\eta_{sp} \cong \sqrt{2}\rho, \\ &\eta_C \cong \frac{1}{(1-\beta_{ph}^{-2})(1-\gamma_0^{-1})}\frac{\beta_\perp^2}{b}, \end{aligned} \tag{6.57}$$

where we have denoted by $\eta_{sp,C}$ the single particle and collective efficiency, respectively [23]. The single particle efficiency, can be written using the analogy in terms of the Pierce parameter as

$$\eta_{sp} \cong \sqrt{2}\rho = \frac{\sqrt{2}\Lambda\Gamma}{4\pi\sqrt{3}}. \tag{6.58}$$

According to the previous identity the saturated power can be cast in the form

$$\begin{aligned} &P_S \cong \frac{\sqrt{2}}{4\pi\sqrt{3}}\frac{\Lambda\Gamma}{(1-\beta_{ph}^{-2})(1-\gamma_0^{-1})}\frac{\beta_\perp^2}{b}P_E, \\ &b = \frac{\beta_\perp^2}{2\beta_z\beta_{ph}\left(1-\dfrac{\beta_z}{\beta_{ph}}\right)}. \end{aligned} \tag{6.59}$$

According to the terminology of [5, 23], b denotes the electron recoil parameter. It accounts for the auto-resonance contribution, including the effect of axial momentum and velocity change with the electron energy loss [29]. Regarding the analogy with U-FEL it can be associated with the undulator tapering parameter [27, 30].

We have recovered all the crucial parameters (gain length and saturated power) to draw the CARM power growth curve, using the logistic equation. However, equation (6.55) accounts only for the exponential growth prior to the saturation and does not contain any lethargic phase. To overcome this problem we replace the exponential term in equation (6.55) with the square modulus of the small signal amplitude derived in the previous section, thus writing

$$\begin{aligned} &P(z) \cong P_0\frac{|\beta(z)|^2}{1+\dfrac{P_0}{P_S}(|\beta(z)|^2-1)}, \\ &\beta(\tau) = \frac{a(\tau(z))}{a_0}. \end{aligned} \tag{6.60}$$

To check the validity of the previous formula we have developed an ad hoc numerical GRAAL (Gyrotron Radiation Amplification Auto-Resonance Laser) code to integrate the CARM equations which will be described in the next section.

6.3.1 The 1D GRAAL code

The dynamical systems accounting for the evolution of CARM devices is described by a set of equations coupling electrons and field [5, 23, 31].

As described previously, the pivotal parameters characterizing the CARM dynamics are summarized by three dimensionless quantities: b, accounting for the auto-resonance (see equation (6.33)), Δ, normalized detuning (see equation (6.31))

$$\Delta = \frac{2\left(1 - \frac{\beta_z}{\beta_{\rm ph}}\right)^2\left(1 - \frac{\omega_r}{\omega}\right)}{\beta_\perp^2(1 - \beta_{\rm ph}^{-2})}, \tag{6.61}$$

and I_g normalized beam current, proportional to the beam current I_b and expressible in terms of ρ parameter as

$$I_g = \left[\frac{4\sqrt[3]{2}\left(1 - \frac{\beta_z}{\beta_{\rm ph}}\right)}{\beta_\perp^2\left(1 - \beta_{\rm ph}^{-2}\right)}\rho\right]^3. \tag{6.62}$$

In terms of these parameters the CARM energy and phase equations, for TE modes interaction, can be cast in the form [23, 31]

$$\begin{aligned}
\frac{du}{d\zeta} &= \frac{[1 - u]^{s/2}}{1 - bu}Re(F_s e^{-i\theta}),\\
\frac{d\theta}{d\zeta} &= \frac{1}{1 - bu}\left[\Delta - u - bQ(\zeta) + \frac{s}{2}[1 - u]^{s/2-1}Re(iF_s e^{-i\theta})\right],\\
\frac{dQ(\zeta)}{d\zeta} &= -\frac{[1 - u]^{s/2}}{1 - bu}Re(i\frac{dF_s}{d\zeta}e^{-i\theta}).
\end{aligned} \tag{6.63}$$

The normalized u, θ variables are associated with the electron energy and the electron–wave phase, respectively, s is the order of the harmonics and F_s accounts for the complex mode field amplitude, whose evolution is fixed by the equation

$$\frac{dF_s}{d\zeta} = I_g\left\langle\frac{[1 - u]^{s/2}}{1 - bu}e^{i\theta}\right\rangle. \tag{6.64}$$

From the mathematical point of view the problem is that of solving a system of nonlinear ordinary differential equations (ODEs), consisting of four differential equations three of which account for the electron motion and the other for the complex field amplitude evolution inside the cavity.

The adopted numerical procedure foresees the use of a Runge–Kutta scheme for the electron dynamics, with the field amplitude kept constant during one discretization step. Furthermore, a finite difference method has been applied to evaluate the differential equation concerning the amplitude wave evolution, in which the crucial

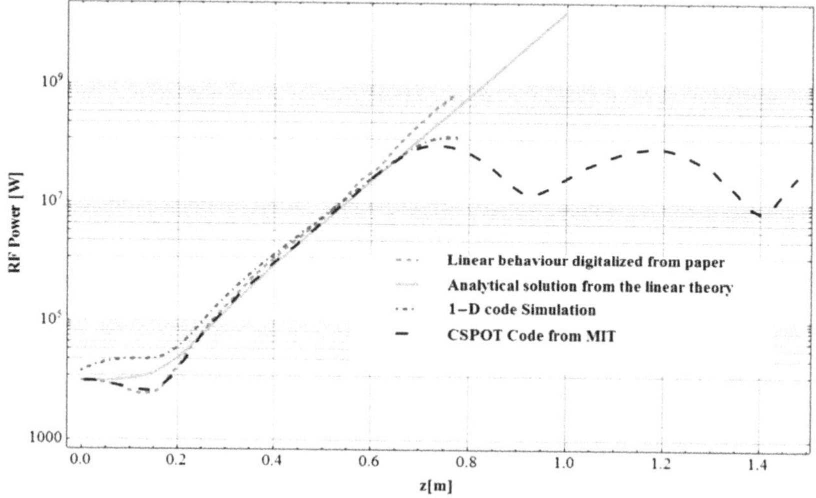

Figure 6.6. The analytical growth linear rate of the signal compared with the numerical simulation.

step is the careful average on the electron phase distribution and on the transverse velocity distribution in order to include correctly the effect of the beam qualities.

To study the effect of the particles velocity spread, starting from a fixed γ_0 beam energy and an α_0 pitch, a Gaussian distribution of the transverse velocity has been generated centered at the initial value of $\beta_{\perp 0}$.

For each particle we considered an ODE system characterized by a $b(\beta^i_{\perp 0})$ and $\Delta(\beta^i_{\perp 0})$ parameter and the average on the electron phase and velocity distribution, has been evaluated by the use of a standard trapezoidal scheme.

Furthermore, the orbital efficiency has been obtained, by averaging the electron motion on the electron phase and velocity distribution, allowing one to evaluate the CARM power growth.

The comparison between equation (6.60) and the power evolution obtained via the numerical implementation are shown in figures 6.6 and 6.7. The two curves compare fairly well; the use of these formulae for fixing the working points of a CARM device is therefore justified.

We have so far shown that a wise application of the CARM theory and U-FEL scaling formulae, developed in the past, may provide a heuristic tool useful for CARM device design. Further 'practical' consequences from our treatment will be drawn in the forthcoming section.

The impact of the beam qualities, demanding for a high performance e-beam which leads to an appropriate modeling of the gun as discussed previously, on the output power will be analyzed in the forthcoming section deriving appropriated scaling laws.

6.4 FEL to CARM scaling law

Before going further, we note that the complexity of the description of the free electron like devices stems from the large number of parameters characterizing these

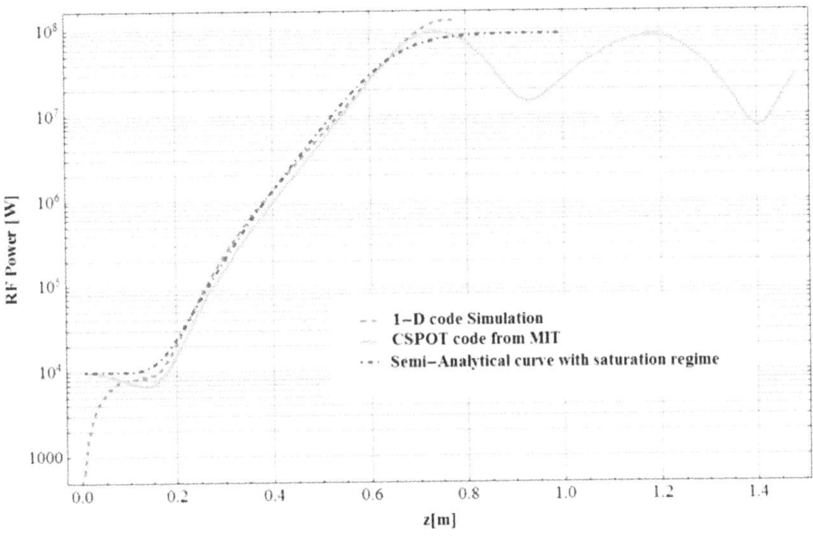

Figure 6.7. The revisited semi-analytical formula from FEL compared with numerical code.

systems. A possible simplification comes from the fact that a few key parameters (as well as an appropriate combination of them) can be selected to express quantities like gain or efficiency in terms of simple formulae.

The CARM dependence of the saturated power on the beam velocity spread can be derived from an accurate analysis of the numerical data and by an extension of an analogous expression obtained in the past for the U-FEL operation [22]. In the case of CARM, one important parameter is the normalized current I_g introduced in [31, 32]. Figure 6.8 shows the CARM efficiency versus the frequency detuning δ for different values of I_g as used by GRAAL code in order to reproduce the data from [25] regarding a CARM operation at 18 GHz and regarding the homogeneously broadened operation (namely with a beam without any significant velocity spread). The procedure that we have developed to get a general formula providing the dependence of the efficiency versus the velocity spread is summarized below.

After fixing the value of δ, maximizing the curves in figure 6.8 for each normalized beam current, we have run the simulation taking into account the velocity spread and evaluated the corresponding efficiency. The results are presented in figure 6.9(a), where we have plotted the efficiency versus the rms value $\sigma_{\beta_\perp}$ of the velocity spread (with a Gaussian distribution of the transverse velocity, centered at $\beta_{\perp 0}$, beam energy γ_0 and a pitch factor $\alpha_0 = v_{\perp 0}/v_{z0}$) for different I_g. A fit of the numerical data with a Lorentzian-like function yields the following expression

$$\eta = \frac{\eta_0}{1 + a\sigma_{\beta_\perp}^2}, \tag{6.65}$$

a being the fit parameter, $\sigma_{\beta_\perp}$ the beam velocity spread and η_0 the efficiency obtained neglecting the velocity spread of the beam electrons.

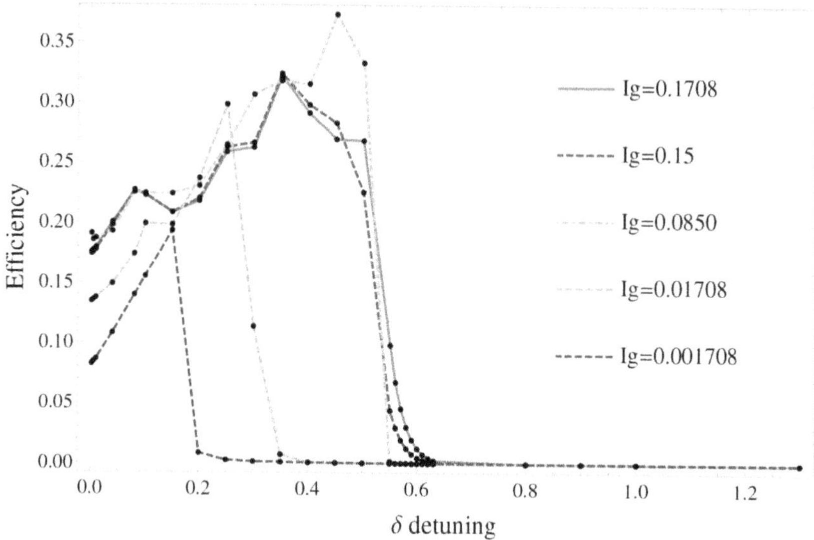

Figure 6.8. The efficiency versus the detuning parameter for a CARM amplifier at 18 GHz with a beam energy of 1.0 MeV and a pitch ($v_{\perp}/v_z$) of 0.5 and an axial magnetic field $B_0 = 4.01$ kG.

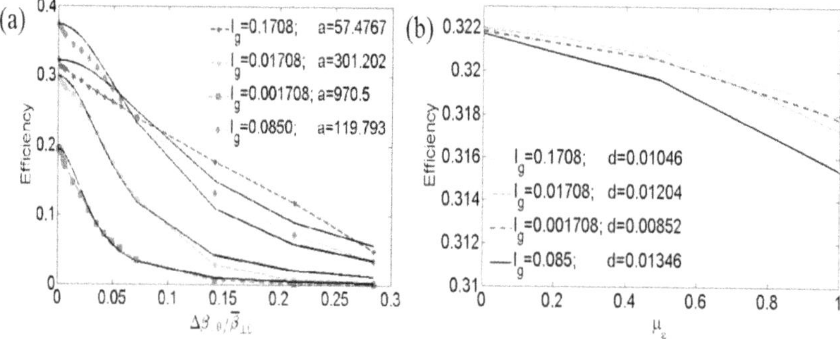

Figure 6.9. On the left side: (a) the system efficiency using the beam parameters of figure 6.8 for different normalized currents (dotted-line) each of which are fitted with a Lorentzian curve (continuous line); on the right side: (b) the system efficiency versus the inhomogeneous broadening parameter.

It has been found that the values of a, derived from the fitting procedure, strongly depend on the normalized current. It is therefore convenient to use a slightly different fitting strategy, involving the use of the inhomogeneous broadening parameters previously used for the study of U-FEL devices.

Taking advantage of the analogy between U-FEL and CARM devices and using the Pierce parameter (ρ) for CARM operation [33], it is possible to derive a 'universal' semi-analytical curve describing the CARM efficiency. The pivotal parameter ruling the effect of velocity spread on CARM performances is completely equivalent to the inhomogeneous broadening parameter, already defined for a FEL device and reads [27]:

$$\mu_\varepsilon = \frac{2\sigma_{\beta_\perp}}{\rho}. \tag{6.66}$$

The efficiency versus μ_ε is plotted in figure 6.9(b). It is evident that the scaling reveals a kind of 'universality' since the observed behavior is well reproduced by the relation

$$\eta = \frac{\eta_0}{1 + d\mu_\varepsilon^2}, \tag{6.67}$$

where the fit parameter d is almost the same for the different I_g. If we choose $d \approx 1.2 \cdot 10^{-2}$ the agreement between the numerical and the fitted values is more than satisfactory. It should, however, be stressed that the effect of beam quality on the CARM saturation is almost negligible. The values we have assumed for the velocity spread (and hence for the corresponding μ_ε) are greatly exaggerated, since in the real CARM devices the values of μ_ε are significantly less than 1.

The problem of finding an appropriate scaling parameter in order to take into account the efficiency deterioration due to an insufficient beam quality has been addressed in [32]. Similar criteria have been exploited in the quoted paper, where the authors have employed an inhomogeneous scaling parameter proportional to the velocity spread through a coefficient depending on $I_g^{-1/2}$. In our case, in order to be consistent with the commonly accepted treatment of the U-FEL devices, we use $\mu_\varepsilon \propto I_g^{-1/3}$, and $\rho = 1/4\pi\chi I_g^{1/3}$ proportional to $I_g^{1/3}$.

The role of μ_ε is, however, manifold and allows the understanding of other parameters of pivotal importance, like the growth rate in a CARM device operating as an amplifier. As is well known, the power growth increases, while the beam is propagating along the longitudinal axis as $P(z) \propto e^{z/L_g}$, where L_g is the gain length [33]. Moreover, a CARM amplifier operating with a beam of poor quality is characterized by larger values of L_g and therefore by a longer saturation length.

We have used a procedure analogous to that exploited for the efficiency to derive the dependence of L_g on $\sigma_{\beta_\perp}$. The analysis of the numerical data supports a quadratic dependence which can be expressed as (see figure 6.10(a))

$$L_g = L_g^0\left[1 + k\sigma_{\beta_\perp}^2\right], \tag{6.68}$$

k being the fitting parameter, strongly dependent on the different values of I_g. On the other hand, the curves acquire a less dispersed behavior when plotted versus μ_ε (see figure 6.10(b)).

This is, however, not the end of the story, because, as shown in figure 6.9(a), the efficiency is extremely sensitive to the beam characteristics. An inspection of the figure shows that if the velocity spread slightly increases we may expect a significant decrement of the efficiency. In order to ensure an operation of the CARM device with a sufficiently large efficiency it is necessary to exploit a beam of electrons with

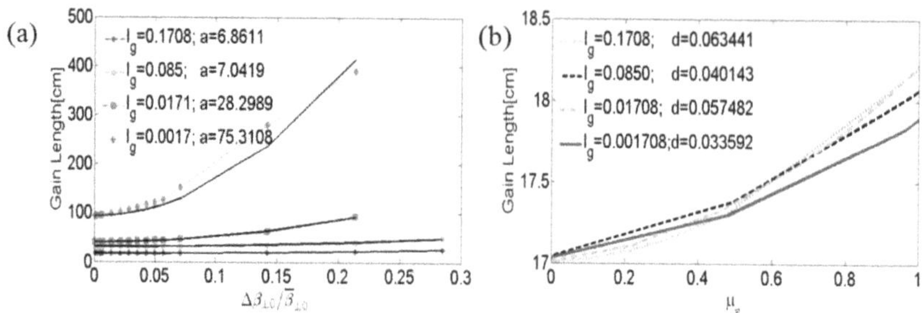

Figure 6.10. On the left side: (a) the gain length obtained from the simulation data taken from the example reported in figure 6.8 for different values of the normalized current (dot-line) each of which is fitted with a parabolic curve (continuous line); on the right side: (b) the gain length versus the inhomogeneous broadening parameter.

'reasonably' small dispersion of the energy and velocity distributions. We can obtain an upper limit to the previously quoted dispersions, by noting that:

1. The inhomogeneous line broadening induced by the longitudinal velocity spread is (see equations (6.28) with $\beta_p \approx 1$)

$$\left\langle \frac{\Delta\omega}{\omega} \right\rangle \cong \frac{\sigma_{\beta_z}}{1 - \beta_z}, \tag{6.69}$$

requiring that it be smaller than its homogeneous counterpart we end up with the following condition on σ_{β_z}

$$\frac{\sigma_{\beta_z}}{1 - \beta_z} < \frac{1}{N}, \tag{6.70}$$

2. From equation (6.21) we end up with

$$\alpha^2 \sigma_{\beta_z} + \sigma_{\beta_\perp} = \frac{1}{\bar{\alpha}} \sigma_\gamma, \tag{6.71}$$

Using therefore $\sigma_{\beta_\perp} \leqslant 10^{-3}$ (as suggested by figure 6.9) and $\sigma_{\beta_z} \leqslant 3 \cdot 10^{-3}$ (as derived from equation (6.70)) we find $\sigma_\gamma \leqslant 0.5\%$.

6.5 Transverse mode selection: operating configuration

The coefficient Ξ reported in equation (6.53) summarizes quite a complicated expression including the transverse mode structure and should indeed be characterized by the indices m, n, l labeling the TE mode coupling and the overlapping integral with the beam itself.

This last quantity defines the filling factor, which, in turn affects the gain coefficient and the gain length as well. The problem becomes more and more serious when higher order modes are considered for lasing. In this case the relevant spatial distribution is not provided by a Gaussian, covering smoothly a transverse surface as for TE_{11} mode, but by a kind of circular corona which, as already stressed, demands for an appropriate shaping of the transverse structure of the e-beam to optimize the coupling. Accordingly the beam transverse distribution should be modeled as a thin circular corona.

In 6.11 we have shown the linear part of the intensity evolution, along the z direction, together with the associated mode distribution. According to the previous remarks, it is not surprising that some collections of mode tend to grow in practically a undistinguishable way.

We must emphasize that, although figure 6.11 put the caveat that many transverse modes can be locked at saturation (thus creating problems of efficiency reduction), it should be stressed that it accounts for the fast growing root only and therefore it might lead to 'pessimistic' conclusions.

The 'degeneration' of the transverse mode evolution can however be removed by analyzing the relevant growth through the inclusion of all the roots of the dispersion equation. The transverse mode power growth is given in figure 6.12 which shows a more complicated growth pattern.

In conclusion, although the cavity is evidently over-moded (see figure 6.13), the use of a convenient shaping of the annular e-beam may allow an efficient tool of mode selection, as further discussed in the following.

Furthermore, the equations system given by the dispersion relation of the resonance condition (equation (6.18)) and the dispersion relation for the modes in a waveguide (equation (6.26)) leads to the following expression for the CARM resonance (ω_+ in figure 6.2)

$$\omega_+ = \frac{\beta_z^{-2}\Omega/\gamma - \sqrt{\omega_{\text{cutoff}}^2(1 - \beta_z^{-2}) + \beta_z^{-2}\Omega^2/\gamma^2}}{\beta_z^{-2} - 1} = \tag{6.72}$$

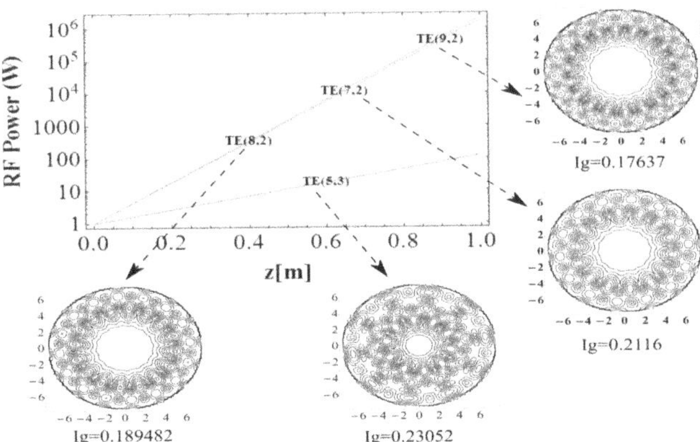

Figure 6.11. Linear growth regime (using the solution from equation (6.42)) for different transverse mode distribution and associated gain length (L_g) for a frequency resonance equal to 250 GHz. The red circle represents the e-beam transverse distribution with radius $R_b = 0.004\,8$ m, which is assumed to stay well inside the mode itself thus maximizing the filling factor.

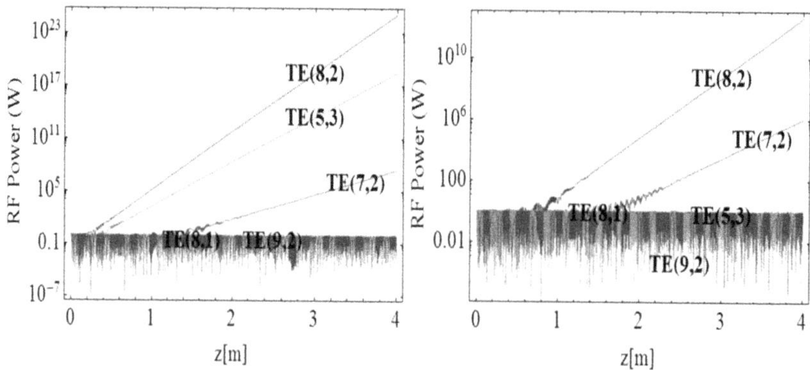

Figure 6.12. Mode intensity evolution including the lethargic part changing the radiation frequency (f) and the annular beam radius (R_b). (a) $f = 258$ GHz, $R_b = 0.004\,8$ m. (b) $f = 250$ GHz, $R_b = 0.004\,2$ m.

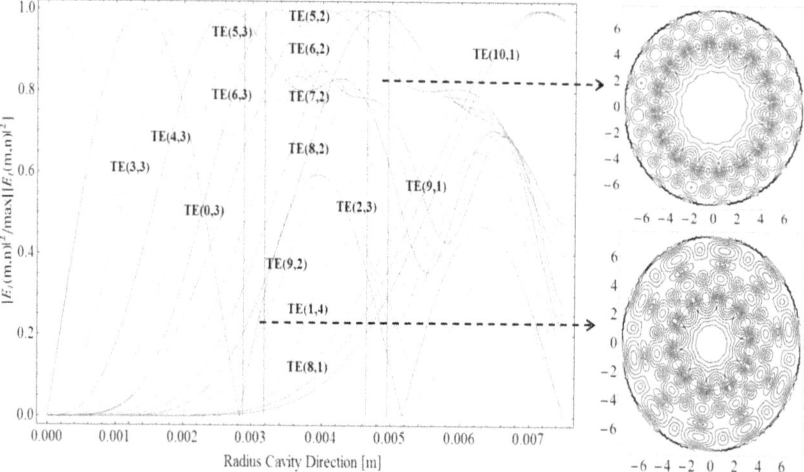

Figure 6.13. Transverse mode distribution versus the cavity radius, the vertical lines refer to beam transverse dimensions interacting with the $TE_{5,3}$ or $TE_{8,2}$ mode; on the right side the transverse mode distribution with the transverse annular beam in red.

$$= \frac{\beta_z^{-2}\Omega\sqrt{1-\beta_z^2(1+\alpha^2)} - \sqrt{\omega_{\text{cutoff}}^2(1-\beta_z^{-2}) + \beta_z^{-2}\Omega^2(1-\beta_z^2(1+\alpha^2))}}{\beta_z^{-2}-1}, \tag{6.73}$$

where $\omega_{\text{cutoff}} = k_\perp/c$ and the pitch $\alpha = v_\perp/v_{||}$.

In figure 6.14 we have reported the ω_+ values for an e-beam with particles having a longitudinal velocity spread of $\delta v_z = 0.5\%$ (with $\gamma = 2.17$ or $E_b = 650$ keV) and pitch value $\alpha = 0.53$) interacting with the mode TE_{53} of a cylindrical waveguide (having a radius $r_w = 7.5$ mm) surrounded by an axial magnetic field of 5.3 T. In this condition and in a cold cavity analysis the beam will interact with only one mode

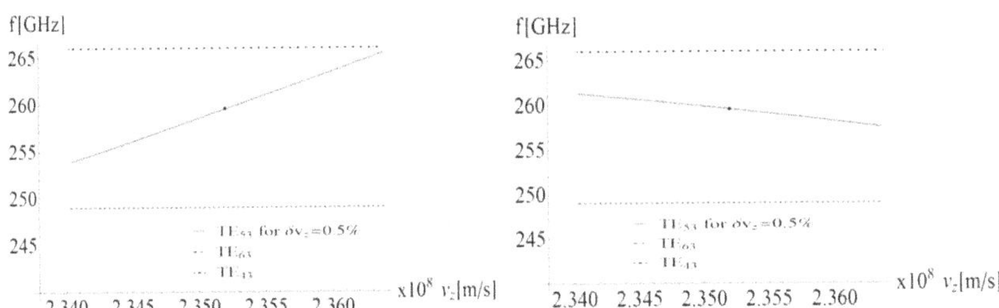

Figure 6.14. The resonance frequency range for a beam line with a velocity longitudinal spread of ($\delta v_z = 0.5\%$) interacting with the mode TE_{53} assuming a constant value for γ (equation (6.72)) (on the left side) and with a γ variation (equation (6.73)) on the right side. The resonances are well separated from those of the competitor modes TE_{63} (dotted line) and TE_{43} (dot-dashed line).

The limit for $\delta\beta_z$ can be derived starting from the approximated CARM resonance condition (equation (6.25)) coupled with the dispersion relation for the modes in a waveguide (equation (6.26)) and assuming a fixed k_z value during the interaction:

$$\delta\beta_z < (1 - \beta_z)^2 \frac{c\delta k}{\Omega} \approx (1 - \beta_z)^2 \frac{\gamma c}{\Omega_c} \frac{\nu_{mn} - \nu_{m'n'}}{R_w}, \tag{6.74}$$

where R_w is the waveguide radius and $(\nu_{mn}/R_w, \nu_{m'n'}/R_w)$ the perpendicular wave vectors of the mode closest to the operating mode.

Formula (6.74) could be refined considering also the induced spread on γ by the particles

$$\delta\beta_z < (1 - \beta_z)^2 \frac{c}{\Omega_c} \frac{\nu_{mn} - \nu_{m'n'}}{R_w} \frac{1}{\dfrac{1}{\gamma} - \gamma(1 - \beta_z)\beta_z(1 + k^2)}, \tag{6.75}$$

in this way for the example of figure 6.14 the limit for $\delta\beta_z$ is 0.55%.

Furthermore, considering the expression for the electron energy

$$E = |e|V + m_e c^2 = \sqrt{(m_e c^2)^2 + c^2(p_z^2 + p_\perp^2)}, \tag{6.76}$$

the relativistic factor is given by

$$\gamma = \frac{|e|V}{m_e c^2} + 1. \tag{6.77}$$

By combining the equations on the frequency selection (6.28) (with $\beta_{\mathrm{ph}} \approx 1$) and the Hamiltonian (6.76) we end up with the following condition on the magnetic field necessary to get the resonance condition

$$B = \frac{m_e \omega_r}{|e|} \frac{1 + \bar{a}^2}{2}, \tag{6.78}$$

where $\bar{a} = \gamma\beta_\perp$ is the corresponding strength of the undulator for an FEL device.

Table 6.2. Preliminary design parameters.

Beam current	5 A
Beam voltage	650 kV
Longitudinal and transverse velocity spread	< 0.5%
Energy spread	< 0.5%
Magnetic field	5 T

Requiring $f_r = \omega_r/(2\pi) \approx 250$ GHz we find a corresponding magnetic field intensity of the main coil $B \approx 5$ T.

We can finally summarize in table 6.2 the design request for the e-beam spread and the magnetic field for a given cavity mode operation and e-beam energy.

The numbers reported in the previous table fix the conditions for a safe operation of the CARM device, but do not specify the form of operation, which can be either an amplifier or an oscillator. Within the present framework, the latter is more convenient for various reasons: it removes the quest for an input source and input couplers, requires a shorter length of the interaction region, with the consequent need of a long high field intensity magnet. Furthermore, the amplifier operation demands for the suppression of the backward-wave instability.

In order to ensure the CARM oscillations at the desired Doppler shifted frequency, it will be necessary to design the system in such a way that the relevant threshold current be less than that of the competing modes. This will be done by an appropriate choice of the beam and cavity parameters, as properly discussed in the next chapter.

6.6 CARM oscillator and cavity design: numerical simulation

In the previous sections we have commented on the high value efficiency as the pivotal requirement to be reached with a CARM device to be a useful tool for magnetic fusion purposes. This request necessarily implies the need of producing and transporting a very good beam in terms of energy spread and longitudinal velocity spread. The transport problem will be treated in a section of chapter 7 with an appropriate gun design since now we would like to pay attention to the oscillator configuration for which at the high quality beam we must add a high selectivity cavity interaction in order to preserve the high system efficiency.

The crucial requirements that must be accomplished by the facility we have in mind is a generation of a millimeter wave operating in the frequency range of ~250 GHz carrying a power of ~1 MW in continuous wave (CW) operation reaching a wall-plug efficiency of ~30%.

The design of a CARM cavity is a very difficult problem, already addressed in [34], where different options have been analyzed and compared. The quasi optical solution is inappropriate for the configuration we are going to analyze, as in order to reach a reasonable power density on the mirror surface (2 kW cm^{-2}) the distance among them should be greater than 2 m and the alignment of three distinct sections (two mirrors with the smooth cylindrical section) becomes an arduous task.

The solution on which we are going to land, consisting of a smooth cylindrical waveguide section delimited by two Bragg reflectors, is close to that used in the last CARM experiment [11] where the upstream Bragg, to save the gun system from radiation, has been substituted by a taper section.

Concerning this last solution, if one side reduces the problems with the machining, the other decreases the cavity selectivity allowing the excitation of the spurious modes. We would like to underscore that with such a configuration, even using a beam with a relative velocity spread of 8%, a considerable level of efficiency 26% was achieved during the experiment [11].

In order to fulfill the requirements mentioned above, in particular for the CW operation to prevent the discharge, the electric field at the surface must be less than 10 kV mm^{-1}. Furthermore, the high efficiency value can be maintained with a power wall load less than 2 kW cm^{-2} avoiding the cavity deformation and the confined power density inside the cavity must be less than 500 kW cm^{-2} [35].

The requirements merged with the previous discussed constraint lead directly to the following cavity parameters: the Doppler shift should be 3–4 times the cyclotron frequency Ω (varying in the range 65–75 GHz) to operate in a region where the mode overlapping is contained (see figure 6.15).

The traveling modes in a cylindrical waveguide with radius r_w are determined by the dispersion relation (6.26)

$$\frac{\omega^2}{c^2} = (\nu_{mn}/r_w)^2 + k_z^2, \tag{6.79}$$

where ν_{mn} is the nth zero of the Bessel function J_m.

In order to limit the operating mode numbers, by considering the confined power density limitation, we get the variation range for the cavity cross section among 1.5–2.4 cm^2.

Using the parameters reported in table 6.3 the system equations of (6.25) and (6.26) lead to two cavity operating modes at ~250 GHz (TE_{53} or TE_{82}) whose choice will be analyzed in the next section; now we focus attention on the numerical support to the cavity design.

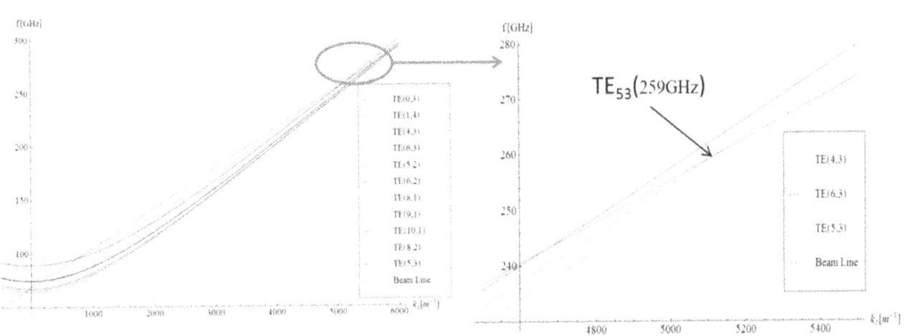

Figure 6.15. Brillouin diagram.

Table 6.3. Oscillator CARM parameters.

Cathode voltage	650 kV
Magnetic field	5.3
Pitch ratio ($v_{\perp}/v_{\parallel}$)	0.53
Cavity radius	7.5 mm

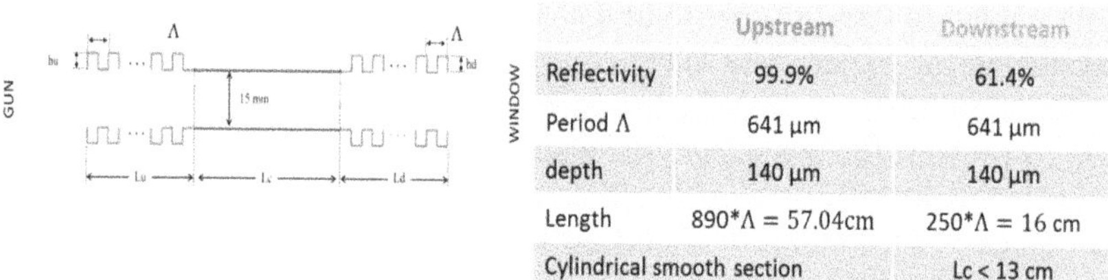

	Upstream	Downstream
Reflectivity	99.9%	61.4%
Period Λ	641 μm	641 μm
depth	140 μm	140 μm
Length	890*Λ = 57.04cm	250*Λ = 16 cm
Cylindrical smooth section		Lc < 13 cm

Figure 6.16. Cavity design with the main parameter values.

The choice of a more compact cavity, with radius of 1.5 cm and a length of ~80 cm, allows a more efficient dissipation of the RF power (see figure 6.16) even if in this case the machining must be supported by an appropriate cold test to verify the goodness of the reflectivity properties for the operating and competitors modes. In fact, the most urgent problem when dealing with the chosen cavity configurations is the competition between the selected operating mode and the neighboring parasitic (spurious) modes.

The radio frequency circuit design must be supported by a numerical modeling allowing one to accurately simulate the intra-cavity beam–wave interaction. The large length of this cavity (80 cm) compared to the wavelength under investigation impairs the effectiveness of the classical particle-in-cell (PIC) codes, due to the huge amount of computer memory and CPU time required by the resulting large mesh sizes. In parallel to the development of the full numerical treatment, envisaged for a 'start to end' simulation, we have undertaken the cavity design by the use of a twofold strategy. We have indeed developed a 'home-made' 1D code GRAAL and extended to the CARM case the 'universal' scaling formulae, already developed for undulator U-FELs [26].

The numerical code, referred to as GRAAL (which stands for Gyrotron Radiation Auto-Resonance Amplified Laser), is based on a self-consistent procedure developed by several authors [4, 23, 24, 36, 37], while the scaling formulae have been derived after a proper comparison between U-FEL and CARM theories [26, 31] as discussed in the previous chapter.

The two procedures have been cross-checked, benchmarked with the numerical predictions available in literature, with the experimental results from the MIT (Massachusetts Institute of Technology)-CARM test facility [25], operating at low frequency (35 GHz) and the commercial PIC code (CST Microwave Studio®) simulation whose parallel version, working with a graphics processing unit (GPU), has been installed at CRESCO HPC facilities. In the case of a CARM amplifier, namely the simplest geometry without Bragg reflectors, experimental data from the Massachusetts Institute of Technology (MIT) [25] have been compared with semi-analytical formulae, GRAAL, the commercial PIC code by CST Microwave Studio® and the CSPOT code by MIT. Measurements are overlapped with predicted curves in figure 6.17 showing a reasonable agreement among the simulation tools and providing some confidence on the reliability of the method.

The approach we are going to describe is based on the assumption that the beam–wave interaction, resulting in the electromagnetic wave generation, takes place mainly in the smooth cylindrical section of the cavity (L_c in figure 6.16). The assumption is also supported by the design of an adiabatic magnetic field compression at the upstream mirror region which allows one to gradually shrink the beam radius, thus inducing a variation in the resonance condition, which prevents the growth of the gyrotron modes, before the constant field profile is reached in the smooth cylindrical section (see figure 6.18).

A PIC simulation has been performed using CST Microwave Studio® considering only the upstream mirror with an appropriate analytical tapered magnetic field which allows one to have a well-defined beam radius $Rw = 2.9$ mm at the entrance of the smooth section 6.18.

In figure 6.19 the results of the simulation show a substantial reduction of the output signal using a tapered magnetic field, in particular the *TE* mode disappears, only the *TM* mode still remains with a low signal level which can be canceled using a slotted cavity, as will be discussed later.

Before concluding the section it is useful to point out the reason why the operating mode TE_{53} has been chosen analyzing the graphs of figure 6.20. The resulting analysis of the beam-wave interaction in a cold cavity, using the parameters of table 6.3, show that the maximum of the coupling coefficient for the co-propagating mode TE_{53} is located in a more inner radial position than mode TE_{82} (see figure 6.20(b)) decreasing the power load on the cavity surface.

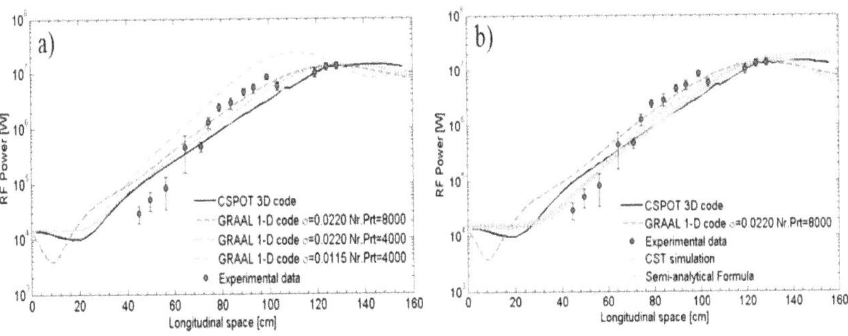

Figure 6.17. Experimental results of a 35-GHz CARM amplifier from MIT compared with: (a) the CSPOT code by MIT and the homemade GRAAL code (varying σ, the velocity spread of the beam particles, and the number of particles (Nr. Prt.) used in the simulation) and (b) the PIC commercial code by CST Microwave Studio and the semi-analytical formulae.

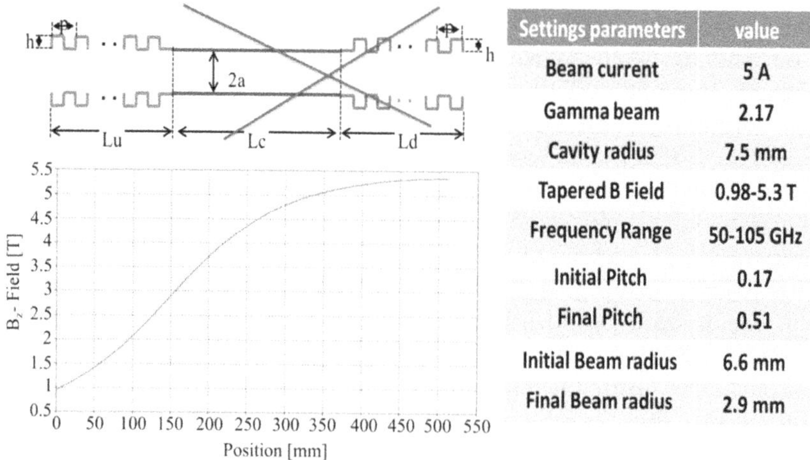

Settings parameters	value
Beam current	5 A
Gamma beam	2.17
Cavity radius	7.5 mm
Tapered B Field	0.98-5.3 T
Frequency Range	50-105 GHz
Initial Pitch	0.17
Final Pitch	0.51
Initial Beam radius	6.6 mm
Final Beam radius	2.9 mm

Figure 6.18. The simulation geometry scheme and setting parameters for a tapered magnetic field.

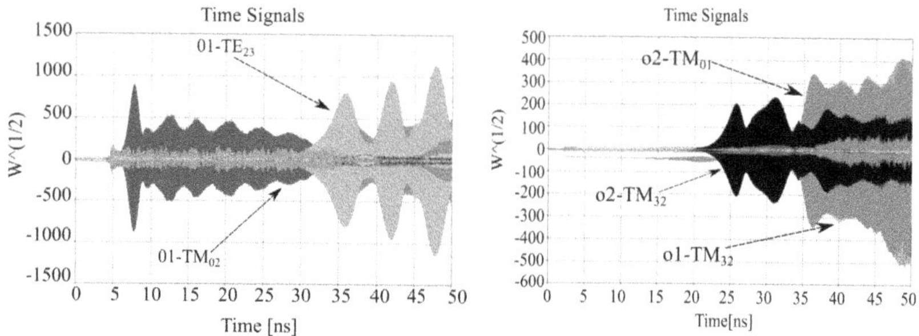

Figure 6.19. The output signal at entrance (01) and at exit (02) of the beam for the first 100 modes propagating inside the cavity using a constant magnetic field (on the left side) and a varying magnetic field (on the right side).

Furthermore, the maximum linear growth signal of an annular beam, with 300 μm of thickness, is at the radial position which maximizes the coupling coefficient, as reported in figure 6.20(c), (d) for mode TE_{53} at the gyrotron (95 GHz) and CARM (259 GHz) resonance, respectively.

The previous discussion leads to the choice of TE_{53} for the operating mode as it allows having more space to play with the tapering magnetic field avoiding the growth signal of the gyrotron mode.

6.7 Operating mode selection: *Q*-factor, starting current and cavity length

In an oversized cavity, like that we are going to analyze, a great number of modes (≈400) with dense spectrum can be excited, as shown in figure 6.21, which presents many possible intersections of their dispersion characteristics (Brillouin diagrams)

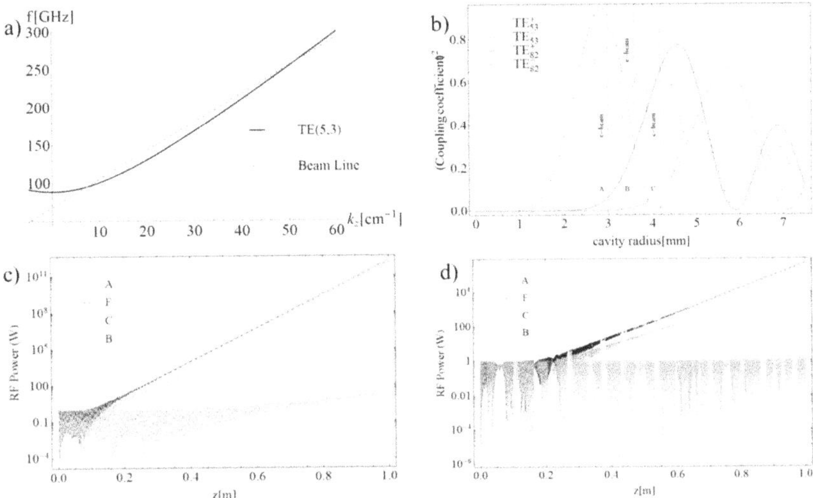

Figure 6.20. In (a) the beam–wave interaction for the parameters of table 6.3 and the coupling coefficient versus the cavity radius for co-counter propagating modes TE_{53} and TE_{82}. The linear growth signal, linearizing the Vlasov–Maxwell equation, for mode TE_{53} at gyrotron resonance (95 GHz) in (c) and at CARM resonance (259 GHz) in (d). To evaluate, the growth signal has been considered an annular beam with 300 μm of thickness centered at different radial positions A, B, C as reported in (b) related to the maximum of the coupling coefficient. F corresponds to the radial position (r_F = 3.1 mm) where the transversal electric field of the mode TE_{53} is maximized.

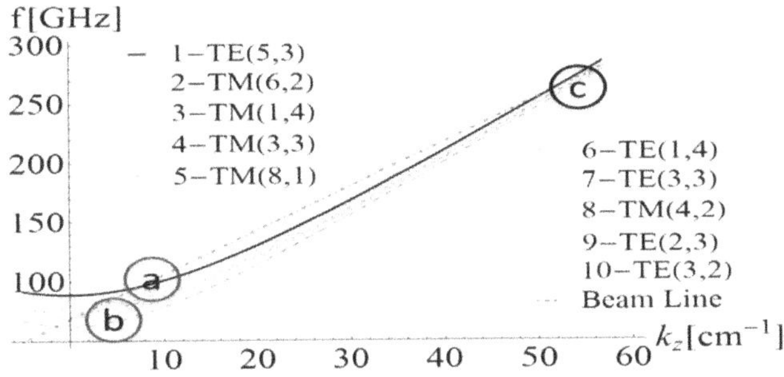

Figure 6.21. Brillouin diagram of a particle beam (γ = 2.17, $v_\perp/v_\parallel$ = 0.53) interaction with modes of a cylindrical cold cavity with radius R_w = 0.75 cm surrounded by an external axial magnetic field B_0 = 5.3 T.

with the beam line. This makes the mode selection and correspondingly the single-mode operation a difficult task. The beam–wave interaction in a cold cavity for a CARM oscillator opens three kinds of problems, as highlighted in figure 6.21:

(a) the suppression of the beam–wave intersection for the operating mode at low frequency;
(b) the suppression of the excitation of the competitor mode near cutoff;
(c) the suppression of the modes competition near the Doppler up-shift interaction.

Starting from point (a) we must note that the signal at low resonance exhibits the undesirable feature of growing faster than the signal at CARM resonance (see figure 6.20(c), (d)). Furthermore, the temporal growth rate decreases monotonically as the resonance frequency increases away from the waveguide cutoff frequency.

This can be easily achieved starting from the dispersion relation (6.36), perturbing the central resonance frequency ω_0 with $\omega = \omega_0 + \Delta\omega$ and deriving the following expression

$$\begin{aligned}&\Delta^4\omega + 2(\Omega_s + k_z v_z)\Delta^3\omega + (\Omega_s - k_z^2 + 2\Omega_s k_z + k_z^2 v_z^2 - k_{mn}^2 + \bar{\epsilon} k_{mn}^2)\Delta^2\omega + \\ &+ 2(\Omega_s + k_z v_z)\bar{\epsilon} k_{mn}^2 \Delta\omega + (\Omega_s^2 - k_z^2 + 2k_z v_z + k_z^2 v_z^2)\bar{\epsilon} k_{mn}^2 = 0,\end{aligned} \tag{6.80}$$

where $\Omega_s = s\Omega$ and $\epsilon_{\rm mnns}^{TE} = \bar{\epsilon}/\beta_\perp$ is the coupling parameter of the beam with the cold-cavity mode TE_{mn}.

The fourth degree algebraic equation (6.80) has two real solutions and two complex conjugate roots whose imaginary part versus k_z is plotted in figure 6.22 (solid red line) together with the dispersion curve of the mode TE_{53} (solid black line) and the dashed beam line. The parameters of the e-beam are those reported in table 6.3.

It is evident from figure 6.22 that the lowest frequency solution exhibits the largest growth rate, therefore, this is the mode competing with the desired CARM operation and, once coupled to the beam, reaches the saturation and suppresses its high frequency counterpart.

In the case of CARM oscillator, a possible solution to prevent the onset of the gyrotron mode is the design of an appropriate cavity with selective losses, inhibiting the growth of the lower-frequency mode. In particular, for a fixed mode the threshold current for the growth of the oscillations can be defined in terms of the parameters of the device including the gain and cavity losses. The energy W of the electromagnetic field stored in the cavity with a quality factor Q is related to the beam power P_b by the equation [1]

$$\eta_a P_b = \omega \frac{W}{Q}, \tag{6.81}$$

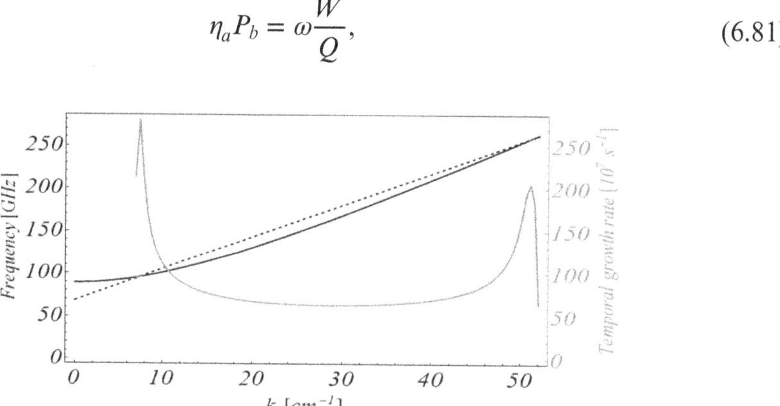

Figure 6.22. The frequencies (left axis) corresponding to the intersection of the dashed beam line with the dispersion curve of the operating mode TE_{53} (solid black line) and the temporal growth rate of the generated signal (right axis).

where η_a is the efficiency of the interaction at an operating frequency ω. The threshold current ensuring the onset of the oscillations and eventually the saturation is accordingly given by (see [5])

$$I_{\text{thr}} = I_A \gamma_0 \beta_z^2 \frac{\pi}{2\lambda} \frac{R^2}{L_c} \frac{1}{QG\tilde{\chi}(\Phi_k)}, \tag{6.82}$$

where I_A is the Alfvén current, G is the beam–wave coupling current coefficient, $\tilde{\chi}(\Phi_k)$ is the gain function, including the beam quality effects, as will be discussed later in this section, and L_c is the interaction length, given by the length of the smooth section. The beam–wave interaction in the upstream mirror region is avoided by an adiabatic magnetic field compression in order to shrink gradually the beam radius, as discussed in the previous section.

The previous identity defines the minimum current, which is necessary in order to overcome the losses and to reach a saturation. It also contains geometrical (L_c) and dynamical factors like the gain function $\tilde{\chi}(\Phi_k)$.

The strategy we will follow is that of modeling the device parameters in order to increase the threshold current of the gyrotron mode and to prevent any competition with the CARM operation.

The underlying optimization procedure is based also on a reasonable compromise between the gain and the efficiency, which exhibits an opposite behavior with increasing the L_c values: a decrease of the gain leads to an increase of the efficiency.

It is furthermore important to stress that, according to [5], the quality factor Q of the cavity can be written as

$$Q = \frac{\omega L_c}{v_{\text{gr}}(1 - R_1 R_2)}, \tag{6.83}$$

R_1, R_2 being the reflection coefficients at both sides of the cavity. In the case of the operation at a gyrotron mode, where $k_z \sim 1/L_c$, we can write [5]

$$Q_G \sim 4\pi \left(\frac{L_c}{\lambda_G} \right)^2, \tag{6.84}$$

which is the minimum diffractive Q_G, obtained neglecting the reflectivity. Equation (6.84) shows that Q_G is a large quantity thus, demanding for a small starting current. Our strategy is therefore that of adjusting the kinematic conditions to shift away from the gyrotron mode the interaction ω^- (see figure 6.2) by increasing of the staring current while preserving an adequate value of the system efficiency.

The optimization procedure, for fixed beam parameters (α, γ, beam current I_0) and magnetic field intensity in cavity B, involves the following steps [38]:

First step: Select an optimal length L_{opt} of the regular cylindrical section of the cavity which optimizes the system efficiency using the following considerations. The CARM efficiency, given by the product of the single-particle efficiency (η_{sp}) and the transverse efficiency ($\eta_\perp$), is close to the η_{sp} if the number of cyclotron turns $N_t = L/(v_z \Lambda_B)$ in the interaction space satisfies the following expression [5]

$$N_{\text{opt}} \sim \frac{2}{\beta_\perp^2}\frac{(1-\beta_z/\beta_{\text{ph}})^2}{(1-\beta_{\text{ph}}^{-2})}, \tag{6.85}$$

from which we derive L_{opt}, using $\Lambda_B = 2\pi c/\Omega$.

Second step: Adjust the beam energy and/or the external magnetic field B in order to shift the ω^- intersection (see figure 6.2) away from the cutoff and to reduce the Q^G value for the gyrotron interaction (in this case the formula given by the (6.84) is not valid) finding a $\bar{B}$ value for which the CARM starting current $I^C_{\text{start}}(\gamma(\bar{B}), L_{\text{opt}})$ is less than the gyrotron starting current $I^G_{\text{start}}(\gamma(\bar{B}), L_{\text{opt}})$. Herein the upper indices G, C stand for gyrotron and CARM modes, respectively. Therefore, for a fixed up-shifted beam–wave interaction (ω^+ in figure 6.2), we change the magnetic field or γ in the range 2–2.4 (which is related to the non-relativistic cyclotron frequency Ω_e by $\Omega_e = \gamma(\omega^+ + k_z^+ v_z)$), in order to increase the oscillation threshold current for the gyrotron mode. In particular, evaluating the starting current according to (6.82) for both of the intersections $\omega^-(I^G_{\text{start}})$ and $\omega^+(I^C_{\text{start}})$ we define the following quantity

$$F(\gamma(B), L_{\text{opt}}) = \frac{I^G_{\text{start}}(\gamma(B), L_{\text{opt}})}{I^C_{\text{start}}(\gamma(B), L_{\text{opt}})} = \frac{\lambda_C Q_C(L_{\text{opt}})\chi_C(\gamma(B), L_{\text{opt}})}{\lambda_G(\gamma(B))Q_G(L_{\text{opt}})\chi_G(\gamma(B), L_{\text{opt}})}. \tag{6.86}$$

Here χ_G and χ_C are the gain in the small-signal approximation for ω^-(G) and ω^+(C), respectively. Then, for a fixed CARM interaction (see the Brillouin diagram in figure 6.2), i.e. for a fixed relativistic cyclotron frequency Ω_c, we change the magnetic field B inside the cavity in order to satisfy the condition $F(\gamma(H), L_{\text{opt}}) > 1$. From it, we can evaluate the maximum value that can be reached for a given ratio $Q_G(\gamma(H), L_{\text{opt}})/Q_C(\gamma(H), L_{\text{opt}})$ since

$$\frac{Q_G(\gamma(B), L_{\text{opt}})}{Q_C(\gamma(B), L_{\text{opt}})} < \frac{\lambda_C(\gamma(B))\chi_C(\gamma(B), L_{\text{opt}})}{\lambda_G \chi_G(\gamma(B), L_{\text{opt}})} = \Gamma(\gamma(B), L_{\text{opt}}). \tag{6.87}$$

Third step: We determine the Q_C, using the formula describing the dynamics of the CARM oscillator reported in [26]. Then, we select the value of $\bar{B}$, which according to (6.87) yields an acceptable Q_G and satisfies the condition $I^C_{\text{start}}(\gamma(\bar{B}), L_{\text{opt}}) < I^G_{\text{start}}(\gamma(\bar{B}), L_{\text{opt}})$.

Starting from $\chi_C(\gamma(B), L_{\text{opt}})$, the saturation intensity I_s and considering the ratio $\zeta = \eta_p/\eta_a$ between the passive η_p and active η_a losses, we can determine the equilibrium intracavity power density I_e given by the relation [26]

$$\begin{aligned} I_e &= C\left(\sqrt{\frac{1-\eta_a(1+\zeta)}{\eta_a(1+\zeta)}\chi_C(\bar{B}, L_{\text{opt}})} - 1\right)I_s, \\ C &= (\sqrt{2}+1), \end{aligned} \tag{6.88}$$

from which we determine the Q_C specifying an output power of $P_{\text{out}} = 1$ MW and using the following equations

$$P_{\text{out}} = \eta_a I_e,\ Q_C = 2\pi\nu \frac{L_{\text{opt}}}{c\eta_a}. \tag{6.89}$$

The underlying optimization is sketched in the block diagram in figure 6.23. The search for an optimum interaction length is based on a compromise between gain and efficiency for a given output CARM oscillator power.

It should be noted that (6.88) is a result of an assumption (only partially supported by the numerical analysis) that the small signal gain of CARM and U-FEL devices exhibits the same dependence on the intra-cavity power.

Major attention must be paid to point (b) of figure 6.21 for which the increase of the starting current is not sufficient for suppressing the modes growth excited by the beam near the cut-off where *TM* modes play a crucial role too.

In figure 6.24 it has been put in evidence how large is the linear growth signal for the TM_{42} at 70 GHz (near cutoff) with respect to the operating mode TE_{53} at 259 GHz. The starting current for TM modes is minimized at the cutoff, since the group velocity of the wave is well separated from the particle velocity, thus increasing the beam–wave energy exchange. The opposite condition occurs at CARM resonance implying an infinity value for the starting current.

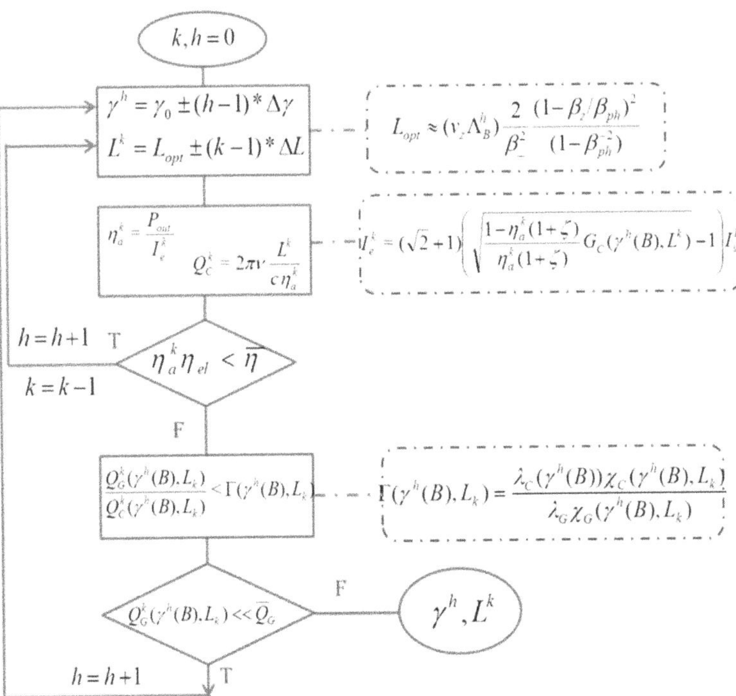

Figure 6.23. Block diagram describing the procedure allowing to fix the beam energy and the smooth cylindrical cavity section length limiting the growing of the lower beam intersection with the operating mode.

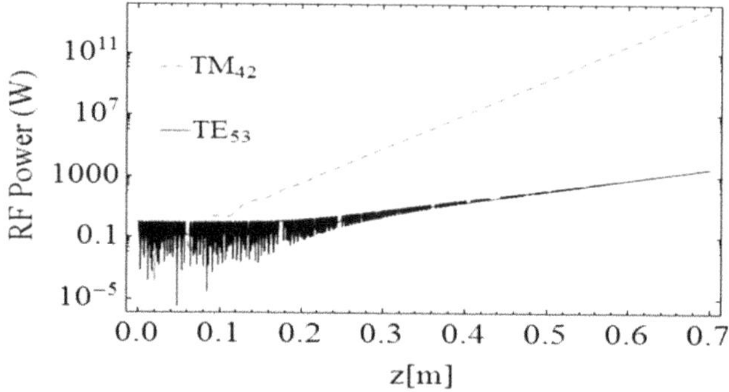

Figure 6.24. The linear growth signal for an e-beam (with $\gamma = 2.17$, $v_{\perp}/v|| = 0.53$, $I = 8\text{A}$) interacting with the modes of a cylindrical cavity of radius $R = 7.5\text{cm}$.

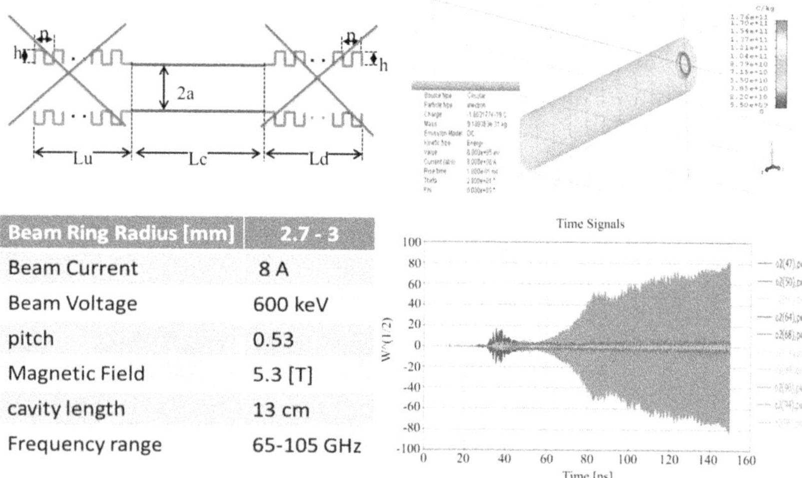

Beam Ring Radius [mm]	2.7 - 3
Beam Current	8 A
Beam Voltage	600 keV
pitch	0.53
Magnetic Field	5.3 [T]
cavity length	13 cm
Frequency range	65-105 GHz

Figure 6.25. PIC simulation results on CRESCO ENEAGRID with two GPUs (upper-left) simulation scheme without the Bragg reflectors, (upper-right) geometry design with the particles e-beam, (lower-left) values settings of the main parameters, (lower-right) largest output modes signals, the most dangerous (in brown) is represented by the mode TM_{42}.

In order to study the cavity mode response at the cutoff frequency range due to the beam–wave interaction a CST simulation has been performed considering only the smooth cylindrical section. This reasonable approximation takes into account the fact that the Braggs has been designed to work close to the CARM resonance and the beam–wave interaction will be drastically reduced tapering the magnetic field inside the Braggs (as previously discussed, in particular see figure 6.19).

In figure 6.25 is shown: in the upper side the layout of the simulation with only the smooth cylindrical section which allows reducing the computational requirement and simulating the structure on CRESCO ENEAGRID and in the lower side the setting parameters with the most significative output modes signals. For each of those modes the cavity quality factor Q and the starting current have been evaluated and are reported in the table of figure 6.26. The *TM* mode, red encircled in figure 6.26, represents the most dangerous competitor mode due to the low value of the starting current and the high growth factor (see figure 6.24). The only way to suppress this *TM* mode is by using a slotted cavity.

A cylindrical cavity with a short longitudinal or transversal cut on the boundary surface will avoid the propagation of the competing modes *TE* or *TM*, respectively (see figure 6.27). The price that we must pay is a reduction of 30% of the system overall efficiency due to the linear polarization of the operating mode induced by the cut.

A further step towards a more definite design of the a CARM oscillator is accomplished by running full-wave simulations of circular waveguides with either transversal or longitudinal slots as depicted in figure 6.27. Some CARM experiments in the past have been impaired by the excitation of gyrotron modes, namely, waveguide modes intersecting the beam line very close to their cutoff frequency. The most dangerous gyrotron mode of the configuration we are considering is the TM_{42}

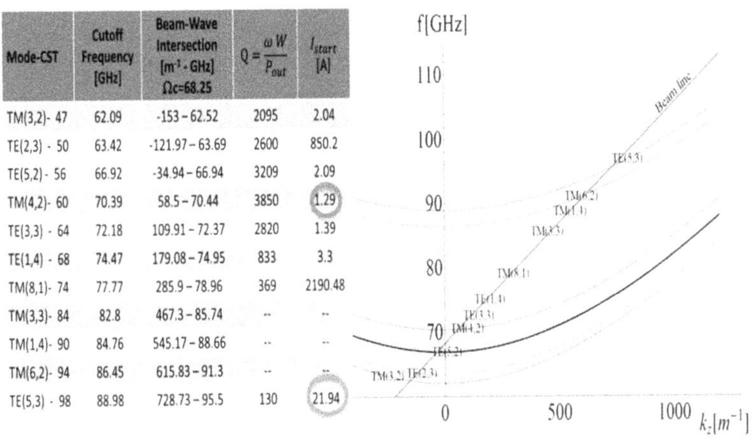

Mode-CST	Cutoff Frequency [GHz]	Beam-Wave Intersection [m^{-1} - GHz] Ωc=68.25	$Q=\frac{\omega W}{P_{out}}$	I_{start} [A]
TM(3,2)- 47	62.09	-153 - 62.52	2095	2.04
TE(2,3) - 50	63.42	-121.97 - 63.69	2600	850.2
TE(5,2)- 56	66.92	-34.94 - 66.94	3209	2.09
TM(4,2)- 60	70.39	58.5 - 70.44	3850	1.29
TE(3,3) - 64	72.18	109.91 - 72.37	2820	1.39
TE(1,4) - 68	74.47	179.08 - 74.95	833	3.3
TM(8,1)- 74	77.77	285.9 - 78.96	369	2190.48
TM(3,3)- 84	82.8	467.3 - 85.74	--	--
TM(1,4)- 90	84.76	545.17 - 88.66	--	--
TM(6,2)- 94	86.45	615.83 - 91.3	--	--
TE(5,3) - 98	88.98	728.73 - 95.5	130	21.94

Figure 6.26. On the left side table is reported the modes with a significant output signal derived by the PIC simulation results reported in figure 6.25; for these modes has been calculated the *Q*-factor, using the transient module of the CST Microwave Studio, and the related starting current from equation (6.82). The chosen system parameters allow increasing the starting current for the low resonance of the operating mode TE_{53} but great attention must be paid to some gyrotron modes, the most dangerous is the TM_{42} which has a very low starting current in this configuration.

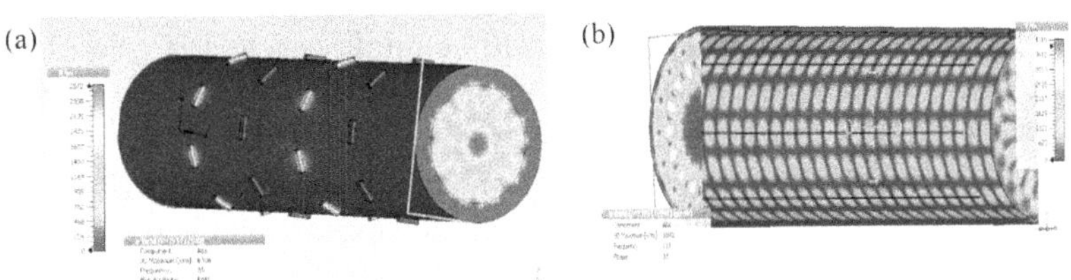

Figure 6.27. Circular waveguide with transversal (a) and longitudinal (b) slots. Light blue identifies vacuum parts, while other colors are used to indicate metals (Courtesy of Dr Gian Luca Ravera.)

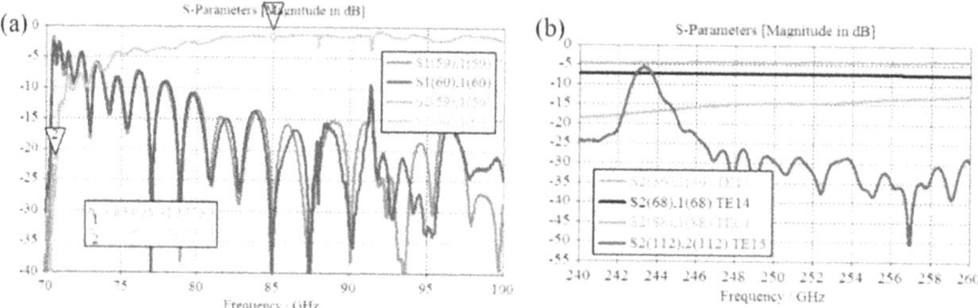

Figure 6.28. (a) Main scattering parameters across the ENEA CARM cavity with five rings of eight transversal slots: both polarization of *TM42* mode, i.e. mode numbers *nn* = 59 and 60, are impaired when transmitted (S2(*nn*),1(*nn*)); adjacent rings are rotated 22.5 degrees. (b) Main transmission parameters of some unwanted modes in a circular waveguide with internal radius of 2.92 mm and five longitudinal slots (Courtesy of Dr Gian Luca Ravera.)

(see figure 6.26), whose propagation can be suppressed by providing waveguide walls with transversal slots. On the other hand, longitudinal slots mostly affect transverse electric modes and can be designed to select a single polarization of the working mode as well as to damp modes with azimuthal index different from the one of the operational mode. Some outcomes of preliminary optimizations carried out for slotted waveguides are reported in figure 6.28.

For the case of transversal slots, the behavior of a waveguide section is studied around the frequency where the dispersion curve of the TM_{42} mode intersects the beam line. This frequency is lower than the operational frequency of the ENEA CARM, leading to a reasonable computation time if the smooth cylindrical section is simulated by means of two GPUs TESLA K40m 2x10 in CRESCO. When moving to 250 GHz, the computational load increases significantly; therefore, a waveguide with smaller diameter has been simulated. Despite this geometry differing from the actual cavity, it equally provides a meaningful, though preliminary, assessment of the effects of longitudinal slots on unwanted modes.

The mode competition of the beam–wave interaction at CARM resonance, third point (c) in figure 6.21, can be cured generating a high quality beam with a limited longitudinal velocity spread whose estimation is given in equation (6.75) (see figure 6.14).

Before concluding this section we must put in evidence that for the configuration under study the spurious dangerous mode could be also the TE_{52} with a low starting current (as reported in figure 6.26) and with the same azimuthal index of the operating mode (TE_{53}) which does not allow suppressing this mode with a slotted cavity. Regardless, the results achieved with a multi-mode PIC simulation performed with CST at frequency near the cutoff show that the output signal for the gyro-backward TE_{52} is much less than output signal for the TM_{42} as reported in figure 6.29. However, if with the simulation including the Braggs and enlarging the frequency range till the CARM resonance we achieve a significant growth of the TE_{52} mode, we can control the effect by a benchmarking of the magnetic field inside the optical cavity. In this way near the cutoff we can drive the interaction of the

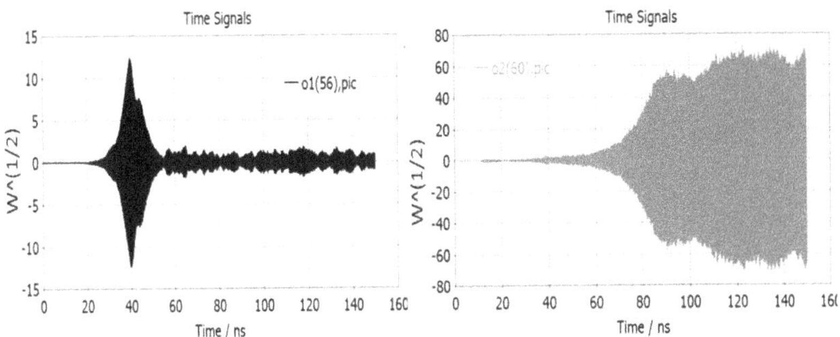

Figure 6.29. The output signal for the gyro-backward mode TE_{52} (on the left) and the TM_{42} mode (on the right): the results have been achieved with a multi-mode PIC simulation for 13 cm of a smooth cylindrical cavity using the parameters set reported in the table of figure 6.25 and considering all the modes having the cutoff between 11–90 GHz .

beam line with a mode having a different azimuthal index of the operating mode and which can be suppressed with an appropriate slotted cavity.

In the following section we will discuss the effect of the energy spread on the starting current.

6.8 Starting current and energy spread

The gain function in the equation (6.82) for a CARM interaction in low gain regime is given by the following expression [5]

$$\begin{aligned}\chi(\Phi_k) = &-\left(1 - \frac{\beta_z}{\beta_{ph}}\right)\phi(\Phi_k) \\ &+ 3\frac{\beta_\perp^2}{2\beta_z\beta_{ph}}[\phi(\Phi_k) + \Phi_k\phi(\Phi_k)'] - \frac{\beta_\perp^2(1 - \beta_{ph}^{-2})Z_{out}}{2}\phi(\Phi_k)',\end{aligned} \tag{6.90}$$

where

$$\phi(\Phi_k) = Re\left[\int_0^1\left(\int_0^\xi e^{i\Phi_k(\xi - x)}dx\right)d\xi\right] = \frac{1 - \cos(\Phi_k)}{\Phi_k^2}, \tag{6.91}$$

characterizing the spectrum of the wave acting on the electrons with constant amplitude in the spatial range $0 < x < Z_{out}$ being $Z_{out} = \omega L/v_z$, $\Phi_k = \delta_0 Z_{out}$, ω the resonance frequency and δ_0 the detuning parameter

$$\delta_0 = 1 - \frac{\beta_z}{\beta_{ph}} - \frac{\Omega_c}{\omega}. \tag{6.92}$$

The inclusion of the gain distortion effects due to the velocity spread can be easily done by convolving the function $\chi(\Phi_k)$ on the velocity distribution reported below.

The relevant procedure is sketched below. The beam velocity spread effect on the detuning parameter δ_0 is taken into account through the ϕ function

$$\phi(\Phi_k + \delta\Phi_k) = Re\left[\int_0^1 \left(\int_0^{\xi} e^{i(\Phi_k - NK\epsilon)(\xi - x)} dx\right) d\xi\right], \tag{6.93}$$

where

$$N = \frac{L}{\Lambda\beta_z}, \; \Lambda = \frac{2\pi c}{\Omega}, \; K = \frac{\omega_R \Lambda}{c\gamma^2\beta_{ph}\beta_z} \text{ and } \epsilon = \frac{\Delta\gamma}{\gamma}.$$

The assumption of a Gaussian energy distribution for the beam velocity spread modifies the ϕ function in the following way

$$\begin{aligned} \phi(\Phi_k^{\sigma_\varepsilon}) &= Re\left[\frac{1}{\sqrt{2\pi}\sigma_\varepsilon}\int_{-\infty}^{\infty} \phi(\Phi_k + \delta\Phi_k) e^{-\frac{\varepsilon^2}{2\sigma_\varepsilon^2}} d\varepsilon\right] = \\ &= Re\left[\int_0^1 (1 - t) e^{i\Phi_k t - \frac{(\pi\tilde{\mu}_\varepsilon)^2 t^2}{2}} dt\right], \end{aligned} \tag{6.94}$$

where μ_ϵ is the inhomogeneous broadening parameter

$$\tilde{\mu}_\epsilon = 4N\sigma_\epsilon \frac{\omega_R}{2\Omega\gamma^2\beta_z\beta_{ph}}. \tag{6.95}$$

In figure 6.30, the gain which takes into account the beam velocity spread has been fitted with a Lorentzian function

$$G(\tilde{\mu}_\epsilon) = \frac{G_0}{1 + a\tilde{\mu}_\epsilon^2}, \tag{6.96}$$

G_0 being the gain without spread. Figure 6.31 shows the plot of $\Gamma(\gamma(B), L_{opt})$ and the starting current values versus the magnetic field B inside the cavity for a beam, with the parameters used in figure 6.26, interacting with the operating mode TE_{53} of a

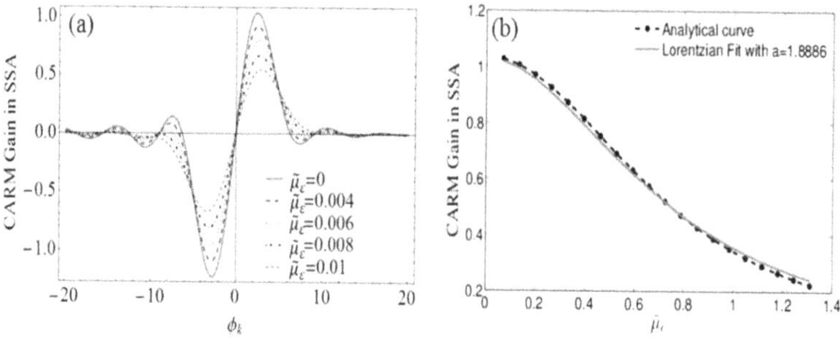

Figure 6.30. (a): The CARM gain in small signal approximation without beam spread (continuous line) and with different values of the beam spread $\tilde{\mu}_\epsilon$ (dashed lines); (b): the maximum value of the CARM gain (dotted line) versus $\tilde{\mu}_\epsilon$.

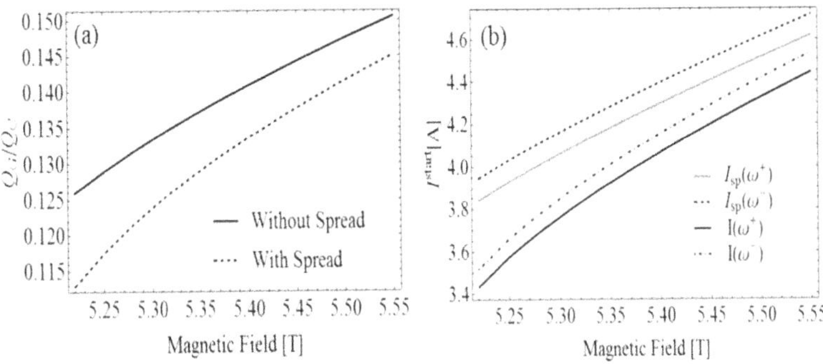

Figure 6.31. (a) the maximum admissible value for the quality factors ratio evaluated at ω^+ and ω_- versus the magnetic field; (b) the starting current values versus the magnetic field in the cavity for a fixed quality cavity factors $Q_C = 3300$ and $Q_G = (\Gamma(\gamma(H), L_{\text{opt}}) - k)Q_C$ with $k = 0.003$. In both cases we show the effect of the velocity spread of the beam electrons considering a Gaussian distribution with a dispersion $\sigma_\varepsilon = 0.003$.

cavity with a radius $R_w = 7.5\,\text{mm}$ and a smooth section $L_{\text{opt}} = 13\,\text{cm}$. The admissible values of the cavity Q-factor ratio (figure 6.31(a)) are constrained within the continuous curve, in the case of a beam with negligible energy spread, or within the dashed counterpart for a more realistic beam.

In this chapter we have provided the base-line for the CARM theoretical foundations and their link with other available sources including U-FEL and gyrotron. The detailed analysis we have devoted to the relevant design has been aimed at underscoring how complex and delicate is their practical conception. Further comments on the design of the electron gun and beam transport will be presented in the forthcoming chapter.

References

[1] Petelin M I 1974 On the theory of ultrarelativistic cyclotron self-resonance maser *Radiophys. Quantum Electron.* **17** 686–90

[2] Botvinnik I E, Bratman V L, Volkov A A, Ginzburg N S, Denisov C G, Kol'chugin D B, Ofitserov M M and Petelin M I 1982 Free-electron masers with Bragg resonators *Pis'ma Zh. Eksp. Teor. Fiz.* **35** 418–20

[3] Bratman V L, Denisov C G, Ginzburg N S and Petelin M I 1983 FEL's with Bragg reflection resonators: cyclotron autoresonance masers versus ubitrons *J. Quantum Electron.* **19** 282–96

[4] Ginzburg N S, Zarnitsyna I G and Nusinovich G S 1981 Theory of relativistic crm amplifiers Radiophys *Quantum Electron.* **24** 331–8

[5] Bratman V L, Ginzburg N S, Nusinovich G S, Petelin M I and Strelkov P S 1981 Relativistic gyrotrons and cyclotron autoresonance masers *Int. J. Electron.* **51** 541–67

[6] Botvinnik I E, Bratman V L, Volkov A A, Denisov C G, Kol'chugin D B and Ofitserov M M 1982 Cyclotron-autoresonance maser with a wavelength of 2.4 mm *Pis'ma Zh. Tekh. Fiz.* **8** 1386–9

[7] Bekefi G, DiRienzo A, Leibovitch C and Danly B G 1989 35 35 GHz cyclotron autoresonance maser amplifier *Appl. Phys. Lett.* **54** 1302–4

[8] Caplan M, Kulke B, Bubp D G, McDermott D and Luhmann N 1991 A 250 GHz CARM oscillator experiment driven by an induction linac *Nucl. Instr. Meth. Phys. Res.* A **304**

[9] Danly B G 1990 High-power CARM experiments *15th International Conference on Infrared and Millimeter Waves* ed R J Temkin (Bellingham, WA: SPIE) pp 692–6

[10] Chong C K, Razeghi M M, McDermott D B, Luhmann N C Jr, Thumm M and Pretterebner J 1991 Bragg reflectors: tapered and untapered *Intense Microwave and Particle Beams II* ed H E Brandt (Bellingham, WA: SPIE) pp 226–33

[11] Bratman V L, Denisov C G, Kol'chugin D B, Samsonov S V and Volkov A B 1995 Experimental demonstration of high-efficiency cyclotron autoresonance maser operation *Phys. Rev. Lett.* **75** 3102–5

[12] Bratman V L, Kalynov K Y, Ofitserov M M, Samsonov S V and Savilov A V 1997 CARMs and relativistic gyrotrons as effective sources of millimeter and submillimeter waves *Digest of the 22nd Int. Conf. on Infrared and Millimeter Waves (Wintergreen, VA)* pp 58–60

[13] Garavaglia S 2018 EU demo EC system preliminary conceptual design *Fusion Eng. Des.* **136** 1173–7

[14] Federici G 2018 Demo design activity in Europe: progress and updates *Fusion Eng. Des.* **136** 729–41

[15] Zhom H 2017 On the size of tokamak fusion power plants *Phil. Trans. R. Soc.* A **322**

[16] Ginzburg V L 1947 On emission of microwaves and their absorption in air *Izv. Akad. Nauk SSSR Fiz.* **11** 165–82

[17] Chen K R, Dawson J M, Lin A T and Katsouleas T 1991 Unified theory and comparative study of cyclotron masers, ion channel lasers and free electron lasers *Phys. Plasmas* **3** 1270

[18] Gapanov A V 1959 Interaction between electron fluxes and electromagnetic waves in waveguides *Izv. Vyssh. Uchebn. Zaved. Radiofiz.* **2** 450–62

[19] Weibel E S 1959 Spontaneously growing transverse waves in a plasma due to an anisotropic velocity distribution *Phys. Rev. Lett.* **2** 83–4

[20] Chu K R 2004 The electron cyclotron maser *Rev. Mod. Phys.* **76** 489–538

[21] Chu K R and Hirshfield J L 1978 Comparative study of the axial and azimuthal bunching mechanism in electromagnetic cyclotron instabilities *Phys. Fluids* **21** 461–6

[22] Dattoli G, Torre A and Renieri A 1993 *Lectures on the Free Electron Laser Theory and Related Topics* (Singapore: World Scientific)

[23] Nusinovich G 2004 *Introduction to the Physics of Gyrotrons* (Baltimore, MD: Johns Hopkins University Press)

[24] Fliflet A W 1986 Linear and non-linear theory of the Doppler-shifted cyclotron resonance maser based on TE and TM waveguide modes *Int. J. Electron.* **61** 1049–86

[25] Chen C and Wurtele S 1991 Linear and non linear theory of cyclotron autoresonance masers with multiple waveguide modes *Phys. Fluids* B **3** 2133

[26] Ceccuzzi S, Dattoli G, Di Palma E, Doria A, Sabia E and Spassovsky I 2015 The high gain integral equation for CARM-FEL devices *J. Quantum Electron.* **51** 1–9

[27] Dattoli G, Ottaviano P L and Pagnutti S 2007 *Booklet for FEL Design: A Collection of Practical Formulae* (Frascati: Edizioni Scientifiche Frascati)

[28] Artioli M, Dattoli G, Ottaviani P L and Pagnutti S 2012 Virtual laboratory and computer aided design for free electron lasers outline and simulation *Energia Ambiente e Innovazione* vol 3 (Rome: ENEA)

[29] Saldin E, Schneidmiller E V and Yurkov M V 2000 *The Physics of Free Electron Lasers* (Berlin: Springer)

[30] Kroll N, Morton P and Rosenbluth M N 1981 Free-electron lasers with variable parameter wigglers IEEE *J. Quantum Electron.* **17** 1436–68

[31] Bratman V L, Ginzburg N S and Petelin M I 1979 Common properties of free electron lasers *Opt. Commun.* **30** 409–12

[32] Nusinovich G S, Latham P E and Li H 1994 Efficiency of frequency up-shifted gyrodevices: cyclotron harmonics versus CARM's *IEEE Trans. Plasma Sci.* **22** 796–803

[33] Di Palma E, Sabia E, Dattoli G, Licciardi S and Spassovsky I 2017 Cyclotron auto resonance maser and free electron laser devices: a unified point of view *J. Plasma Phys.* **83** 1

[34] Bratman V L and Denisov G G 1992 Cyclotron autoresonance maser–recent experiments and prospects *Int. J. Electron.* **72** 969–81

[35] Gilden M and Gould L 1964 *Handbook on High Power Capabilities of Waveguide Systems* (Burlington, MA: Microwave Associates Inc.)

[36] Gaponov A V, Petelin M I and Yulpatov K V 1967 The induced radiation of excited classical oscillators and its use in high-frequency electronics Radiophys *Quantum Electron.* **10** 794–813

[37] Yulpatov V K 1967 Nonlinear theory of the interaction between a periodic electron beam and an electromagnetic wave *Radiophys. Quantum Electron.* **10** 846–56

[38] Di Palma E, Dattoli G, Sabia E, Sabchevski S and Spassovsky I 2017 Beam-wave interaction from FEL to CARM and associated scaling laws *IEEE Trans. Electron. Devices* **99** 1–8

IOP Publishing

High Frequency Sources of Coherent Radiation for Fusion Plasmas

G Dattoli, E Di Palma, S P Sabchevski and I P Spassovsky

Chapter 7

Plasma heating with coherent FEL-like sources

7.1 U-FEL and fusion applications

In the last three chapters, we have reported the physical and technological features of free electron laser (FEL)-type coherent generators of radiation.

In figure 7.1 we have comprised, in a single chart, different devices in terms of frequency and wavelength and the spectral region useful for plasma physics, including inertial/magnetic fusion and diagnostic purposes.

The first items we have examined, namely the U-FEL devices, are characterized by a low intrinsic efficiency and, therefore, their use for plasma heating is doubtful. Significant efforts were pursued during the 1980s towards the design of FEL, with enhanced efficiency, dedicated to this purpose (see e.g. [1]). For the theoretical foundations of FEL plasma heating see [2, 3].

In those times, the number of existing or proposed FELs was not so crowded as nowadays. A relevant scenario is reported in the Letardi-chart of figure 7.2. Most of the device were at the level of proposals and a non-negligible proportion of them had been designed to provide radiation in the sub-mm/mm region. A view of an FEL-Tokamak arrangement is shown in figure 7.3.

The first attempt, in the direction of exploiting U-FELs for fusion purposes, had been performed at Livermore on Alcator-C [4]. In this experiment an FEL pulse with peak power levels up to 0.2 GW and pulse lengths up to 10 ns at 140 GHz has been employed. Before entering into further details, we should clarify the proposals to overcome the inadequacy of the relevant intrinsic efficiency.

We have already commented that low gain FELs have efficiency of the order of $1/4N$, where N is the number of undulator periods. Therefore, at best, we can foresee values not exceeding the level of a few percent.

Regarding the high gain regime, the efficiency is proportional to the Pierce parameter, which, for x-ray devices, is limited to values not exceeding 0.1%.

doi:10.1088/978-0-7503-2464-9ch7

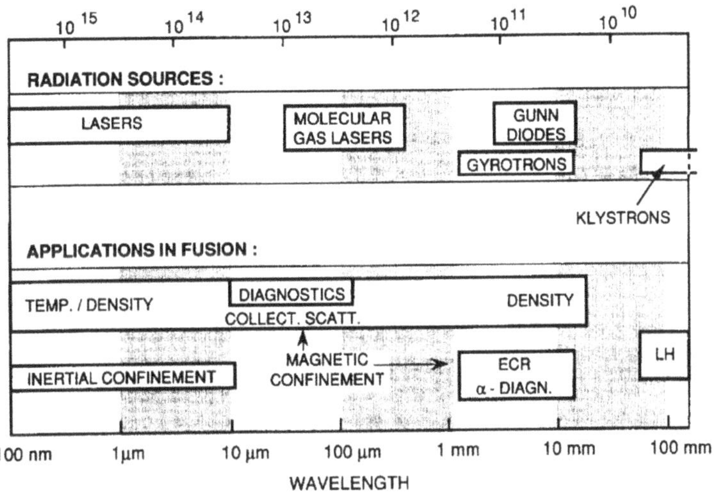

Figure 7.1. Frequency and wavelength chart of coherent electromagnetic (em) sources and regions of interest for applications.

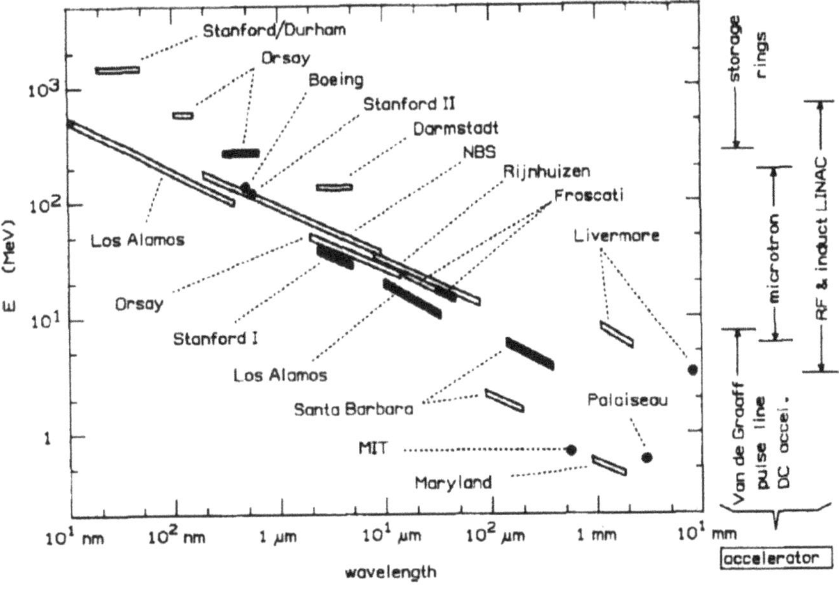

Figure 7.2. Energy–wavelength chart of FEL device designed or proposed at the end of the Twentieth Century. On the right side axis we have marked the employed accelerator.

Regarding long wavelengths, larger values of the Pierce parameter can be obtained (see figures 4.13 and 7.4), but still not sufficient for plasma heating purposes.

This is not a drawback for the foreseen applications of FEL oscillators or fourth generation radiation sources, but certainly for plasma heating applications, demanding efficiencies larger by at least one or two orders of magnitude.

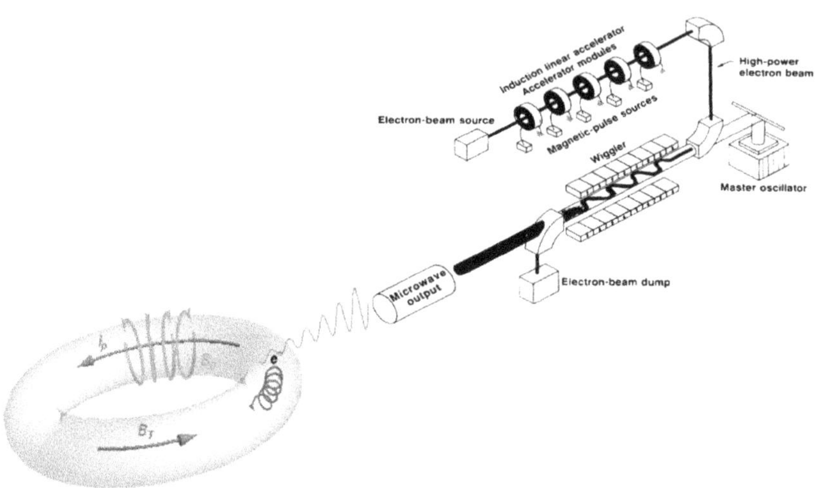

Figure 7.3. Schematic of an FEL wave injection inside a Tokamak.

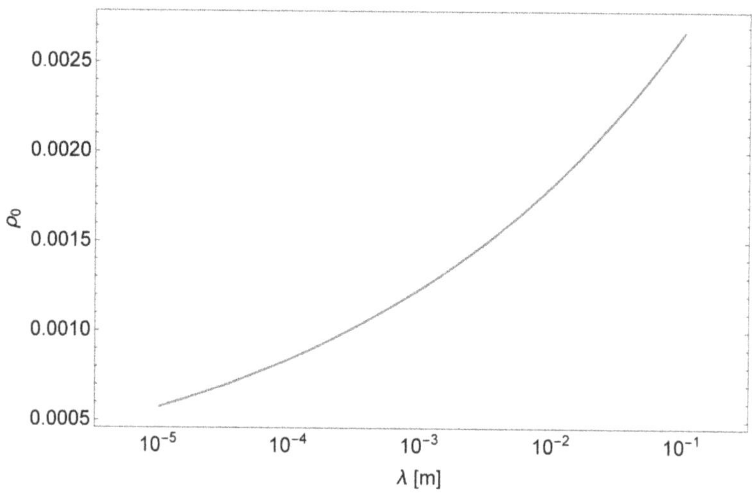

Figure 7.4. Pierce parameter for a high gain FEL operating in the wavelength range 1 mm–0.1 m with a beam current of 1 kA.

It is worth stressing that the possibility of overcoming the problems associated with beam diffraction (dominant at large wavelengths) are made possible by the effect of the high gain, inducing '***wave guiding***' [5]. It constrains the radiation within the beam transverse dimension, in turn enhancing the gain since the filling factor (namely the ratio between electron and radiation cross sections) remains large.

What we have so far described are FELs operating in the ***Compton regime***, in which the electrons interact via the em field. When the e-beam current increases, namely when it is above the kA threshold (at relatively low energies), the direct Coulomb interaction cannot be considered irrelevant. The occurrence of these effects determines the ***Raman-FEL*** regime.

According to the discussion of the first part of this book, these effects can quantified through the electron plasma frequency, which, including the relativistic mass effect, reads

$$\tilde{\Omega}_p = \frac{\Omega_p}{\sqrt{\gamma}}. \tag{7.1}$$

The plasma oscillations effects are important if the relevant period is much larger than the beam transit time inside the undulator, the condition expressing this regime can be written as

$$\frac{\Omega_p}{\sqrt{\gamma}} \frac{L_u}{\gamma c} \gg 1. \tag{7.2}$$

The efficiency of Raman FEL devices is written as [6, 7]

$$\begin{aligned} \eta &\approx \frac{1}{4N} \Delta_R, \\ \Delta_R &= \left(\frac{2}{\pi} \frac{\Omega_p}{\sqrt{\gamma}} \frac{L_u}{\gamma c} \right), \end{aligned} \tag{7.3}$$

with Δ_R being the Raman efficiency enhancement factor. A slightly more suggestive form is given below ($L_u = N\lambda_u$)

$$\eta \approx \frac{2\sqrt{\gamma}}{1 + \frac{K^2}{2}} \frac{\lambda}{\lambda_p}. \tag{7.4}$$

The electron number density is linked to the amplification factor by the identity

$$\begin{aligned} n_e &= \frac{\pi}{16} \frac{\gamma^3 \Delta_R^2}{r_0 L_u^2}, \\ r_0 &\equiv \text{electron radius}. \end{aligned} \tag{7.5}$$

Requiring an amplification factor of 30 and an undulator of the order of tens of meters and a modestly relativistic energy ($\gamma \cong 6$), we obtain densities in excess of $10^{18} \mathrm{m}^{-3}$, which are rather difficult to be transported.

Excluding Raman-FEL and considering conventional U-FEL only, two mechanisms allowing efficiency enhancement can be foreseen:

1. e-beam energy recovery [8];
2. undulator tapering [9] (for a gentle introduction see [6]).

Regarding the first point, we note that in conventional systems the beam is dumped after any interaction, even though only something like a few % of its energy has been lost. This is evidently a large amount of wastage. It is, therefore, natural to look for energy recovery. A possible solution, in this direction, is sketched in figure 7.5. In these devices the energy FEL's wall-plug efficiency can be significantly increased by recovering the e-beam energy by re-circulating it, after the FEL interaction, through

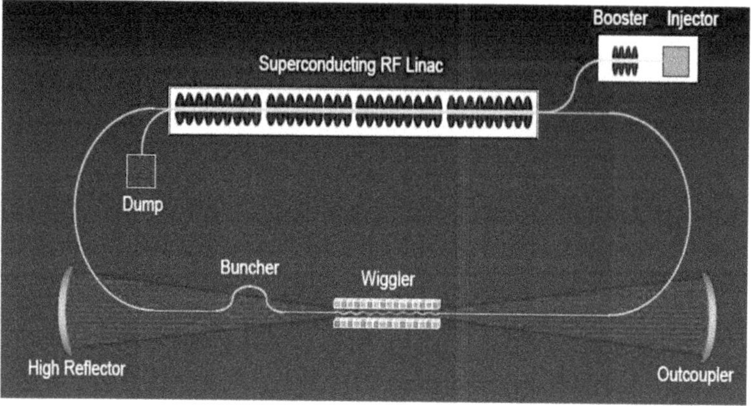

Figure 7.5. Representative schematic of an ERL FEL. The electrons accelerated to relativistic energies in the injector and linac sections enter the undulator and trigger FEL oscillator operation. They are successively injected inside the superconducting cavities and finally spent in the beam dump. Reproduced from [6] with permission.

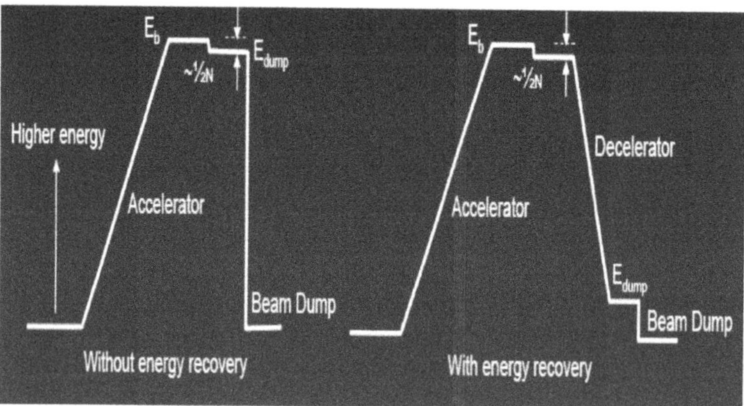

Figure 7.6. Energy diagram without and with the energy recovery mechanisms. Reproduced from [6] with permission.

the superconducting RF linac at the deceleration phase. The electrons give back the power, acquired during the accelerating phase, to the linac cavities.

The beam is eventually dumped at lower energy.

FELs employing this device are recognized as ERL FEL.

The energy diagrams in figure 7.6 clarify that, without recovery, the energy of the beam lost in the beam dump is almost the same as that of the original beam, while it is much less, in the other case.

The overall efficiency can therefore be written as

$$\begin{aligned} \eta_{\mathrm{ERL}} &\approx \frac{1}{4N}\Delta_{\mathrm{ERL}}, \\ \Delta_{\mathrm{ERL}} &\cong 1.67\frac{E_b}{E_{\mathrm{dump}}}. \end{aligned} \tag{7.6}$$

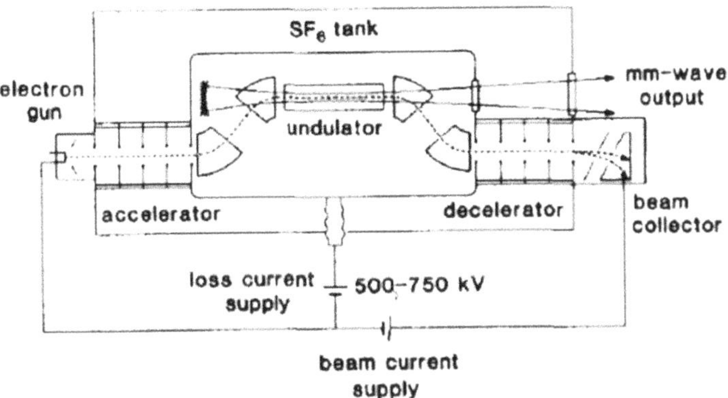

Figure 7.7. FEL operating with an electrostatic accelerator and a charge recovery system. Reprinted from [10] as reported in the text below, copyright (1987), with permission from Elsevier.

A different flavor of energy recovery device is reported in figure 7.7 and sketches an electrostatic accelerating device [10]. The conceptual content is the same as that of figure 7.5; the e-beam after feeding a mm-wave oscillator is injected into a decelerator, and then sent to a collector. In this scheme what is recovered is the charge, which then restores the necessary accelerating voltage, with a small external current contribution, compensating the charge recovery losses.

Let us now comment on the ***undulator tapering*** and the way it contributes to the efficiency enhancement [6, 9].

We have mentioned, in chapter 6, that the distinctive feature of the CARM device is the property of compensating the effect of the electron energy losses, thus keeping the electrons in resonance, for longer time, before the system reaches the saturation. This is the natural mechanism yielding a larger efficiency, with respect to U-FEL.

According to the discussion of chapter 4, when the electrons move inside the undulator they lose energy. The emission process becomes progressively out of resonance. Suitable changes of the undulator parameters could therefore be capable of ensuring the resonance condition.

We have sketched below the steps to be followed to ensure synchronism while electrons are progressing through the undulator

$$\begin{aligned}&\text{Resonance condition at } z_o\\ &\lambda = \frac{\lambda_u}{2\,\gamma^2}\left(1 + \frac{K^2}{2}\right)\\ &\text{Resonance condition at } z_o + \Delta z\\ &\lambda = \frac{\lambda_u}{2\,(\gamma - \Delta\gamma)^2}\left(1 + \frac{(K - \Delta K)^2}{2}\right)\\ &\text{Rate of energy change versus } z\\ &\frac{d}{dz}\left(\frac{\Delta\gamma}{\gamma}\right) = \frac{1}{2}\frac{K^2}{1 + \dfrac{K^2}{2}}\frac{d}{dz}\left(\frac{\Delta K}{K}\right).\end{aligned} \tag{7.7}$$

An effective variation of the K-parameter is obtained by changing the magnetic field using fixed prescription, e.g.

$$B(z) \cong B_0\left(1 - \frac{z - z_o}{L_u}\right). \tag{7.8}$$

As displayed in figure 7.8 and 7.9 the tapering produces a reduction of the phase-space area and a reduction of the particles trapped in a bucket.

The quantity used to express the amount of tapering is given below and is linked to the fractional K variation $\Delta K/K$ by

$$\delta = \sin(\phi_R) = \left(\frac{K}{1 + \frac{K^2}{2}}\right)^2 \frac{1}{|a|}\frac{d}{dz}\left(\frac{\Delta K}{K}\right), \tag{7.9}$$

where $|a|$ is the dimensionless field amplitude (see chapter 4).

In figure 7.10 we have shown how the FEL phase space plot is modified by the presence of undulator tapering.

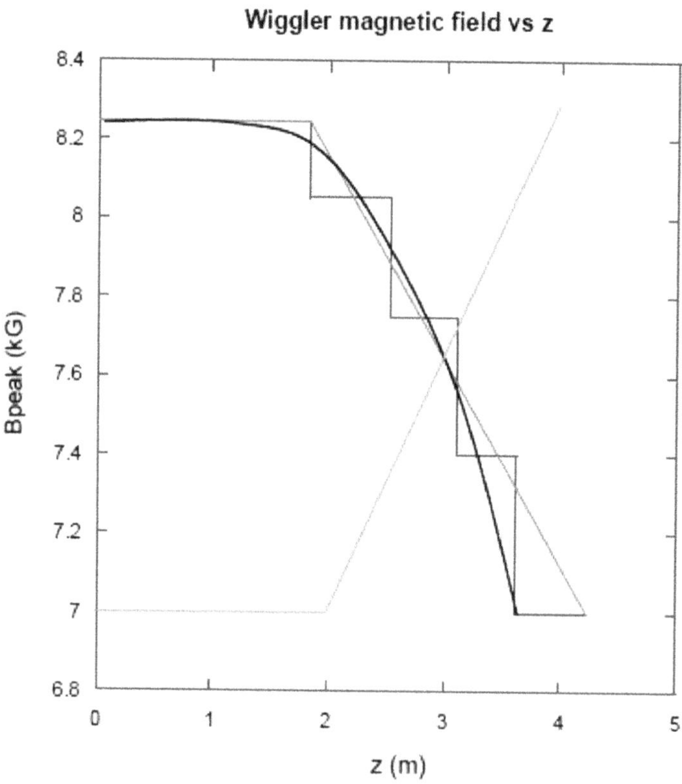

Figure 7.8. Undulator tapering (quadratic and linear) and linear anti-tapering (green). Reproduced from [6] with permission.

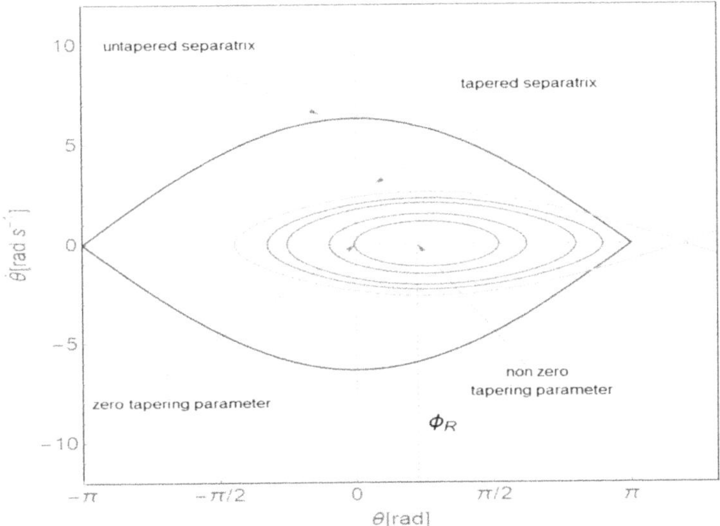

Figure 7.9. FEL phase space plot in presence of tapering effects. We have marked tapered and untapered separatrices and zero and non-zero tapering parameters.

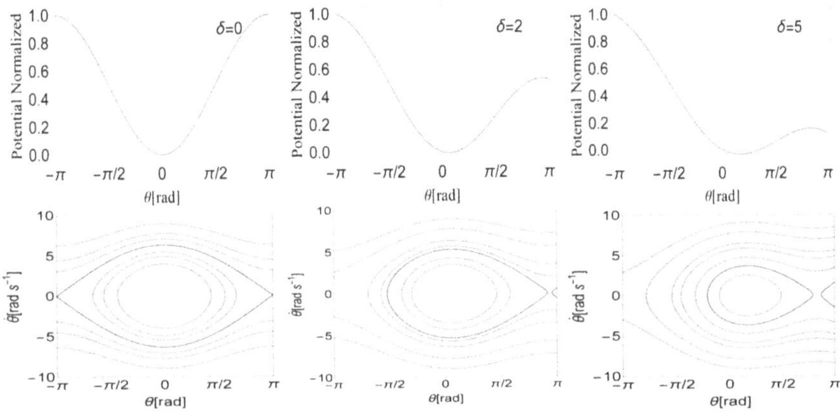

Figure 7.10. Modification of FEL phase space with the inclusion of undulator tapering effects and increasing values of the tapering δ parameter.

An idea of the relevant effect on the output power is eventually given in figure 7.11, where we have reported the power growth of an FEL amplifier versus z. The tapered section is inserted at the onset of the saturation for the section with $\delta = 0$.

In figure 7.11 we report the effect of the undulator tapering on the power evolution for a tapered undulator. In figure 7.11(b) the comparison with simulation is from [11].

According to the discussion of the first part of the book, U-FELs might have different directions of impact in the field of magnetic fusion.

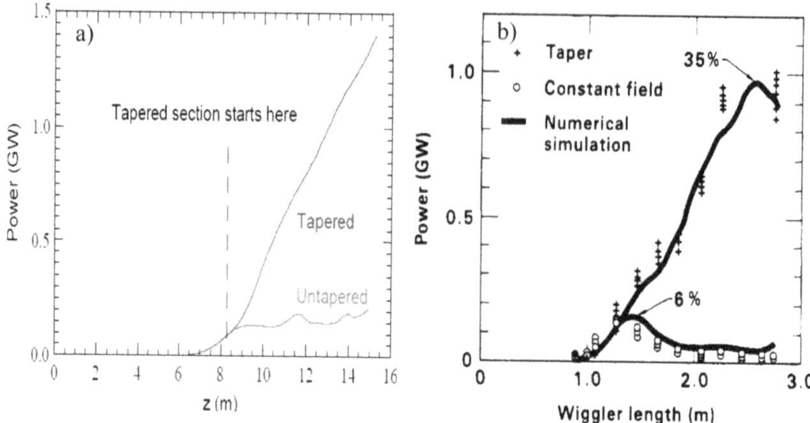

Figure 7.11. Efficiency enhancement in FEL operating with tapered undulators. The tapered section is set at the point in which the system has reached saturation with constant parameters undulator. The plots in (a) and (b) have been taken with permission from [6] and [11], respectively.

The combined use of tapering and energy recovery, along with the other properties characterizing the U-FELs (like the tunability), makes these devices appealing (in principle) for fusion purposes.

Leaving apart the application for diagnostics, we note that heating and current drive control, in high field Tokamaks, could involve FELs operating above 200 GHz, where high-power gyrotrons appear to reach a limit. This is a greatly challenging task. For these applications, long-pulse or continuous wave (CW) operation is required, with units delivering at least 1 MW power.

An experiment in this direction had been performed at the Alcator-C Tokamak, on electron cyclotron resonance heating (ECRH) under conditions of high density and high B [2]. In this preliminary experiment an FEL operating at 140 GHz, 0.2 GW (untapered) 0.4 GW (tapered) and 10 ns pulses has been used, with and undulator of 4 m total length with a period of 9.8 cm. An idea of the interplay between magnetic field tapering and laser power growth is offered in figure 7.12 taken from [12].

The experimental data have reported ECRH absorption, at the cutoff frequency. The shortness of the pulse did not allow other plasma specific measurements on plasma heating effects. Similar types of FEL, aimed at analogous proposals, were proposed by the ENEA Frascati laboratories. The FEL-facilities involved here are very large and complex devices, while the output characteristics—GW pulses at low duty cycle—are different from what one would conventionally expect to use for ECRH. To gain by making use of non-linear phenomena, it would seem that on a longer timescale true CW FELs have clear advantages; the present design effort indicates much more compact and simpler-to-operate devices, with efficiencies in the 50% range. Finally, the requirements for inertial confinement should be mentioned. For inertial fusion process, lasers in the sub-micron region with power at petawatt level and pulse duration of tens of nanoseconds are required. FELs satisfying these requirements, are well beyond the present state of the art. In particular, much further work on fully employing the potential benefits of the optical guiding principle is needed.

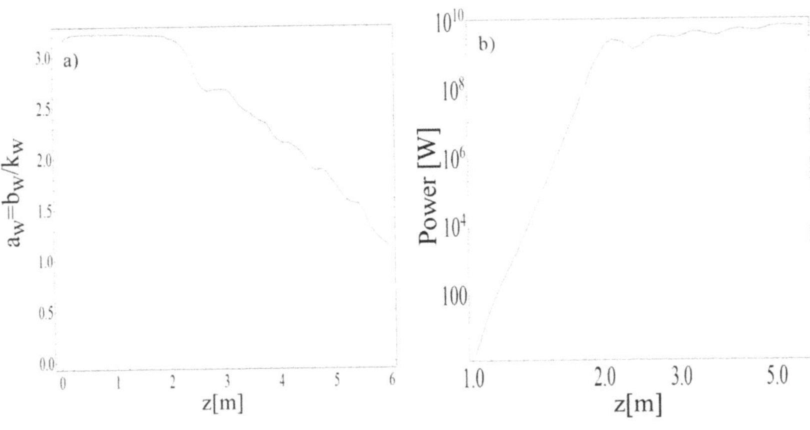

Figure 7.12. (a) Axial profile of the wiggler magnetic field for a sample calculation for a 140 GHz FEL system with a 6 m long wiggler and a 7 MeV electron beam. Run quantities: $P_{\text{laser}} = 10$ W, $P_{\text{out}} = 9 \cdot 10^9$ W, $a_{w0,\text{init}} = 3.2$, $a_{w0,\text{final}} = 1.2$ rms, $B_{w0,\text{init}} = 6.1$ kG, $B_{w0,\text{final}} = 2.2$ kG, $P_{\text{beam,init}}4 = 2.1 \cdot 10^{1}0$ W, extraction = 43.6%, $E_{\text{beam}} = 7$ MeV, wiggler length = 6 m, $\phi_R = 0.06$. (b) TE_{01} output power for the 140 GHz FEL system sample calculation, plotted as a function of the wiggler length. Reprinted from [12], copyright (1989), with permission from Elsevier.

7.2 Gyrotron for fusion and current status

Gyrotrons, as has already been mentioned in chapter 5, are being utilized as high-power radiation sources in many scientific and technological fields. By far the most notable among them, however, is fusion research, where gyrotrons are used for additional heating of magnetically confined plasma in various reactors for controlled thermonuclear fusion. Besides for ECRH and current drive (ECRCD) they are used also for plasma ignition, plasma control (suppression of the plasma instabilities due to the neoclassical tearing modes) as well as for plasma diagnostics based on the Thomson collective scattering (TCS). Although all gyrotrons have a common structure, gyrotrons for fusion have several distinguishing features that are related to the currently demonstrated unprecedented levels of the output power [13, 14]. These megawatt-class tubes with advanced designs necessarily include (i) internal (built-in) mode converter with a Denisov-type launcher; (ii) diamond output window, and (iii) depressed collector. A schematic of the design of a typical gyrotron for fusion is shown in figure 7.13. This illustration presents Japan's 170 GHz/1 MW tube [15].

This gyrotron has a triode-type magnetron injection gun, a cylindrical resonator working at 170 GHz with $TE_{31,8}$ mode, a water-cooled diamond window, and a depressed collector. After the demonstration of a performance that meets the requirements of the ITER project, the gyrotron has been operated for three years and recorded ~200 GJ of total output energy. The next tube developed by JAEA was a gyrotron that oscillates in a higher-order resonator mode, $TE_{31,12}$, and was used to study the long pulse oscillation at output power higher than 1 MW. It has

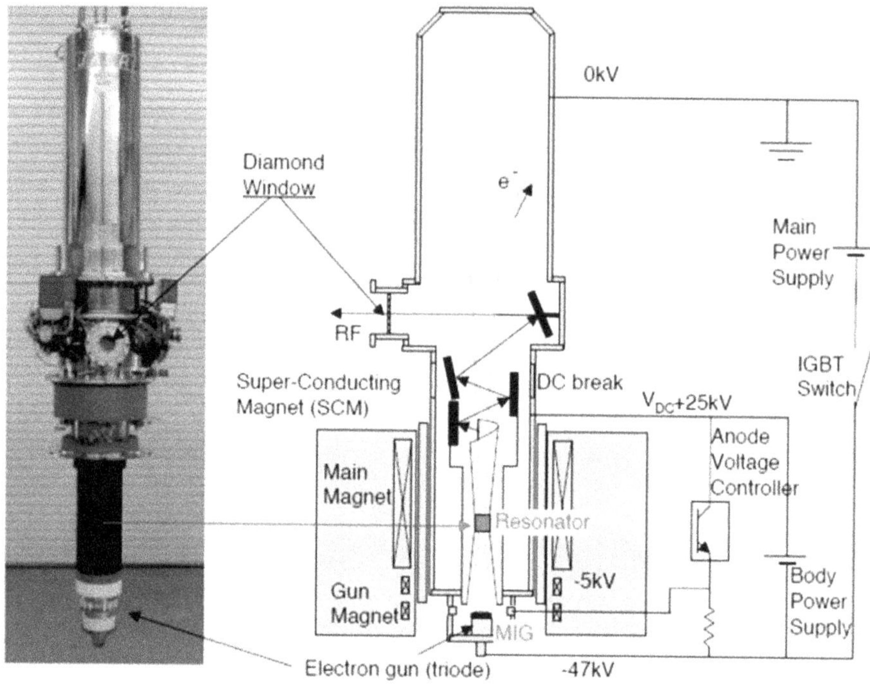

Figure 7.13. JAEA 1 MW 1709 GHz gyrotron. Reproduced from [15], copyright 2009 IAEA, Vienna.

been recognized a long time ago that at such enormous powers the long pulse and CW operation is a great challenge and a difficult task.

The electron cyclotron heating and current drive (ECH&CD) system of ITER requires RF power of 20 MW to be injected in the plasma vessel of the Tokamak. It will be delivered by 24 gyrotrons. The Japanese Domestic Agency (JADA) of the ITER project QST (former JEEA) will supply eight gyrotron units, each of which produces 170 GHz/1 MW. The specification of ITER gyrotron is 170 GHz/1 MW output/50% efficiency/3600s pulse duration. Recently, the final design of Japan's ITER gyrotron has been accomplished and the manufacturing of two tubes has been completed. During the tests at QST the first ITER gyrotron achieved an output power of 1.05 MW operating for 300 s with 51% efficiency. Additionally, operation at 5 kHz modulation has been demonstrated up to 200 s with power higher than 0.8 MW at the flat top of the pulses. The operation in pulses of 300 s at 1 MW has been repeated in 20 shots, of which 19 successfully demonstrated in such a way more than 95% reliability [16]. The first prototype-gyrotron which operated at $TE_{31,8}$ mode has demonstrated the following output parameters: (i) 170 GHz oscillation with the power of 1.0 MW, efficiency 55%, pulse duration 800 s, and (ii) 0.8 MW, 57%, and 3600 s. The studies aiming an optimization of the operational performance have revealed that the cavity for $TE_{31,8}$ mode has a heat load capacity close to the acceptable limit at 1 MW and, therefore, is susceptible to fatigue deformation for long-term operation. That is why this operating mode has been replaced by $TE_{31,11}$. The newly selected mode can be excited in a cavity with a larger radius, which reduces the thermal load of its wall. Additionally, careful optimization of the start-up scenario of $TE_{31,11}$ mode using a special anode voltage control has allowed avoiding the competition between the design mode and several counter-rotating modes. After such improvements, the prototype gyrotron has achieved 1.24 MW output power and demonstrated sufficient thermal capability of the cavity for long-term (3600 s) operation with an efficiency of 57% [16]. The internal mode converter consists of a quasi-optical launcher and a system of four mirrors. It forms a Gaussian beam with a mode purity of 95% and 2% RF losses. A diamond disk with a thickness of 1.86 mm and edge cooling is used as an output window. The main operational parameters of the gyrotron are presented in table 7.1 and the summary of the acceptance test criteria in table 7.2. These results show that the first ITER gyrotron developed at QST satisfies all the FAT (factory acceptance test) criteria and has the necessary performance for EC H&CD system operation of ITER.

The Russian prototype gyrotron complex with the necessary parameters for ITER developed in Nizhny Novgorod (Russia) by scientists of the Institute of Applied Physics of the Russian Academy of Sciences (IAP-RAS) and the Kurchatov Institute, Moscow in cooperation with the Scientific and Production Enterprise GYCOM and the company RTSoft in 2015 was the first among all countries participating in the project [17]. In May 2018 the FATs were successfully carried out

Table 7.1. Gyrotron operation parameters.

Cathode voltage	–46 kV
Anode voltage	–3 kV
CPD voltage	29 kV
Beam current	46 A
Cavity magnetic field	6.65 T
Frequency	169.85 GHz
Output power at the window	1.05 MW electric efficiency
Electronic efficiency	51%
CRM efficiency	31%

Table 7.2. Summary of acceptance test criteria and results of the first ITER gyrotron.

Item	Acceptance test criteria	Test results
Power	>0.96 MW at MOU output	1.02 MW at MOU output 1.04 MW at output window
Frequency	170 ± 0.3 GHz	169.85 GHz
Pulse duration	300 s	300 s
Duty cycle	25%	25% (5 min pulse/15 min interval)
Reliability	>95%	95% at 20 shots
Modulation	1, 3, 5 kHz 100% power modulation with 60 s/> 0.8 MW	1, 3, 5 kHz 100% power modulation 200 s, 0.87–0.90 MW

Figure 7.14. The IAP RAS/GYCOM (Russia) gyrotron for ITER during the acceptance tests in GYCOM (image courtesy E Tai).

on the second and third gyrotrons of the Russian procurement program by specialists at the IAP and GYCOM (figure 7.14). These ITER gyrotron systems have demonstrated reliable operation in 1000 s pulses at megawatt power levels and an efficiency exceeding 50% [18]. The produced microwave beam is coupled to a corrugated HE11 waveguide with a diameter of 50 mm. The main results of the recent tests are summarized in figure 7.15.

Megawatt output power in very long pulses (300–1000 s) has been demonstrated by several other gyrotrons operating at frequencies of 105 GHz and 140 GHz and developed for the EC systems of EAST (China) and KSTAR (Korea) superconducting Tokamaks [18]. Megawatt class gyrotrons with moderate pulse duration from 2 to 10 s have been developed also for TCV, HL-2A, and ASDEX Upgrade Tokamaks. The gyrotrons for KSTAR and ASDEX Upgrade are capable of operating at both frequencies.

Work on the next generation of fusion gyrotrons for DEMO, which will follow the ITER project, is also underway. Compared with ITER, it will require a higher frequency (230 GHz instead of 170 GHz) and an output power (1.5–2.0 MW instead of 1 MW), higher efficiency (higher than 60% instead of 50%), and multi-frequency operation, which is needed in order to avoid a wide-angle scanning of the wave beams in plasma. In preparation for reaching these challenging goals, the first experimental tests of a powerful 250 GHz gyrotron for future fusion research and collective Thomson scattering diagnostics have been carried out recently [20]. Similar efforts are being pursued also in the framework of EUROfusion [21] and in Japan [22].

Next, we present briefly the current state of the art [14, 23] of gyrotrons for fusion that are being developed by the European Gyrotron Consortium, EGYC, (which includes KIT—Germany, SPC—Switzerland, HELLAS—Greece, CNR and ENEA—Italy) in collaboration with Thales Electron Devices (TED) as an industrial partner.

Gyrotron Parameters	1st serial gyrotron	Gyrotron for F4E	2nd serial gyrotron
Frequency, GHz	169.9±03	170	169.9±03
RF output power, MW	0.97±0.05	0.92	0.96±0.05
Output radiation - HE_{11} mode content, %	97.3±0.5	96.8	Waiting for measurement
Efficiency (with depressed collector), %	56±3	56.3	53±3
Pulse duration, sec	1000	400/1000	1000
Beam voltage, kV	42.5 ± 0.4	42.5 ± 0.4	44.1± 0.4
Depression voltage, kV	27.5 ± 0.3	28.5 ± 0.3	26.5± 0.3
Beam current, A	42 ± 0.2	37.5 ± 0.2	42.4± 0.2
Body current, mA	10 ± 0.1	< 20	14 ± 0.1
Reliability, %	> 95	> 80	> 95

Figure 7.15. Test results of the Russian ITER gyrotrons. Reproduced from [19], copyright 2018 the authors, published by EDP Sciences.

The European 1 MW CW, 170 GHz industrial prototype ITER gyrotron utilizes a conventional cylindrical cavity optimized for operation at $TE_{32,9}$ mode. Its design is based on the previously developed modular short-pulse (SP) prototype and the 1 MW, 140 GHz gyrotron development for the stellarator W7-X [14]. The electron-optical system (EOS) of the tube uses a diode-type MIG and a conical beam tunnel assembled from a stack of BeO/SiC (60/40) ceramic damping rings and indented copper rings. Two differences of the design with respect to both the Russian and the Japanese tubes should be mentioned. The first distinction is that the body insulation is located in the lower part of the tube close to the electron gun. The second difference is the absence of an adjustable last reflector in the advanced three-mirror quasi-optical output coupler. The collector uses 7 Hz VFSS (vertical magnetic field sweeping systems) combined with 50 Hz transversal sweeping with six coils arranged around a stainless steel collector section [14]. During the measurements at the KIT test facility, the gyrotron has demonstrated an output power of 1.0 MW at 170.1 GHz with 30% efficiency operating without a depressed collector. The optimal operational parameters that had been determined during the short-pulse operation have been used for conditioning the tube while increasing the pulse length. Finally, the tube has been operated at power levels of 0.8 MW in pulses up to 180 s, which is a limit imposed by the used HV power supply. The maximum efficiency reached in depressed collector operation has been 38%. The Gaussian mode content has been estimated to be at least 97% [14]. Figure 7.16 shows the parameters and summarizes

	JA	RF	EU
Operating cavity mode	$TE_{31,8}/TE_{31,11}$	$TE_{25,10}$	$TE_{32,9}$
Radial mode eigenvalue	63.76750/74.32574	63.31966	68.56314
Cavity radius	17.9 mm/20.87 mm	17.77 mm	19.24 mm
Electron beam radius	9.13 mm	7.4 mm	9.5 mm
Launcher angle	0.2 deg.	0.2 deg.	0.23 deg.
Output mode purity	96%	97%	97%
Output power	1 MW (0.8 MW)/1 MW (1.2 MW)	1 MW (1.2 MW)	0.8 MW
Pulse duration	800 s (3600 s)/300 s (2 s)	1000 s (100 s)	180 s
Overall efficiency	55% (57%)/46% (47%)	54% (53%)	38%
Magnetron injection gun	Triode	Diode	Diode
Electron beam energy	72 keV	70 keV	72 keV
Depression voltage	30 kV	27.5 kV	25 kV
Beam current	43 A	42 A	45 A
Mean emitter radius	46.5 mm	41.5 mm	53.5 mm
Magnet bore hole	240 mm	160 mm	220 mm

Figure 7.16. Parameters and typical results (up to the end of 2018) of the Japanese, Russian and European gyrotrons for ITER. Reproduced from [14], copyright 2019 EURATOM.

the typical results (up to the end of 2018) of the Japanese and Russian 170 GHz ITER gyrotrons and of the first European prototype tube developed for ITER.

Although as a whole both the operational and the output parameters are similar, there is a notable difference in one of the most critical parameters, namely the pulse duration. In this respect, as can be seen in table 7.16, the best performance has been achieved by the Japanese tube followed by the Russian one. The same applies also to the maximum efficiency.

The European and US (CPI) gyrotrons for W7-X stellarator (at IPP Greifswald) and EU Tore Supra gyrotron are megawatt-class 140 GHz long-pulse gyrotrons with CVD diamond output window and single-stage depressed collector. They were developed by CPI and the EU team (KIT, SPC, THALES, Max Planck Institute for Plasma Physics in Greifswald (IPP Greifswald), IPF University of Stuttgart), independently, for the 10 MW ECH&CD system of the W7-X stellarator at IPP Greifswald [14]. Currently, the ECH&CD system of W7-X utilizes nine TED tubes and one CPI gyrotron.

The concept of the coaxial gyrotron has many attractive advantageous features [24, 25]. Among them, the most significant are the following [24]. By special profiling of the inner rod it is possible to provide effective electrodynamic mode selection in the radial indices. Additionally, the profiling of the insert together with an appropriate corrugation of the cavity wall allows improving further the mode selectivity in both the radial and azimuthal indices. The presence of an insert eliminates the problem of beam voltage depression (due to the space charge) and thus allows using volume operating modes that have very low ohmic losses.

Currently, an advanced coaxial gyrotron for ITER is being developed at IHM-KIT. The initial experiments obtained with the 170 GHz 2 MW short-pulse coaxial-cavity pre-prototype (figure 7.17) at pulse length of a few milliseconds (ms) have shown the potential of the coaxial-cavity concept in the multi-MW operation regime [26]. Currently, the efforts are focused on the verification of the feasibility of the operation at longer pulses up to 1 s. With this aim in view all main components of the tube, namely the beam tunnel, the cavity, the launcher, the mirrors of the quasi-optical

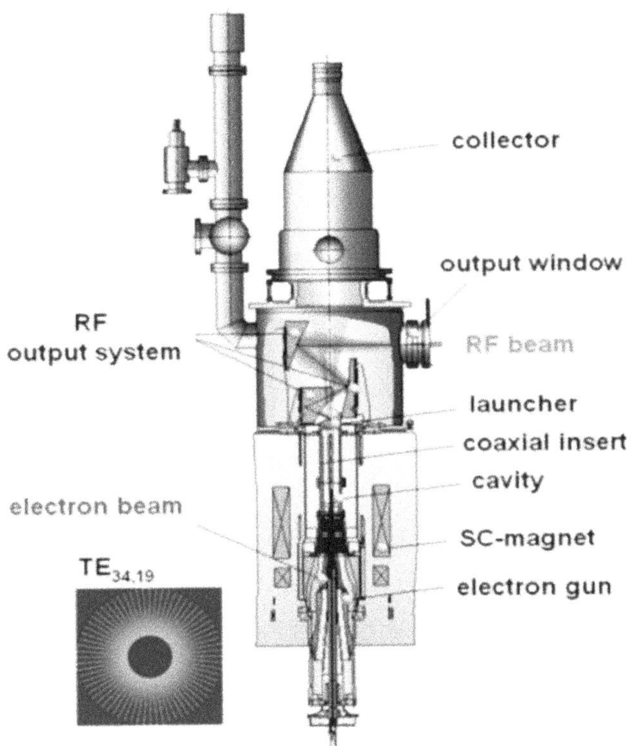

Figure 7.17. Schematics of the 2 MW/170 GHz coaxial-cavity gyrotron [26].

system, the CVD) diamond RF output window, and the collector have been equipped with an efficient cooling system. The final improved design is shown in figure 7.18. In order to preserve the modularity of the tube, each component is supplied with an individual cooling system. The gyrotron utilizes an inverted type triode magnetron injection gun (IMIG) which forms a high-quality helical electron beam with a current of 75 A, voltage of 90 kV and pitch factor of 1.3 (figure 7.19). Since in the IMIG the cathode and the emitter are placed at the outer side of the gun, the outer surfaces of the cathode assembly can be directly cooled by passing oil. The cathode and anode pairs are made of materials with high thermal conductivity, mainly CuCrZr (in figure 7.18 in orange color) and molybdenum (marked by the dark blue color). Such construction of the gun decreases the temperature of the thermally loaded elements by 40% (in comparison with the conventional MIG) down to ~150 ° C.

The cavity is optimized for operation at $TE_{34,19}$ mode. The launcher has a helix cooling structure which allows the thermal loading to be kept below the limit. A similar scheme is used also for cooling the cavity.

The presented examples that illustrate the current state-of-the-art of the gyrotrons for fusion are a manifestation of the remarkable progress in their development and manufacturing. The continuous improvements of the technology and the design solutions lead to better operational performance which satisfies the ITER criteria.

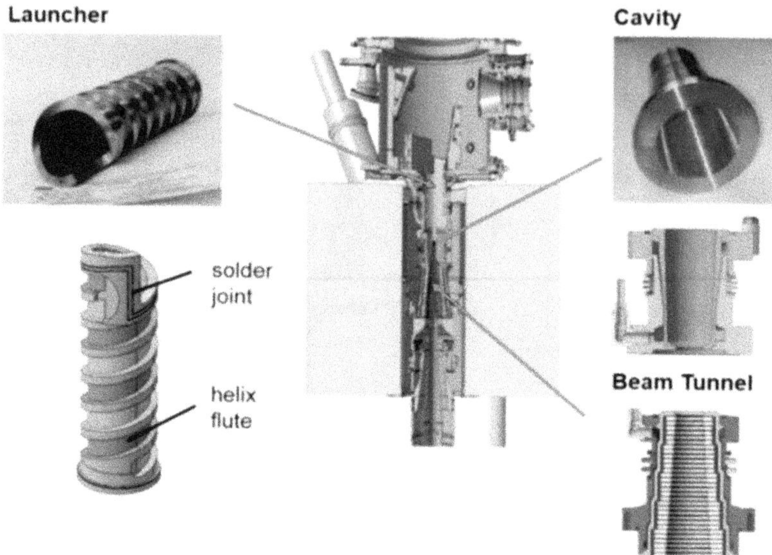

Figure 7.18. Main components of the KIT 170 GHz 2 MW coaxial-cavity longer pulse gyrotron [26]. Copyright Cambridge University Press and the European Microwave Association 2018. CC BY 4.0.

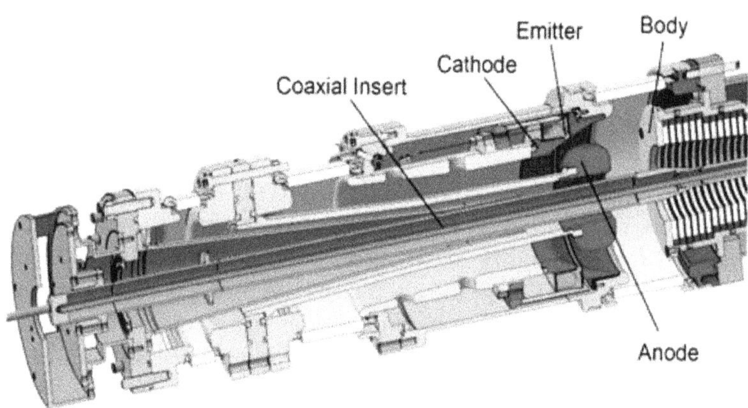

Figure 7.19. Schematics of the inverted MIG of the coaxial-cavity gyrotron.

However, the required output parameters (higher frequency, output power, and efficiency) for DEMO are a big challenge. In this respect, we believe that another promising gyro-device, namely CARM may prove to be a more advantageous candidate as a powerful radiation source at frequencies in the range 230–250 GHz envisaged for the next generation of fusion devices. We present the theory of CARM and discuss its feasibility as a prospective source for plasma heating in the next section of this book.

We end this chapter with some apologetic remarks. As mentioned in the Introduction, nowadays gyrotrons are being used in many other scientific and technological fields besides fusion research. Due to both the limited volume and the specialization of the book we omitted them here but strongly recommend the reader to study the vast literature on these subjects since it presents another class of gyrotrons that is characterized not by enormous output power (as with gyrotrons for fusion) but by record-high frequencies reaching the THz region of the electromagnetic spectrum. Most of them operate not on a single mode but rather on a sequence of modes and therefore on different frequencies. Some of these tubes (e.g. for spectroscopy) have demonstrated continuous frequency tunability in a broad band, frequency and amplitude modulation, etc. For further reading, we refer the interested reader to the review papers [27–34] and the references therein.

7.3 The CARM design for fusion application

In chapter 6 we discussed the possibility to use the CARM as an external additional plasma heating for the future Tokamak plant, like DEMO. The constraints that must be satisfied by such a device, following the Euro-Fusion study (see references in chapter 6), are fixed in the following: 1 MW of delivered power at 250 GHz with an efficiency of 30% .

In this section we dwell in particular on the issues associated with the necessity of generating and transporting a high quality beam (with low energy and velocity spread) necessary to get suitable efficiency.

In figure 7.20 we have reported the layout of the CARM RF source which we have foreseen for fusion purposes. Its main elements, framed within a red dashed line, are listed below:

(a) High voltage modulator,
(b) Electron gun,
(c) RF circuit,
(d) Magnetic circuit and the relevant ancillary components, outside the dashed frame, are

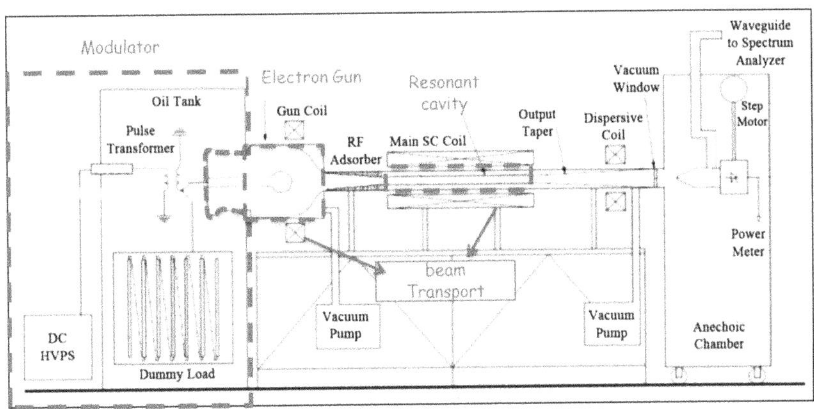

Figure 7.20. Layout of the CARM source with the main elements (in red) and including all the ancillary devices.

(e) Beam dump,
(f) Vacuum system,
(g) Supporting structure,
(h) Diagnostic tools,
(i) Cold test facility,
(j) Experimental room,
(k) Control room and control system.

We briefly describe the main task which must be accomplished by each element and the critical issues.

(a) ***High voltage modulator***
The high voltage (HV) modulator is one of the most crucial, complex and expensive parts of the whole CARM system. The relevant complexity comes from the noteworthy required stability of its electrical parameters, comparable to that of the modulators for high power klystrons.

The design and realization complexity has suggested the strategy of developing the apparatus in two steps.

The first step is mainly aimed at the realization of a CARM device to be tested in pulsed operation, with a maximum pulse length of 50 μs, at low repetition rate. With these tests the CARM electrical parameters will be optimized for maximizing the output power at the nominal output frequency.

The second step is instead aimed at developing a long pulse to CW CARM (necessary for plasma fusion application) prototype equipped with a depressed collector for beam energy recovery, and a Vlasov-like output mode converter to conveniently extract the RF power.

Due to the power losses into the RF circuit, an intensive, forced water, cooling circuit is necessary.

The main difference between the two steps is in the pulse length, even if rise and fall times and flat top accuracy of the pulses will be the same for both the steps. A different design approach has been provided for the HV modulator .

The single shot operation is the easiest way to test the single components and parts of the CARM and the CARM assembly as a whole. Apart from the modulator, all the other CARM components and units are designed for CW operation. The pulsed modulator for the first step will cover the pulse length range from 1 to 50 μs with a voltage tunability of 0.1% at the maximum output voltage of 700 kV. The other two important modulator parameters are the voltage ripple and the flat top smoothness, which should be in the range of 0.1% too.

The modulator will be fully immersed in a metallic oil tank together with a dummy load and the cathode holder of the electron gun.

(b) ***Electron gun***
The electron beam is emitted and formed inside a diode-type electron gun and then transported through a drift tube immersed in a direct high magnetic field. Most of the resulting helical beam properties are strongly dependent on the gun design. In general, the electron beam is very sensitive to small changes of the emitter

dimensions, emitter surface roughness and to its chemical properties. The road map followed during the preliminary design of the electron gun is the limitation of both the maximum surface electric field and the initial electrons velocity spread.

In order to avoid discharge problems, for a gun designed for CW operation, the surface electric field at any point inside the gun region is maintained at a level less than 10 kV mm.

The electron velocity spread is determined by the geometrical shape of the electrodes, their surface roughness, the emitter temperature and its uniformity over the whole emitter surface.

The space charge effect is proportional to the beam current density and is known as an emitter current load. In the preliminary design of the emitter the current density has been limited to 3 A cm^{-2}, a value that ensures it a long lifetime (>10 000 h). The emitter operational temperature is limited to 1300 °C to minimize the electron beam initial thermal velocity spread and to increase the emitter resistance to poisoning. These values are lower than the present technological limits and will guarantee the highest possible beam quality.

(c) ***RF circuit***

The CARM RF circuit is the assembly of the RF cavity, the RF beam expander provided by the output tapered bringing the radiation to the vacuum window. Any component is designed for CW operation.

Within the RF resonant cavity of a CARM, usually oscillating in a TE_{mnp} mode, the electron beam energy is partially transferred to the high frequency electric field. The energy transfer efficiency depends on the electron beam quality, mainly determined by a high performance gun design. The efficiency level also depends on the appropriate coupling between the electron beam and the cavity TE_{mnp} mode. The RF cavity must therefore guarantee an efficient beam interaction with the generated electromagnetic wave, characterized by a Doppler up-shifted frequency, and the suppression of the down-shifted (cutoff) counterpart. The mode selection is therefore of crucial importance for the CARM efficiency.

An oversized cylindrical cavity, whose dimensions are set by the acceptable RF power dissipation limits on the cavity wall and by the electric breakdown in vacuum, will be used. In general it is extremely difficult to design a stable high Q cavity operating far from the cutoff. In addition, the electron beam geometry along the CARM longitudinal axis has to be accurately studied because any interaction with the cavity walls, mainly during CW operations, must be absolutely avoided.

In this frame, the use of a quasi-optical resonant cavity, even though it is a fairly natural solution from the conceptual point of view in this frequency range, has been *a priori* discarded because of the thermal loads induced by the RF on the reflecting mirrors and also because of the difficulties with the electron beam transport.

The solution adopted is a cylindrical smooth cavity delimited by Bragg reflectors, which, although complicated from the mechanical point of view, provides distributed feedback and does not exhibit crucial drawbacks on the cavity cooling and on the electron beam transport.

The resonant cavity is connected to the larger CARM circular output waveguide through an accurately designed taper able to avoid any accidental mode conversion.

The electron beam waste energy is dissipated on the CARM collector while the microwave radiation is launched into an anechoic chamber through a short circular transmission line. A CVD (diamond vacuum window) circular vacuum window, axially brazed on the collector wall, separates the evacuated CARM device from the transmission line. The optimum vacuum level is ensured by a suitable pumping system assisted by an accurate mechanical design of the CARM RF inner components.

The CARM prototype object of the second design step, will be completed with an output mode converter, transforming the TE_{mn} operational mode into a Gaussian RF beam, and with a depressed collector in order to increase the overall efficiency of the device.

(d) ***Magnetic transport system***

The magnetic channel provides the correct electron beam formation before the injection into the cavity. It consists of a gun coil, a large cavity coil and a kicker coil. On the occurrence an additional correcting coil, positioned before the cavity coil, will be used to properly shape the desired magnetic field topology along the CARM longitudinal axis. All the components of the CARM assembly must be aligned very accurately along the CARM horizontal axis in order to realize an efficient beam transport and an optimal beam–RF coupling into the resonant cavity.

The gun coil encircles the gun region of the CARM and provides the necessary magnetic field for an appropriate electron beam emission and transport inside the diode.

The design of this coil has been done with the help of the CST Microwave Studio® (tracking module) to obtain the proper magnetic field intensity, which, in synergy with the static electric field inside the diode, allows the most appropriate electron beam kinematic conditions at the input of the RF cavity for the optimum CARM operations.

The relatively weak magnetic field in the gun region can be obtained by a short, water-cooled, coil. Its diameter must be larger than the grounded gun electrode. Its large cross section, short length and significant weight require a sophisticated supporting structure and an accurate alignment procedure.

The cavity magnetic field, due to its relatively high intensity, will be provided by a superconducting magnetic coil.

Both length and field intensity profile of the cavity coil have been evaluated by following the induced electron motion trajectory tracking along the system.

The design of additional coils for an accurate field profile correction is, however, foreseen to optimize the beam–wave interaction.

(d.2) ***Kicker coil***

The magnetic system includes also a kicker coil that generates a field with a nominal value in the range 0.08–0.1 T, perpendicular to the longitudinal CARM axis. This coil is aimed at removing any stray electrons by forcing them into the grounded wall.

(e) ***Beam dump***

Once it has left the resonant cavity region, the exhaust electron beam is uniformly spread on the collector walls, where its waste energy is finally dissipated.

The impact of the high-energy electron beams (700 keV) on the metallic collector walls generates hence a significant amount of x-ray radiation by bremsstrahlung. Thus the beam dump has to be carefully designed for absorbing the x-ray flux. Usually a multi-layered lead screen encloses this part to protect the surrounding area from the stray x-ray radiation. The beam impact also heats the CARM collector due to both scattering and ohmic losses. Therefore, the beam dump is cooled through suitable water pipes brazed on the outer part of its wall.

(f) ***Vacuum system***
An efficient vacuum system is of paramount importance for the correct operation of any high-power microwave tube. This system is designed to maintain inside the CARM an extremely low pressure at high pumping rate. In general, a bad vacuum level is responsible for two extremely dangerous phenomena. The first one is a vacuum breakdown due to a high electric field across the electrode gaps. The second is a surface breakdown due to a high surface electric field. There are rules regulating the limiting values for both cases that must be strictly respected during the design.

The CARM region under vacuum is about 2 m long with a minimum cross-section radius of 15 mm. The very first CARM prototype will operate at pulse lengths up to 50 μs. Thus to maintain a pulse repetition rate of 10 Hz, double side pumping is required. Most of the components in the evacuated region cannot be slotted or drilled for better pumping performance. These components need, therefore, to be carefully designed from the mechanical point of view in order to ensure a suitable vacuum level around them.

A low vacuum level is also important for the emitter just to avoid a poisoning possibility. A special chemical treatment of the components assembled in the vacuum region before the final assembling has therefore been considered.

(g–h) ***Supporting structure and diagnostics***
An anechoic chamber is the most important diagnostic tool for characterizing a CARM device. It is a large metallic box, whose inner walls are lined with microwave absorbing material, pyramidally shaped for reducing the RF reflections. This chamber in practice simulates the free space propagation. The microwave pattern generated by the CARM output waveguide inside the anechoic chamber is sampled with a horn pickup, an open-ended rectangular waveguide with a cutoff frequency of about 170 GHz, externally dressed by RF absorbing epoxy foam. The microwave signal picked up by the horn is split into two. One half of this signal is directly sent to a power meter, the second half is instead sent to a frequency spectrum analyzer in order to have a complete characterization of the RF power launched in the vacuum chamber. The horn is supported by a dielectric rod and moved inside the anechoic room by a remotely controlled motor. The total emitted RF power can be measured by mapping the radial pattern of the microwave radiation and then by integrating it over the whole chamber volume.

(i) ***Cold test facility***
The cold test facility is an important section of any microwave laboratory aimed either at the development of new devices or at the characterization of the existing ones. The optimum performance of any microwave device in fact depends on the

accuracy of the cold tests. In particular for this device, working with an oversized cavity with dense spectrum, it is essential to provide an appropriate cold test.

The frequency range under investigation is a not completely explored sector so that many microwave components must be expressly designed and developed. In particular a complex mode converter for feeding the CARM cylindrical resonant cavity in the TE_{53} mode, starting from the TE_{10} mode generated by a network analyzer in rectangular waveguide, is presently investigated.

A preliminary list of an essential outfit for our cold test facility is given in the following:

1. Network analyzer with output frequency up to 300 GHz;
2. Mode converters from the TE_{10} mode in rectangular waveguide (*WR 3* or *WR 4*) to the cavity mode (i.e. the TE_{53}) in circular waveguide;
3. Circular tapers to connect the previous mode converter to the oversized circular waveguide of the resonant cavity;
4. Splitters, bends, attenuators, phase shifters, directional couplers and so on.

7.3.1 Gun design and e-beam qualities

According to the discussion reported in the chapter 6, the generation of an electron beam with appropriate qualities is the prerequisite to achieve the desired CARM performance.

A thermionic gun has been designed and proven to be a suitable tool for the production of the electron beam with the foreseen velocity and angular spread. In table 7.3 we have summarized the design parameters of the electron gun system.

The considered gun is essentially a diode, which, unlike the triode gun used for gyrotron, is a non-adiabatic device. The relevant constituents: the cathode and the anode, are shown in figure 7.21.

The gun parameters of table 7.3 have been figured out on the basis of a simple argument, which takes the heating power as a pivoting reference. RF and electron beam power are linked by

$$P_{e-b} = \frac{P}{\hat{\eta}} \tag{7.10}$$

Furthermore, since P_{e-b} is given by the product of the current time and the accelerating voltage

$$P_{e-b} = IV \tag{7.11}$$

Table 7.3. Gun properties.

Cathode voltage	$500 \div 700$ kV
Relativistic factor γ	$2 \div 2.4$
Beam current	$1 \div 10$ A
Pitch ratio ($v_\perp/v_\parallel$)	γ^{-1}
Axial and transverse velocity spread	<0.5%
Electric field at the cathode surface	<10 kV mm^{-1}

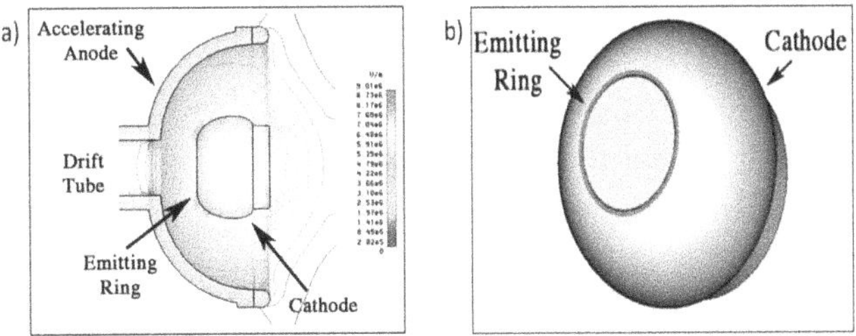

Figure 7.21. (a) Diode-like gun, (b) Emitter ring.

we obtain

$$I = \frac{P}{\hat{\eta}\, V}. \tag{7.12}$$

If we require a total electron efficiency around 30% and demand a CARM RF power of about 1 MW we find for the corresponding e-beam power $P_{e-b} \cong 3$ MW, which can be achieved by imposing suitable constraints on the accelerating voltage and beam current.

The efficiency of 30% can be obtained if the beam energy spread is suitably small, therefore the most convenient design solution is that of as high as possible accelerating voltage to increase the e-beam kinetic energy and a low current to avoid additional velocity spread, induced by the space charge effects. By choosing, e.g. an accelerating voltage of 600 kV, the resulting beam current is about 4.5 A.

A non-secondary issue in the design of the gun is the necessity of producing an electron beam with a ring shape to be positioned on a selected radial RF field maximum of the desired cavity mode, allowing the optimum coupling conditions. Usually, those devices use the whispering gallery modes for operation, which determines the beam profile (see figure 7.21). The electrons emitted from the ring are pushed ahead by the strong electrostatic field given by the accelerating tension and are guided by a longitudinal magnetostatic field generated by a gun coil.

Having specified the value of the accelerating voltage we can provide a first idea of the relevant geometrical dimensions by starting from the following relation between the cathode and anode radius $R_{\text{cath,an}}$ in the case of a spherical shape

$$R_{\text{cath}} = \frac{R_{\text{an}}}{2}\left(1 - \sqrt{1 - 4\frac{V}{E_{\text{surf}} R_{\text{an}}}}\right) \tag{7.13}$$

which implies an equilibrium condition between the accelerating field V and the electric field E_{surf} at the cathode surface. The condition to be fulfilled to ensure the positivity of the argument of the square root in the previous equation is

$$R_{\text{an}} \leqslant 4\frac{V}{E_{\text{surf}}} \tag{7.14}$$

the use of the threshold value of $E_{\text{surf}} \cong 10^7 \frac{V}{m}$, ensuring no cold field emission effects, implies $R_{\text{an}} \cong 30$ cm.

The electron beam generation and transport can be roughly divided into three parts:

(a) The electrons are emitted from a circular corona at the cathode emitting surface, which is kept flat to ensure an homogeneous longitudinal velocity. At the exit of the corona the electrons are captured by the gun accelerating (longitudinal) field and guided by a superimposed static magnetic field, focusing the beam by compensating the transverse components of the velocity, induced by a transverse electric field having a cusp at the transition edge of the diode and drift tube (see figure 7.22).

(b) Near the drift tube the electrons receive a transversal kick induced by the transverse electric field acting as a beam defocusing lens.

(c) Inside the drift tube the electron beam undergoes the combined effect of the fringing fields of the gun coil and the cavity coils (figures 7.22 and 7.23).

The longitudinal and transverse velocities (in m s^{-1}) along the gun (axis) in mm are shown in figure 7.23(a) and (b), respectively.

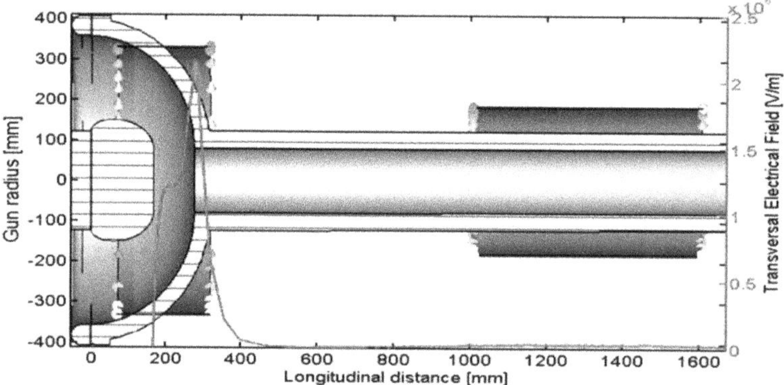

Figure 7.22. Gun geometry with the transverse electric field induced by the anode edge whose values are reported on right side of the *y*-axes.

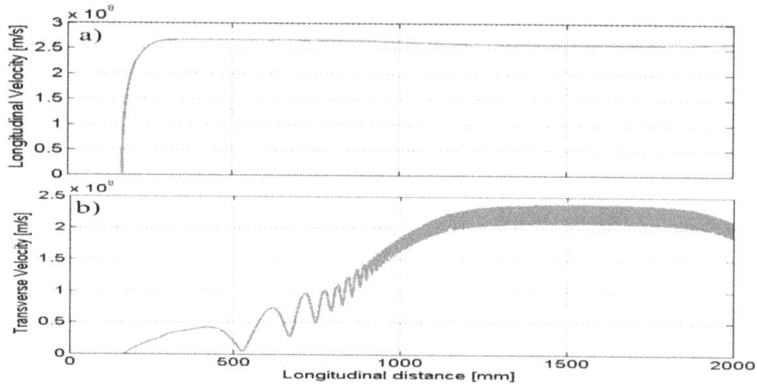

Figure 7.23. Longitudinal (a) and transverse (b) velocity versus longitudinal direction of the gun axial symmetry.

The longitudinal and transverse relative velocity spread are reported in figure 7.24(a) and (b), respectively, while the pitch ratio is shown in figure 7.25. The results provided by the simulation are compatible with the homogeneity requests quoted in the previous sections.

The proposed gun design is based on a non-adiabatic electric field solution, with a transverse component in correspondence of the ending edge of the diode (see figure 7.22). The magnitude of the effect can be easily understood, with the use of standard formulae for the Pierce gun design, such that we write the electric field at anode

$$E_A \cong \frac{V}{|R_{\text{cath}} - R_{\text{an}}|} \frac{R_{\text{cath}}}{R_{\text{an}}} \tag{7.15}$$

thus getting, for our design parameters an anode field value on the order of $10^6 \frac{V}{m}$, which agrees with the results provided by the simulation.

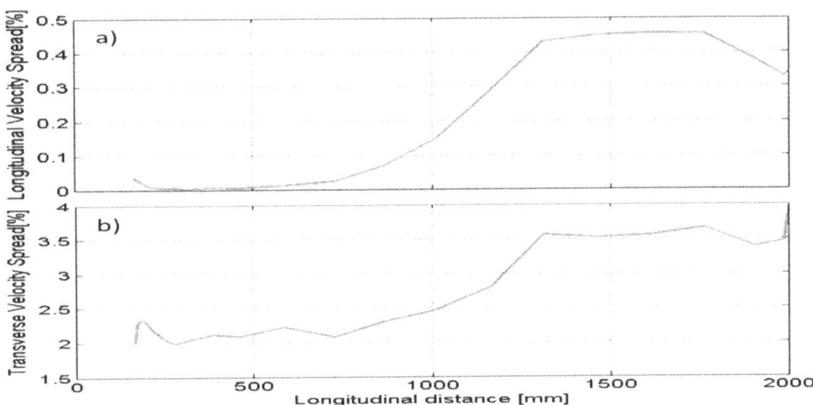

Figure 7.24. Longitudinal (a) and transverse (b) particles velocity spread versus longitudinal direction of the gun axial symmetry.

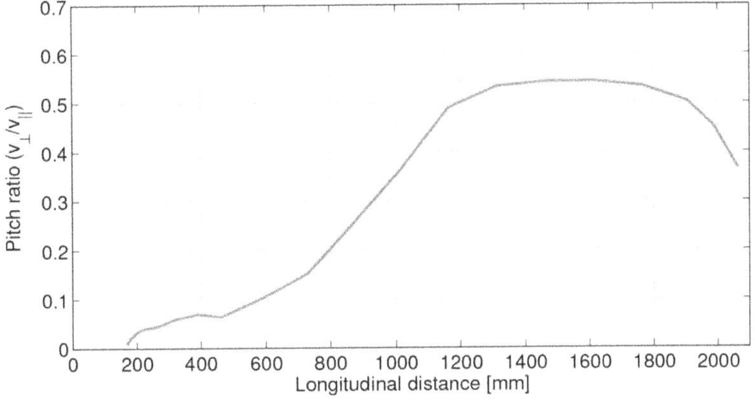

Figure 7.25. Pitch ratio versus longitudinal direction of the gun axial symmetry.

Owing to the cusp, the effect of the field on the electrons can be viewed as a defocusing with a focal length specified by

$$f = -4\frac{V}{E} \cong -4\frac{\rho_c}{\rho_a}(\rho_c - \rho_a) \tag{7.16}$$

An idea of the relevant effect on the particle dynamics is given in figure 7.26, where the effect of the field edge on the transverse velocity is emphasized.

As already stressed, the electrons are produced in a ring-shaped form to fulfill the optimum overlapping with the chosen operating mode inside the CARM cavity. The beam transport conditions ensuring the safe transport of the annular beam, along the line from the cathode to the interaction region, will be discussed in the next section.

In figure 7.27 we have reported the beam transverse section at different positions inside the channel from the cathode to the CARM cavity; the electrons are distributed along a circular corona ring and crucial parameters are

1. The average beam radius versus z (see figure 7.28).
2. The beam corona section versus z (see figure 7.29).

The beam is captured on a ring orbit with smaller radius inside the cavity, owing to the large strength of the CARM magnetic field. The radius stability is related to stability of the magnetic field produced by the set of the solenoids. The width of the corona undergoes a kind of strong focusing effect yielding a minimum at the center of the cavity and is then defocused. The importance of the beam radius control comes from the coupling conditions with the CARM operating modes.

The preliminary considerations, developed in this section, show that the e-beam can be transported preserving its shape, which can be controlled to ensure the desired electron beam electromagnetic mode coupling in the resonant cavity. The specific design details are reported in the next section.

The beam shaping accuracy, with particular reference to its annular structure, is very important because strictly associated with the interaction between electrons and electric field modes in the resonant cavity, that critically depends on the beam shape

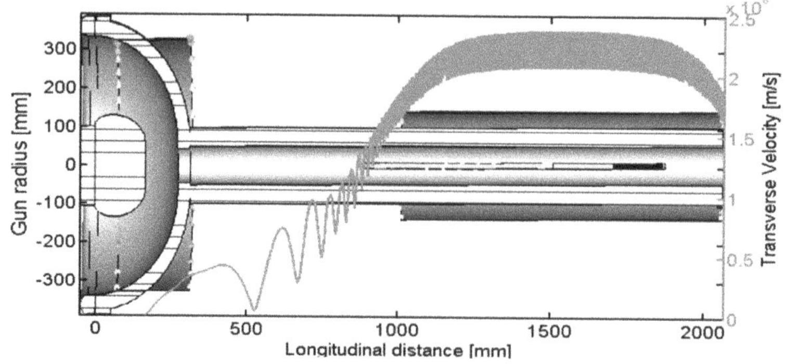

Figure 7.26. Transverse velocity from the cathode to the drift tube.

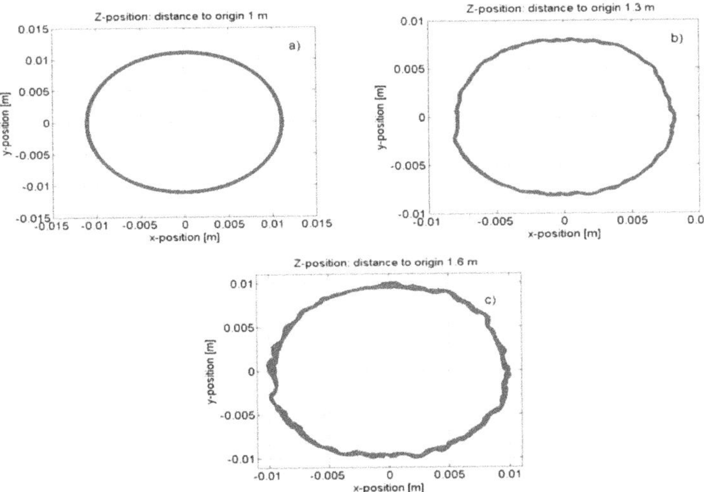

Figure 7.27. The electron beam ring at different positions along the transport line: (a) at the cavity entrance, (b) at the cavity middle, (c) at the cavity output.

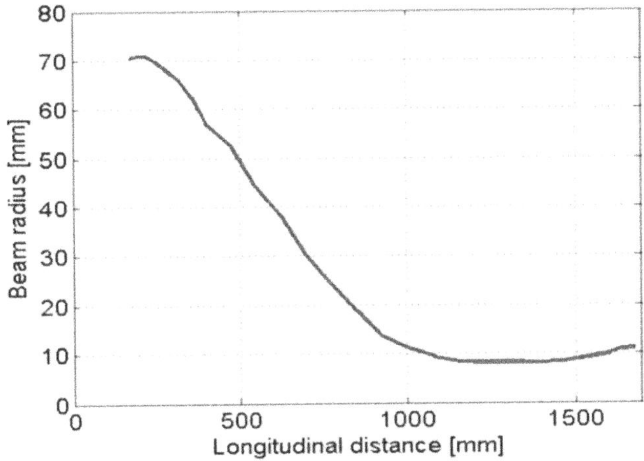

Figure 7.28. Average radius of the electron beam versus longitudinal coordinate z [mm].

and thickness along the cavity itself. It is evident that the mode excitation crucially depends on the portion of beam overlapping the structure of the transverse eigenmodes. This mechanism is crucial for determining the number of excited modes, which can simultaneously grow and eventually induce a reduction of the CARM efficiency. The matching between electron beam and electric field in the resonant cavity will be therefore one of the pivotal topics of the next section.

7.3.2 The e-beam transport line modeling

In figure 7.30 a detailed field profile with the expected intensity values is shown superimposed on the core part of the device from the cathode to the region in which

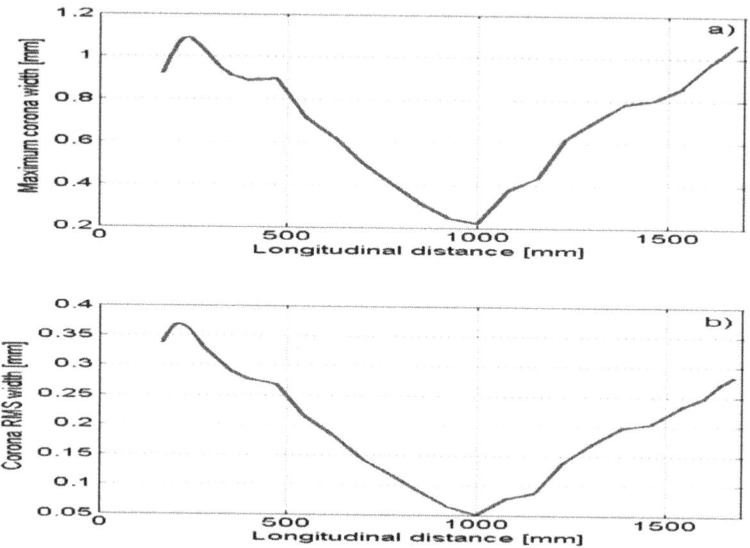

Figure 7.29. Average corona thickness versus z.

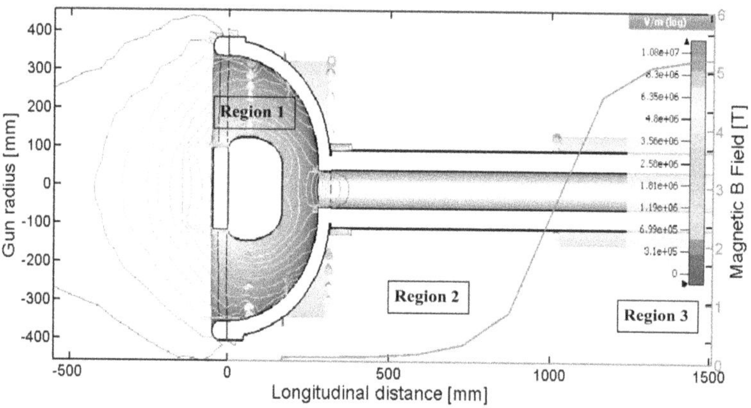

Figure 7.30. Magnetic field intensity profile from the cathode to the interaction region with the three operating regions: (1) acceleration and beam forming region; (2) adiabatic tapering; (3) high field interaction region.

the interaction between the electron beam and the high intensity magnetic field determines the CARM coherent emission.

As already stressed previously, the transport line from the cathode to the cavity region, is of paramount importance to obtain an e-beam with suitable characteristics for the CARM operation. The relevant design strategy can be summarized as follows (see figure 7.30):

(a) The first part segment (namely the transition from region (1) to (2) has the role of bringing the electrons to a kinetic energy of 0.6 MeV, furthermore, it is supposed to induce a kick due to the electric field of the edge cusp (see below) leading to a transverse component velocity, yielding an appropriate pitch angle of the e-beam providing a suitably large coupling with the cavity modes.

(b) The 'drift' section before region (3), where the combined magnetic field of the principal and correcting coil is present, has a manifold role.

(c) To control the transverse motion components, by adjusting the pitch to an optimum value.

(d) To control the radius of the annular beam and eventually the thickness of the corona.

(e) To match the section of the beam to the cavity mode to be enhanced by the CARM interaction. The matching should be achieved in an adiabatic way by eliminating (through a gentle tapering of the e-beam ring radius) the coupling to the unwanted mode in the region of the cavity containing the Bragg reflector (see figure 7.30 where we have superimposed the field intensity tapering to the cavity region).

In figure 7.31 we have reported the pitch ratio evolution along the longitudinal coordinate and we have underscored the region where the kick occurs and the successive concurring effect of the magnetic field, which, among other things, counteracts the growth of the transverse motion.

Figure 7.32 reports the evolution along the transport line of the spread of the longitudinal velocity, which grows with increasing transverse velocity component, but remains, in the interaction region well within the limits for a safe CARM operation.

Figure 7.33 shows the beam radius versus the transport coordinate and it is evident that in the interaction region the beam has the suitable dimensions to couple to the cavity modes.

A more appropriate idea is offered by figure 7.34, yielding the transverse profile of the beam at different positions along the transport line. It is evident that the control of both radius and the thickness of the circular corona may become problematic.

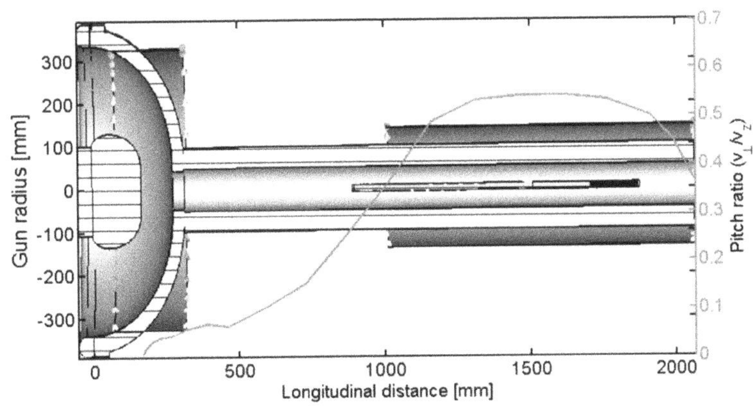

Figure 7.31. Pitch evolution $\left(\frac{v_\perp}{v_z}\right)$ versus the longitudinal direction.

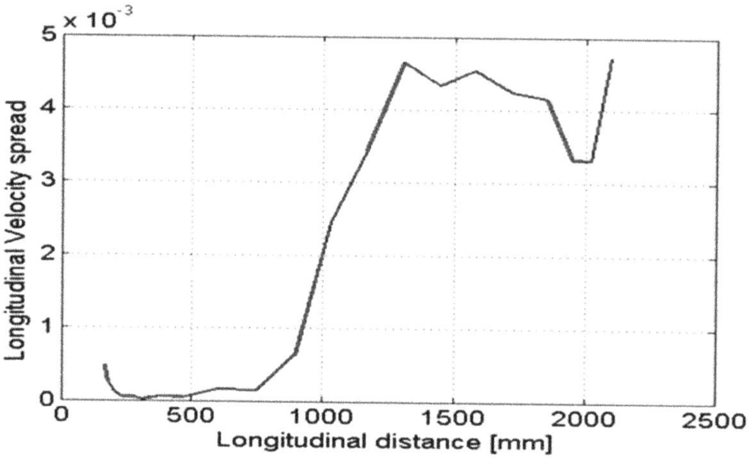

Figure 7.32. Relative longitudinal velocity spread $\left\langle \frac{\Delta v_z}{v_z} \right\rangle_{\text{rms}}$ along the transport line (coordinate z).

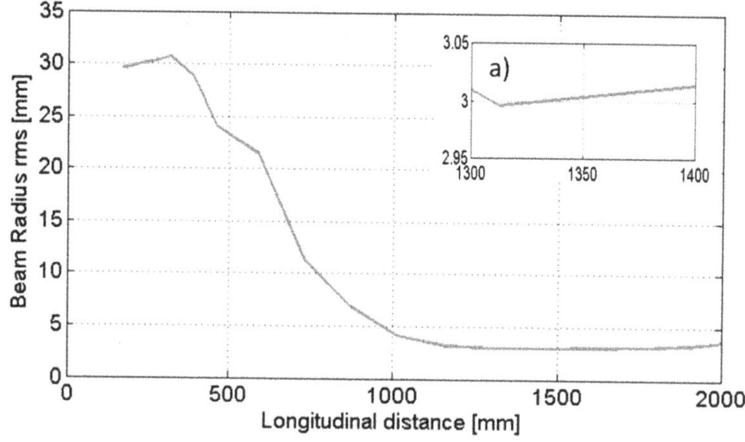

Figure 7.33. Annular beam radius versus the longitudinal coordinate optimized for the interaction with the waveguide mode TE_{53}. In inset (a) is evidenced the beam radius value at the entrance of the smooth cylindrical section of the cavity where the coupling coefficient assumes the maximum value, as was discussed in the last chapter (see figure 6.20).

The study we have developed so far is not based on a very refined theory to optimize the transport detail. We have indeed merely used a ray-tracing procedure to study the particle distribution and not an elaborate technique, as in the case of strong focusing, based on a formalism à la Courant and Snyder. Within such a context concepts like emittance and Twiss parameters could provide substantive help too. We must, however, emphasize that we are dealing with the transport of an e-beam with unusual characteristics. In standard acceleration and transport problems the beam is solid (without any hole) so that the concept of quantities like emittance and phase space domain are easily defined. To overcome such a

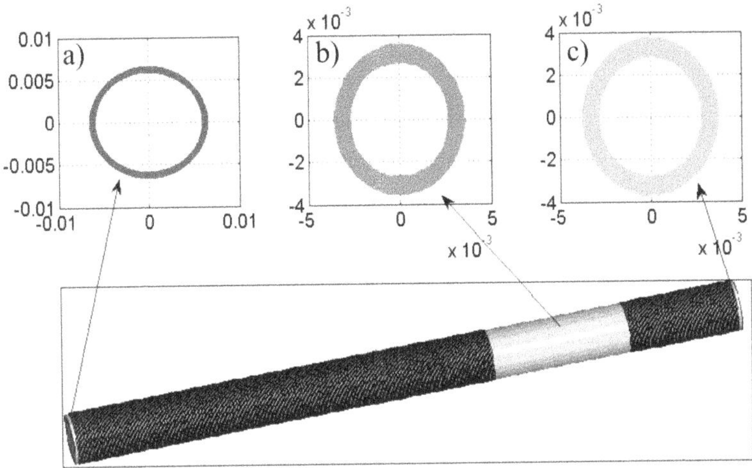

Figure 7.34. Geometrical distribution of the beam at different positions along the longitudinal axis, (a) $z = 900$ mm, (b) $z = 1613$ mm, (c) $z = 1873$ mm.

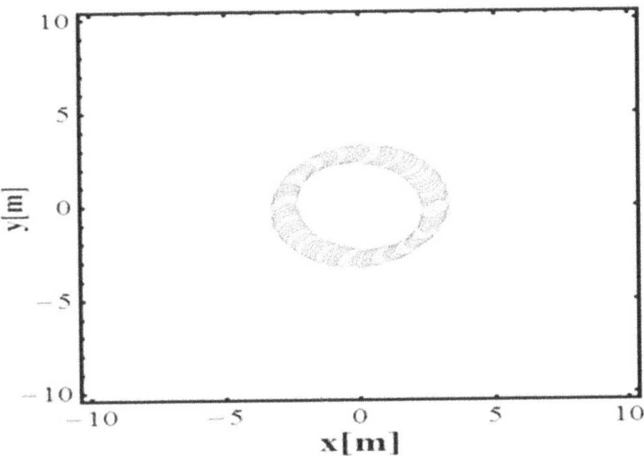

Figure 7.35. Beamlet realization of the beam circular corona.

drawback, we have developed a different point of view, namely we have visualized the beam as a collection of beamlets around the corona, as illustrated in figure 7.35. The use of such a point of view is helpful to check the reliability of the previous analysis and reconcile our procedure with a more conventional point of view.

According to the usual procedure we define the rms of the x and y position and of the relevant velocity components, defined as

$$\xi' = \frac{d\xi}{dz}, \qquad (7.17)$$
$$\xi = x, y$$

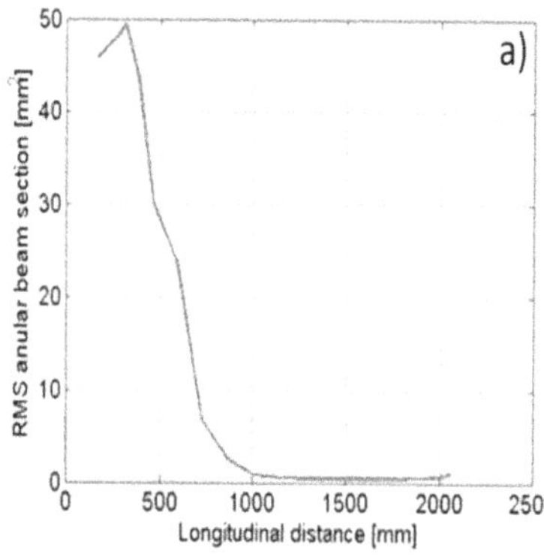

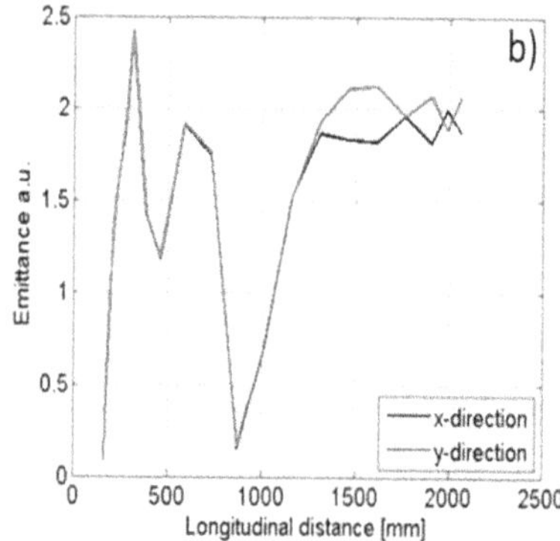

Figure 7.36. (a) Annular beam rms area $\left(\sqrt{|\hat{C}_{x,y}|}\right)$ versus longitudinal coordinate; (b) emittance (arbitrary units) $\left(\sqrt{|\hat{C}_{x,x'}|}, \sqrt{|\hat{C}_{y,y'}|}\right)$ versus the longitudinal coordinate (in current units the a.u. units should be multiplied by $10 \cdot \text{mm} \cdot \text{mrad}$).

We then introduce the 2×2 covariance matrices

$$\begin{aligned} &\hat{C}_{x,x'}, \hat{C}_{y,y'}, \hat{C}_{x,y}, \\ &\hat{C}_{\xi-\xi'} = \begin{pmatrix} \sigma_{\xi\xi} & \sigma_{\xi\xi'} \\ \sigma_{\xi'\xi} & \sigma_{\xi'\xi'} \end{pmatrix}, \hat{C}_{x,y} = \begin{pmatrix} \sigma_{xx} & \sigma_{xy} \\ \sigma_{yx} & \sigma_{yy} \end{pmatrix}, \\ &\sigma_{\xi\xi} = \sqrt{\langle \xi^2 \rangle - \langle \xi \rangle^2}, \\ &\sigma_{\xi'\xi'} = \sqrt{\langle \xi'^2 \rangle - \langle \xi' \rangle^2}, \\ &\sigma_{\xi\xi'} = \sqrt{\langle \xi'\xi \rangle - \langle \xi' \rangle \langle \xi \rangle}. \end{aligned} \tag{7.18}$$

To give an idea of how these quantities evolve, we have reported in figure 7.36 the evolution along the longitudinal direction of the square root of the associated determinants. In figure 7.36(a) we have reported the behavior of $\sqrt{|\hat{C}_{x,y}|}$ which represents the rms area of the annular beam. The evolution of $\sqrt{|\hat{C}_{x,x'}|}, \sqrt{|\hat{C}_{y,y'}|}$ is shown in the plots of figure 7.36(b). These quantities could be interpreted as the longitudinal and vertical emittances[1]. A general conclusion which can be drawn from the previous plots is that the condition of minimizing either $|\hat{C}_{x,y}|$ and $|\hat{C}_{x,x'}|, |\hat{C}_{y,y'}|$ in the interaction region cannot be achieved; the 'matching' condition

[1] There are technical issues regarding the definition of emittance suggested in this text. The quantity we have mentioned does not, strictly speaking, represent an emittance, it refers to the particle beam phase space area and is a conserved quantity during the transport. Its rigorous definition requires the use of canonical variables to properly treat the phase space and the relevant transport. Regarding our case, we have used the transverse velocities, which, for the type of fields and forces involved, are kinetic and not canonical variables. Furthermore, the quantities we have reported are not a transport invariant for different reasons, including the fact that we are not dealing with a Liouvillian transport along the entire line and that, inside the main solenoid, the $x - y$ motion is coupled and therefore the real invariant is associated with the 6-D phase space.

consists in a gentle compromise, to be achieved by relaxing the request on the surface minima.

A general comment and a caveat should be added; in designing the transport line of the CARM device we have been faced with a number of design problems, hereafter specified:

(a) The absence of an appropriate theoretical framework to handle this type of beam transport as in the usual Courant–Snyder (C–S) theory, allowing the definition of quad-like lenses to drive the beam.
(b) The difficulties associated with the control of both annular beam radius and corona thickness.
(c) The extreme sensitivity of these quantities to the magnet field distribution inside and outside the cavity interaction region.
(d) The importance of the emitting cathode surface roughness, only partially accounted for.

The results we have obtained show that

(i) A C–S-like theory can be partially recovered and some of the relevant paradigms can be exploited within the present context.
(ii) The annular beam geometry and shaping can in principle be controlled, but an analysis of the relevant criticity has not been done.
(iii) A global matching of the beam (including the adequate condition on velocity spread) can be obtained, but optimization criteria has not been found yet.

In this section we have drawn the paradigmatic elements for the development of a 'modern' CARM fusion device. We have put in evidence the structure of the design tools and how challenging is the embedding of the various composing elements. As we have mentioned in chapter 6, the development of CARM sources has been hampered in the past by the impossibility of reaching high quality beams and an adequate control of the whole system. The technological developments after thirty years will certainly provide the elements to fill such gaps.

7.4 A hint to the development of future technologies

7.4.1 RF undulators and wigglers

Coherent and quasi coherent radiation from FEL and synchrotron radiation sources is obtained by exploiting either undulators or wigglers. They are called insertion devices (IDs) [35–40], since they are placed (inserted) into the accelerator track of the electron beam. The wiggler is a different flavor of undulator (or vice versa) and the distinctive difference between the two devices is the value of the strength parameter, which, for typical wigglers, is significantly larger than unit. Thus allowing a significant emission at higher harmonics [41–44].

In the following, we will see how laser and other sources of radiation of the type discussed in the second part of this book can be exploited instead of undulators or wigglers.

In chapter 4 we have underscored that magnetic undulators constitute a crucial component of FELs. We have mentioned that it is a static magnetic device, nowadays realized with permanent magnet undulators.

From the point of view of a relativistic electron, moving inside the undulator, its field is 'seen' not dissimilarly from an electromagnetic wave. On the other side, an intense wave may play the same role of a magnetic undulator, namely that of inducing the electron transverse motion component, necessary to couple to an external wave, thus giving rise to the FEL process, according to the lines discussed in chapter 4.

The idea of replacing magnetic undulators with an intense laser wave arose quite early in the FEL community [45–52]. With respect to this, we would like to mention that one of the first suggestions of FEL operation was based on the stimulated electron Compton scattering of a CO_2 laser [53].

The interest in the wave undulators (WUs) is justified by the fact that their use offers the possibility of achieving shorter wavelengths with lower energy beams and, more in general, access to smaller size and cheaper FEL devices. There are evident technological issues which hampered their use in the past, as, e.g. the wave power which is required to be large to reach values of the K parameter enough to ensure sufficient gain. These handicaps have in part been solved and will be overcome in the not too distant future.

The role of a WU may be played by powerful lasers but also by any RF wave, produced by the devices we have discussed in the previous chapters.

Before entering the relevant details of construction and operation, we premise a few important points.

The dynamics of an electron passing through an undulator of period λ_u is not different from that of an electron moving into the field of an electromagnetic wave. According to the Weizsacker–Williams approximation [42, 54, 55] the undulator is viewed by the relativistic electrons as an electromagnetic field with a wavelength $2\lambda_u$. In turn, the laser can be treated as an undulator with an equivalent period $\lambda_L/2$. This is easily understood from the condition (see equation (4.4))

$$\frac{1-\beta}{\beta}\lambda_u = \lambda_L, \tag{7.19}$$

which can be written as

$$\frac{1}{\beta}\lambda_u = \lambda_L - \lambda_u \rightarrow \lambda_u = \frac{\beta}{1+\beta}\lambda_L, \tag{7.20}$$

and, for relativistic electrons $\beta \cong 1 - \frac{1}{2\gamma^2}$, yields the aforementioned correspondence.

The radiation emitted by electrons in an electromagnetic WU, through inverse Compton scattering (ICS), exhibits much shorter wavelengths, than that radiated in a conventional static undulator (whose period is typically much longer than the wavelength corresponding to the wave field).

In order to better fix the previous concepts, we emphasize that

(a) The undulator is seen by the electrons as an ensemble of photons;

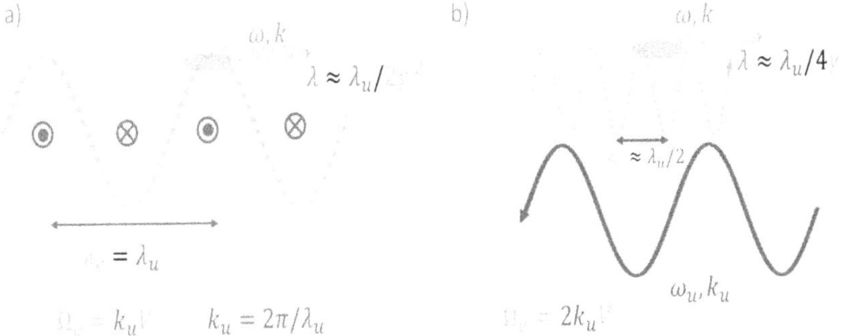

Figure 7.37. Analogy between (a) permanent magnet undulator and (b) and RF undulator.

(b) A laser wave is viewed by the electrons as a dynamic undulator, where the electrons experience an alternating magnetic field, such that they perform transverse oscillations and the mechanism ensuring the emission of radiation is provided by the Compton backscattered photons (see figure 7.37)

Despite these analogies, we cannot state a full equivalence between the two radiator tools.

The means of the comparison in our hands are of kinematic nature only, using the Compton scattering formulae yields for the ICS radiation wavelength λ_S the expression

$$\lambda_L = \frac{\lambda_L}{4\gamma^2}\left[1 + (\gamma\vartheta)^2 + 4\frac{\hbar\omega_L\gamma}{m_e c^2}\right], \tag{7.21}$$

which is equivalent to that already derived in the case of the undulator, with the significant exception of the last term, containing the Planck constant. Such a contribution cannot be recovered within the classical approximation adopted so far and will be neglected[2] (unless we consider electrons with a relativistic factor $\gamma \gg \frac{1}{4}\frac{m_e c^2}{\hbar\omega_L}$). The missing piece is the K-dependent shift term, which can be included within the context of the present analogy, by noting that the undulator is equivalent to a high intensity electromagnetic wave. The electrons, moving in the field of such a wave, experience a deflection analogous to that induced by the undulator. The induced deflection is responsible for a wavelength shift [56]

$$\frac{\Delta\lambda_s}{\lambda_s} = (1 + K_L^2), \tag{7.22}$$

with

$$K_L^2 = \frac{n_L}{(2\pi)^2} r_e \lambda_e \lambda_L, \tag{7.23}$$

[2] Strictly speaking we should refer to Thomson scattering and inverse Thomson scattering. For completeness sake we recall that the ICS frequency is $\hbar\omega_S = \frac{\hbar\omega_L(1+\beta)}{1 - \beta\cos\theta + \frac{\hbar\omega_L}{\gamma m_e c^2}(1+\cos\theta)}$ which yields (7.21) in the energy limit.

where r_e is the electron classical radius, λ_e the electron Compton wavelength and n_L the photon number density carried by the wave. Having established that, from the point of view of the electrons, the undulator is an ensemble of photons, and we can easily derive the equivalent density [55, 57, 58], which reads

$$n_U = (2\pi)^2 \frac{K_U^2}{r_e \lambda_e \lambda_u}. \tag{7.24}$$

Hence, according to equation (7.22), we re-gain the FEL K-dependent shift.

The rate of backscattered photons by an electron bunch is linked to the Thomson cross section and reads [59]

$$\begin{aligned} \frac{dn_s}{dt} &= f_r \frac{N_e}{2\pi\, \sigma_x \sigma_y} n_L \sigma_T \\ \sigma_T &= \frac{8}{3} \pi\, r_e^2, \end{aligned} \tag{7.25}$$

with f_r being the scattering rate and N_e the number of electrons inside the bunch.

The laser wave intensity and magnetic field are linked by the relationships

$$\begin{aligned} I_L &= \frac{P_L}{2\pi \sigma_x \sigma_y} = c \frac{B^2}{\mu_0}, \\ P_L &= \frac{n_L \hbar \omega_L}{\tau_L}, \end{aligned} \tag{7.26}$$

where P_L is the laser power and τ_L is the duration of the laser pulse. If we denote by N^* the number of equivalent periods contained in a pulse length, we end up with

$$\begin{aligned} n_S &\simeq \frac{4\pi}{3} \alpha_{\rm em} N^* K^2 \\ N^* &= \frac{c\tau_L}{\lambda_L}. \end{aligned} \tag{7.27}$$

According to the discussion developed so far, we are allowed to proceed by considering the radiation process, in an undulator, as an elementary electron–photon interaction (and vice versa). The description in these terms does not allow the inclusion of the transverse field in-homogeneities, associated with the static magnetic nature of the undulator. However, the essential features of the emission mechanism are correctly captured and we will not further comment on these points.

The possibility of accessing to electron beams with unprecedented large brightness and to high power lasers with intensities of the order 10^{28-30} W m^{-2} has made feasible the realization of x-ray FEL with short undulators and low energy e-beams.

The crucial quantities for the FEL operation are the undulator period and the associated K-parameter. In the case of laser-wave undulators we find

$$\begin{aligned} K_L &\cong 0.85\, \lambda[\mathrm{m}] \sqrt{I_L[\mathrm{W\,m^{-2}}]}, \\ \lambda_{U,L} &= \frac{\lambda_L}{2} \end{aligned} \tag{7.28}$$

A CO_2 laser with an intensity corresponding to 10^{19} W m^{-2} is capable of providing a strength K_L sufficiently large to support a SASE-FEL operation, ensured if the Pierce parameter is sufficiently large.

The advantage, in terms of undulator length, is provided by the following identities

$$L_{g,L} = \frac{\lambda_{U,L}}{4\,\pi\,\sqrt{3}\rho} = \frac{\lambda_L}{8\pi\,\sqrt{3}\rho},$$
$$L_{s,L} \cong \frac{L_{g,L}}{2\rho}, \tag{7.29}$$

yielding gain and saturation lengths. Equation (7.29) for $\rho \cong 10^{-4}$ and $\lambda_L \cong 10\,\mu$m yields as $L_{s,L}$ of a few cm.

In figure 7.38, we have reported the gain growth of an FEL operating at 1.34 nm with the parameters shown in table 7.4. It should be noted that the laser energy of 250 J is quite a large value and a peak current of 5 kA is not easy to be transported at low energy.

We have underscored the possibility of getting x-ray FELs with a small undulator and a small accelerator although we have been unfair in the comparison, because we did not include the dimensions of the powerful laser ensuring the WU parameters.

For a laser with characteristics not too far from those reported in table 7.4 see [50, 51].

How far are we from the possibility of getting a high power laser useful to sustain a FEL SASE operation? The plot reported in figure 7.39 shows the technological evolution of PW (P = Peta = 10^{15}) laser devices. A reasonable prediction could be 10–15 years starting from 2020.

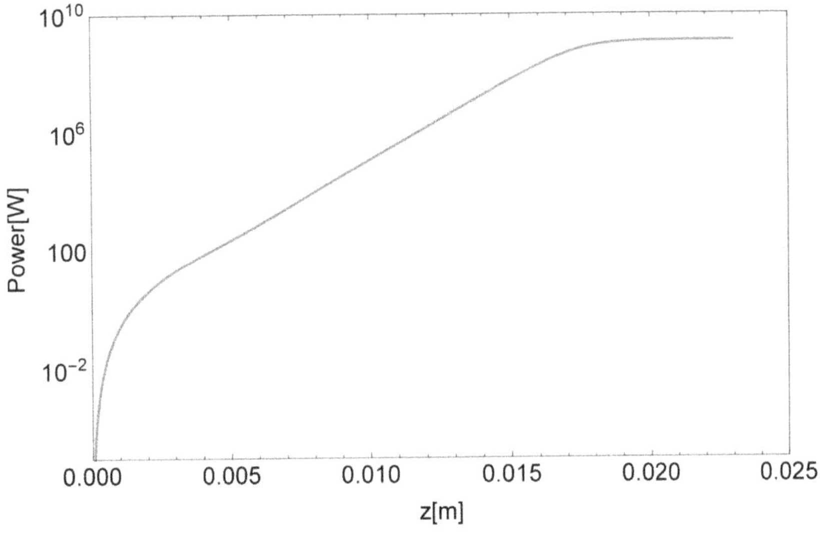

Figure 7.38. Power growth of a SASE FEL operating at 1.34 nm with the parameters of table 7.4.

Table 7.4. Laser and beam parameters for wave undulator FEL-SASE operation.

$E \cong 25\text{MeV}$
$I[\text{MW m}^{-2}] \cong 8 \cdot 10^{19} \Rightarrow K \cong 0.76 \Rightarrow \lambda \cong 1.34\text{nm}$
$J[\text{A m}^{-2}] \cong 10^{11} \Rightarrow \rho \cong 3 \cdot 10^{-4} \Rightarrow L_g \cong 7.7\text{mm}$
$Z_R \cong 5 \cdot 10^{-3}\text{m} \Rightarrow L_s \cong 1 - 2\text{cm}, E_L[J] \cong 260$!!!
$I[A] \cong 5 \cdot 10^3$

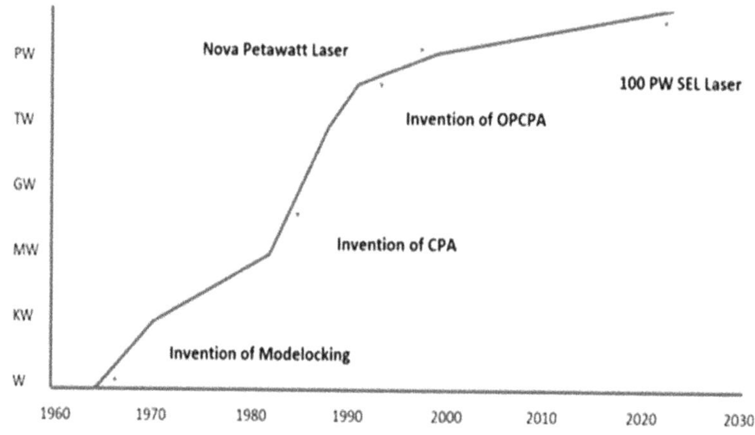

Figure 7.39. Evolution of PW laser power versus the year. CPA = chirped pulse amplification, OPCPA = optical parametric chirped pulse amplification. Reproduced from [60], copyright the authors. CC BY 4.0.

The parameters we have quoted for the laser wave undulator operation, are very tough.

The RF technology provides more relaxed parameters and wavelengths of the order of few mm would guarantee X-FEL SASE operation with beam energies $\sqrt{10}$ less than the operation with conventional tools.

In figure 7.40 we have reported a general comparison between the FEL performances in the x-ray region between laser wave, RF and magnetic undulators.

There are a number of requirements to be satisfied by an RF undulator (RFU) to become an effective tool for a SASE-FEL operation, namely:

1. High microwave fields (~1 Tesla at the wavelength of ~1 cm) are required to provide high enough ($K \sim 1$) undulator parameter. Therefore, powerful RF sources are needed.
2. The high field carried by the RF wave causes problems of breakdown in the cavity where it is propagating.
3. The analysis of the electron motion inside the RF undulator field is not as straightforward as for the ordinary magnetic devices and the evaluation of the 'spontaneously' emitted radiation requires extreme care.

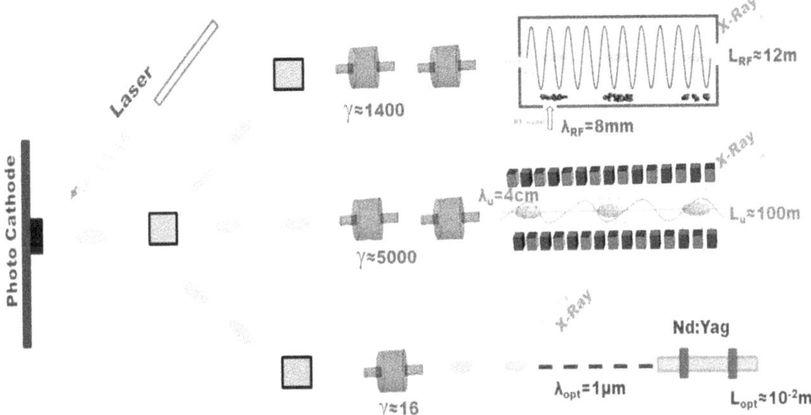

Figure 7.40. Comparison between laser wave (or optical), magnetic and RF undulators from right-top to right-bottom, respectively, for an output of x-ray radiation with $\lambda = 1\text{nm}$. The figure reports an electron beam emitted from the photocathode to the 'undulator', after a suitable acceleration. The expected undulator lengths and the e-beam energies, necessary to reach the SASE-FEL performances, are underlined too.

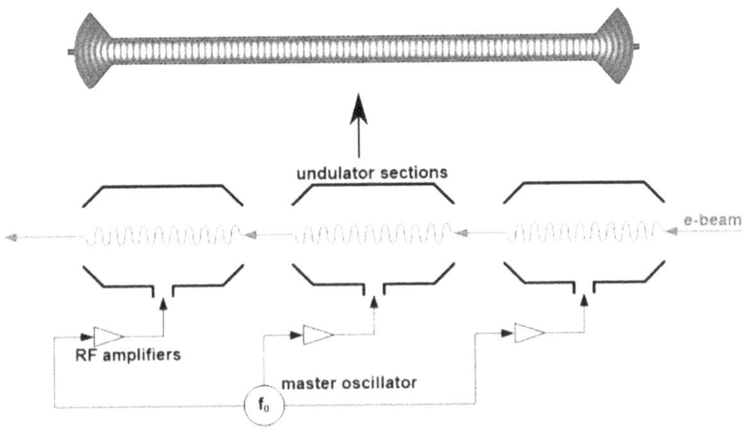

Figure 7.41. RF undulator and different sections locked in phase.

4. In the case of long saturation length, different sections of synchronized RF sources should be used

A proposal in this direction is sketched in figure 7.41, where we show the suggested scheme (experimentally tested) by Tantawi and collaborators [61, 62], in which a high power RF feeds into a high Q cavity to form an undulator section to be connected in phase with other sections.

We have previously naively stated that the equivalent period of an RF undulator is just 1/2 of the field wavelength, and did not specify how the K parameter should be defined, except that it should be linked to the square root of the field intensity.

Regarding the first point, we note that the wavelength of the field inside the waveguide is fixed by

$$\lambda_{U,RF} = \frac{2\pi}{h + k}, \tag{7.30}$$

where h is the propagation constant in the waveguide[3] and $k = \omega/c$ is the field wave vector. Under the assumption that $h = k$ we get

$$\lambda_{U,RF} \cong \frac{\lambda_{\mathrm{RF}}}{2}. \tag{7.31}$$

The beam deflection inside the RFU depends on the electric and magnetic components of the field and reads

$$K_{\mathrm{RF}} = \frac{|e|\,(E_{\perp} + cB_{\perp})}{mc^2(h + k)}. \tag{7.32}$$

The previous relations can be exploited to start a quick analysis of an FEL-SASE design and find optimization criteria, which can be found in [63] and will not be reported here.

The assumption that $K_{\mathrm{RF}} \cong 1$ implies for the required field amplitudes

$$E_{\perp} + cB_{\perp} \cong \frac{mc^2}{e}(h + k) \cong \frac{mc^2}{e}\frac{2\pi}{\lambda_{\mathrm{RF}}}. \tag{7.33}$$

Accordingly, in an RFU fed by RF radiation field of 1 cm the required amplitudes are of the order of hundreds of MW m^{-1}, which should be reduced to avoid cavity discharges.

Such a power level can be ensured by the use of cavities with high Q-factors, powered by RF sources like klystrons or gyro-klystrons, capable of delivering power levels of megawatts.

A further drawback emerges from the fact that, as already noted, the RFU should be composed by many, mutually phased sections (each 1 m long) and each of them should be fed by its own RF source, therefore an RFU, with tens of these sections, might be even more expensive than a UM.

Considerations based on the optimization criteria show that K_{RF} with values significantly below 1, can be exploited to run soft x-ray operating SASE-FELs [63]. Low K_{RF} values implies less input power and therefore lower costs.

Advanced microwave undulators and electron generators for new-generation FEL are discussed in [64]. In particular, the possibilities of using high-power millimeter-wave radiation pulses for electron pitching in the operating space of the laser (in a microwave undulator), as well as for cooling and focusing of electron bunches, are considered.

[3] We recall that the field propagation inside a waveguide is described by the product $e^{i\omega t}e^{-i\alpha z}$ where α is a complex quantity called a propagation constant.

A design of an RF undulator at 91.392 GHz that has a period of 1.75 mm and a minimum beam aperture (as seen by the beam) 2.375 mm has been proposed in [62]. To confine the fields inside the undulator a corrugated waveguide is connected through a matching section to a linear taper and a mirror. After the mirror, a Bragg reflector and a matching section are used to reflect back all the fields leaking out of the mirror opening. This undulator requires approximately 1.4 MW for sub-microsecond pulses to generate an equivalent K value of 0.1. For transmission of such enormous power at mm-wave frequencies a highly overmoded corrugated waveguide is used. It is coupled to the undulator cavity through the beam pipe without disturbing the undulator fields. This system of the RF undulator and its coupling scheme allows for fast dynamic control of the polarization of the emitted light.

One of the advanced concepts is the so-called flying radio frequency undulator [65]. It aims to produce coherent x-ray radiation by means of a relatively low-energy electron beam and pulsed mm-wavelength radiation.

In figure 7.42 we have shown a comparison between normal UM, RF and the further possibility offered by the flying undulator concept.

It should be noted that regarding UM and RF, the e-beam is counter-propagating with respect to the field, thus determining a backscattering and a Doppler up-shift scaling with γ^2.

The use of the flying undulator scheme (see figure 7.43), namely a short pulse of radiation co-propagating in a waveguide, is a solution to avoid the already mentioned problems of breakdown. The experiment has shown that ns RF pulses, up to GW level, can go through the propagating structure without creating significant discharge troubles [66–68].

Sources capable of providing powerful fields of this type are the backward wave oscillators (BWOs) [69, 70], and can deliver pulses with the desired level of power (with high repetition rate), in both X and Ka band[4]. The possibility of mutual phase

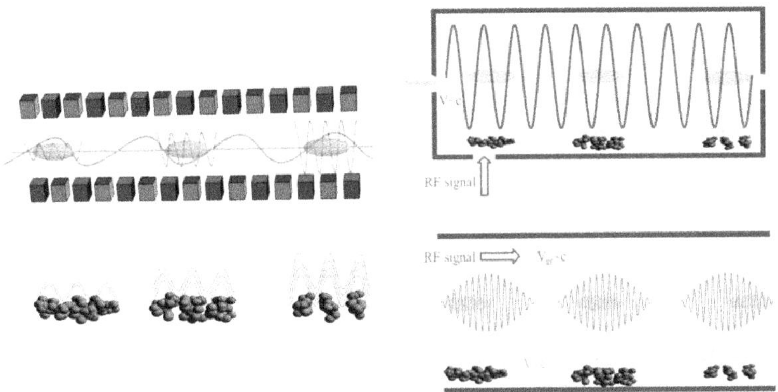

Figure 7.42. On the left: normal UM undulator. On the right-top: RF undulator. On the right-bottom: flying undulator.

[4] We are referring to the microwave frequency band Ka between 27 – 40 GHz and $X8 - 2$ GHz.

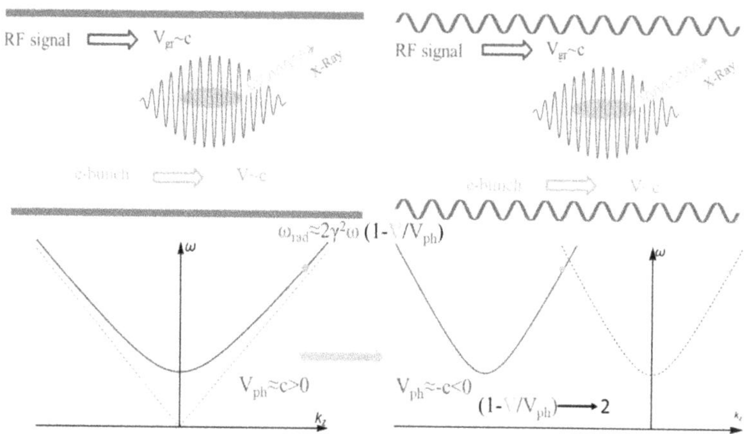

Figure 7.43. Flying undulator scheme for which the condition of $V_{\text{ph}} \approx -c$ enhances the Doppler shift of the radiated frequency wave.

locking between the different sections of U-flying has also been experimentally checked [70].

Let us now try to understand what are the advantages of this type of undulator configuration, apart from the possibility of reducing the damage due to breakdown discharges.

To begin with, we consider the case of a U-flying counter-propagating with respect to the electrons. If we denote by τ its duration and v_g its group velocity, one finds that the effective undulator length experienced during the interaction is [65]

$$L_{U,-} = \frac{v_g \tau}{1 + \dfrac{v_g}{c}}. \tag{7.34}$$

Assuming $\tau \cong 20$ ns and $v_g \cong c$, we find $L_{U,-} \cong 1.5$ m, which is not sufficient to sustain an x-ray SASE FEL. In contrast, for the co-propagating configuration we get

$$L_{U,+} = \frac{v_g \tau}{1 - \dfrac{v_g}{c}}. \tag{7.35}$$

Expressing the group velocity in terms of $L_{U,+}$ we find

$$v_g = \frac{L_{U,+}}{L_{U,+} + c\tau} c, \tag{7.36}$$

keeping for $L_{U,+}$ 50 m, we obtain the following condition for the group velocity $v_g \cong 0.893c$.

The corresponding undulator period and K parameter can be evaluated as for the RFU counterpart, so that we find also

$$L_{U,+} \cong \frac{\lambda_{\text{RF}}}{2\rho_{U,+}}, \tag{7.37}$$

which yields

$$\rho_{U,+} = \frac{\lambda_{\rm RF}}{2} \frac{1 - \frac{v_g}{c}}{v_g \tau}. \tag{7.38}$$

An idea of the behavior of $\rho_{U,+}$ versus v_g is reported in figure 7.44, which shows that a larger Pierce parameter demands lower group velocities.

Regarding the wavelength selection, we just note that the wavelength of the radiation emitted by the electrons moving along with the flying undulator, is determined by the dispersion relation

$$\omega_+ - k_+ v \cong \omega_{\rm RF} - k_{||,\rm RF} v, \tag{7.39}$$

where

$$k_+ = \frac{\omega_+}{c}, \; k_{||,RF} = \frac{\omega}{v_{\rm ph}}, \tag{7.40}$$

with v the electron velocity and $v_{\rm ph}$ the phase velocity, which eventually yields

$$\omega_+ \cong 2\omega_{\rm RF}\gamma^2\left(1 - \frac{v}{v_p}\right). \tag{7.41}$$

The practical realization of a flying undulator can be achieved using different schemes, two of which are reported in figure 7.45, where we have shown:

1. the periodic mirror transfer line in which the electrons intercept in multiple positions after many reflections between suitably positioned mirrors;
2. the corrugated circular waveguide in which the electrons travel with the retarded radiation feed inside the cavity by an external RF.

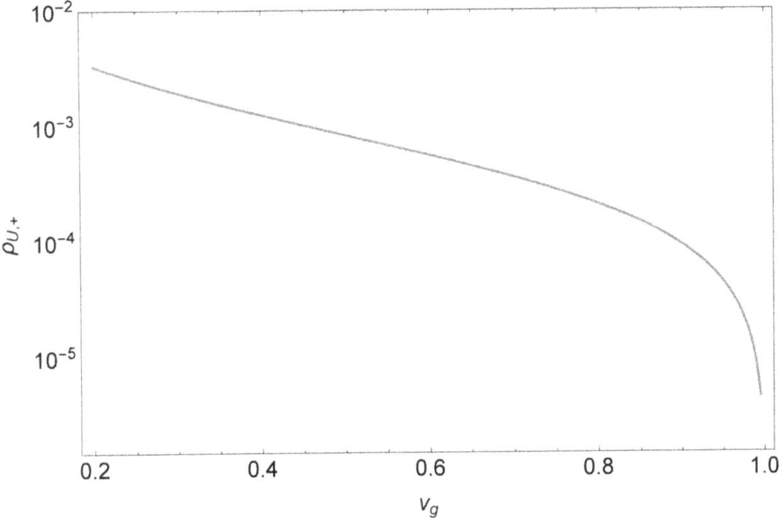

Figure 7.44. Pierce parameter versus group velocity.

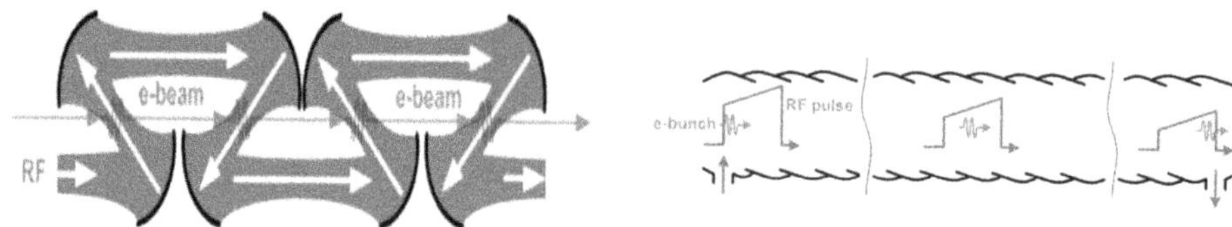

Figure 7.45. On the left: periodic mirror transfer line. On the right: corrugated helical wave guide.

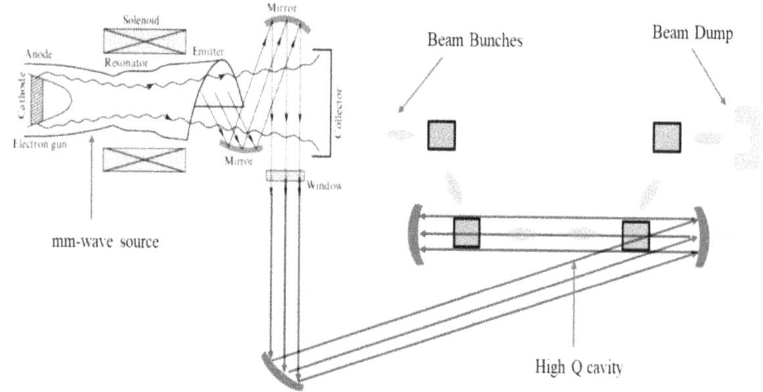

Figure 7.46. One of the possible configurations of FEL based on gyrotron powered electromagnetic wiggler proposed in [71].

For further comments we invite the reader to check the quoted literature at the end of the chapter.

Finally, a few words on the development of microwave wigglers, most of which are utilizing powerful electromagnetic waves produced by gyrotrons. One of the earliest works where the principles of gyrotron powered electromagnetic wigglers for FELs was proposed and discussed is [71]. The theoretical analysis presented there has demonstrated that the usage of short wavelength electromagnetic wigglers in waveguides and resonant cavities can significantly reduce the required electron beam voltages, resulting in compact FELs. The presented gain calculations, in the low- and high-gain Compton regime, take into account the effects of emittance, transverse wiggler gradient, and electron energy spread. Interestingly enough, gain scaling laws of the type discussed in chapter 4 are also presented.

One of the proposed configurations for FELs (operating in the infrared region using moderate energy <10 MeV electron beams) with electromagnetic wigglers powered by millimeter wavelength gyrotrons is shown in figure 7.46.

In this configuration both the gyrotron and FEL interaction take place in the same closed cavity having two sections. The gyrotron interaction occurs in the region where the wave number of the excited mode is almost entirely in the transverse direction (i.e. close to cutoff) and where there is the necessary axial magnetic field. The radiation produced in the gyrotron section is trapped inside the high-Q cavity and acts as a wiggler for the FEL interaction there. The high voltage electron beam can be brought into the cavity through tapered openings which are cut off to the trapped mode or through slots in the resonator walls. In an alternative configuration for an electromagnetic wiggler, the necessary field could be a traveling wave produced by a gyrotron or CARM. FELs with electromagnetic standing wave wigglers have been proposed also in [72], where a detailed analysis of the electromagnetic standing wave wiggler has been conducted for both circular and linear wiggler polarizations, following a single-particle approach. The presented results provide clear understanding of the resonance conditions and the coupling strength associated with each resonance of this type of FEL. In particular, the authors have pointed out a striking feature revealed by the observation that the electromagnetic standing wave wiggler FEL, under certain circumstances, exhibits a rich harmonic content caused by the presence of both the forward and backward wave components of the standing wave wiggler field. The presented nonlinear self-consistent equations for this type of FEL are permitting further investigation of it by the theoretical techniques and numerical codes developed for conventional FELs.

Other schemes for infrared free-electron lasers based on gyrotron-powered electromagnetic wigglers have been proposed by Fliflet *et al* in [73, 74]. They consider the design of a 10 μm electromagnetic wiggler FEL for a proof-of-principle experiment performed using a 6 MeV electrostatic accelerator with beam currents 2 and 5 A. One of these schemes utilizes a waveguide and the other a quasi-optical wiggler resonator. Both configurations appear capable of producing kilowatt peak powers at 10.6 μm. The design of a far infrared FEL with an electromagnetic wiggler based on a quasi-optical gyrotron presented in [75] is characterized by the following parameters: (i) output wavelength 195 μm; (ii) wiggler wavelength 3.2 mm; (iii) electron beam energy 600 keV; (iv) effective beam current 100 A; (v) wiggler circulating power 50 MW; (vi) output power 115 kW.

One of the most remarkable demonstrations of an FEL with an electromagnetic wiggler has been reported by Granatstein *et al* [76]. In their experiment, in order reduce the beam energy requirements a two-stage device has been used where the same intense relativistic electron beam first generates 500 MW radiation of 12.5 GHz through a BWO mechanism and then uses the BWO radiation as a wiggler for an FEL operating at 200 GHz.

Finally, for further reading on the microwave wigglers and an in-depth study of their theoretical foundations the reader is referred to the book chapter by Freund and Antonsen [77].

References

[1] Prater R 1985 Review of electron cyclotron heating experiments in toroidal devices *Presented at the Course and Workshop on Applications of RF Waves To Tokamak Devices, Varenna, Italy, 5–14 Sep.*

[2] Thomassen K I 1986 *Free-electron Laser Experiments in Alcator C* Technical Report LLL-PROP-00202 (Lawrence Livermore National Lab, CA, USA)

[3] Cohen B I, Cohen R H, Nevins W M and Rognlien T D 1991 Theory of free-electron-laser heating and current drive in magnetized plasmas *Rev. Mod. Phys.* **63** 949–90

[4] https://lasers.llnl.gov/multimedia/publications/pdfs/etr/1986_12_6.pdf

[5] Scharlemann E T, Sessler A M and Wurtele J S 1985 Optical guiding in a free electron laser *Nucl. Instrum. Methods Phys. Res.* A **239** 29–35

[6] Nguyen D, Russell S and Moody N 2009 *Theory and Practice of Free Electron Lasers U.S. Particle Accelerator School* (Albuquerque, NM: University of New Mexico)

[7] Marshall T C 1985 *Free Electron Lasers* (New York: Macmillan)

[8] Sprangle P, Penano J, Hafizi B and Ben-Zvi I 2009 *Wall-plug Efficiencies of High-power Free Electron Lasers Employing Energy Recovery Linacs* Technical Report NRL/MR/6790-09-9187 (Naval Research Laboratory)

[9] Kroll N, Morton P and Rosenbluth M 1981 Free-electron lasers with variable parameter wigglers IEEE *J. Quantum Electron.* **17** 1436–68

[10] van der Wiel M J and van Amersfoort P W 1989 A role for free electron lasers in fusion? *Fusion Eng. Des.* **11** 245–53

[11] Prosnitz D 1988 The physics of free electron lasers and applications to electron cyclotron heating *Plasma Phys. Control. Fusion* **30** 1523–34

[12] Jong R A and Scharlemann E T 1987 High gain free electron laser for heating and current drive in the ALCATOR-c tokamak *Nucl. Instrum. Methods Phys. Res.* A **259** 254–8

[13] Litvak A, Sakamoto K and Thumm M 2011 Innovation on high-power long-pulse gyrotrons *Plasma Phys. Control. Fusion* **53** 124002

[14] Thumm M K A, Denisov G G, Sakamoto K and Tran M Q 2019 High-power gyrotrons for electron cyclotron heating and current drive *Nucl. Fusion* **59** 073001

[15] Sakamoto K, Kasugai A, Kajiwara K, Takahashi K, Oda Y, Hayashi K and Kobayashi N 2009 Progress of high power 170 GHz gyrotron in JAEA *Nucl. Fusion* **49** 095019

[16] Oda Y, Ikeda R, Kajiwara K, Kobayashi T, Hayashi K, Takahashi K, Moriyama S, Sakamoto K, Eguchi T and Kawakami Y 2019 Development of the first ITER gyrotron in QST *Nucl. Fusion* **59** 086014

[17] The world's First Serial Production Gyrotron for ITER Was Developed by Nizhny Novgorod Scientists https://scientificrussia.ru/partners/institut-prikladnoj-fiziki-ran/pervyj-v-mire-serijnyj-girotron-dlya-iter

[18] Denisov G, Litvak A, Sokolov E, Chirkov A, Eremeev A, Tai E, Soluyanova E, Myasnikov V and Popov L 2019 Development of megawatt gyrotrons in IAP/GYCOM *Proc. of the Int. Vacuum Electronics Conf. (IVEC) (Busan, Korea (South))* 1–2

[19] Popov L *et al* 2018 Status of the gyrotron complex for ITER: composition of the complex, manufacturing, obtained parameters, delivery conditions *EPJ Web of Conferences* **187** 01016

[20] Denisov G G *et al* 2018 First experimental tests of powerful 250 GHz gyrotron for future fusion research and collective Thomson scattering diagnostics *Rev. Sci. Instrum.* **89** 084702

[21] Jelonnek J, Aiello G and Alberti S *et al* 2017 Design considerations for future DEMO gyrotrons: a review on related gyrotron activities within EUROfusion *Fusion Eng. Des.* **123** 241–6

[22] Ikeda R, Oda Y, Kobayashi T, Terakado M, Kajiwara K, Takahashi K, Moriyama S and Sakamoto K 2016 Development of multi-frequency gyrotron for ITER and DEMO at QST *Proc. of the 41st Int. Conf. on Infrared, Millimeter, and Terahertz Waves (Copenhagen)*

[23] Thumm M 2020 State-of-the-art of high power gyro-devices and free electron masers *J. Infrared Millim. Terahertz Waves* **41** 1–140

[24] Flyagin V A, Khizhnyak V I, Manuilov V N, Moiseev M A, Pavelyev A B, Zapevalov V E and Zavolsky N A 2003 Investigation of advanced coaxial gyrotrons at IAP RAS *Int. J. Infrared Millim. Waves* **24** 1–17

[25] Dumbrajs O and Nusinovich G S 2004 Coaxial gyrotrons: past, present, and future *IEEE Trans. Plasma Sci.* 32 934–46

[26] Ruess S *et al* 2018 KIT coaxial gyrotron development: from ITER toward DEMO *Int. J. Microw. Wirel. Technol* **10** 547–55

[27] Idehara T and Sabchevski S P 2018 Development and application of gyrotrons at FIR UF *IEEE Trans. Plasma Sci.* **46** 2452–9

[28] Kartikeyan M V, Borie E and Thumm M K A 2003 *GYROTRONS High Power Microwave and Millimeter Wave Technology* (Berlin: Springer)

[29] Nusinovich G 2004 *Introduction to the Physics of Gyrotrons* (Baltimore, MD: Johns Hopkins University Press)

[30] Chu K R 2004 The electron cyclotron maser *Rev. Mod. Phys.* **6** 489

[31] Du C-H and Liu P-K 2014 *Millimeter Wave Gyrotron Traveling Wave Tube Amplifiers* (Berlin: Springer)

[32] Tsimring S E 2007 *Electron Beams and Microwave Vacuum Electronics* (New York: Wiley-Interscience)

[33] Gilmour A S 2011 *Klystrons, Traveling Wave Tubes, Magnetrons, Crossed-Field Amplifiers, and Gyrotrons* (Boston, MA: Artech House)

[34] Barker R J, Luhmann N C, Booske J H and Nusinovich G S 2005 *Modern Microwave and Millimeter-Wave Power Electronics* (New York: Wiley-VCH)

[35] Ciocci F, Dattoli G, Torre A and Renieri A 2000 *Insertion Devices for Synchrotron Radiation and Free Electron LaserSeries on Synchrotron Radiation Techniques and Applications* vol 6 (Singapore: World Scientific)

[36] Clarke J A 2004 *The Science and Technology of Undulators and Wigglers* (Oxford Series on Synchrotron Radiation) (Oxford: Oxford University Press)

[37] Onuki H and Elleaume P 2003 *Undulators, Wigglers and their Applications* (London: Taylor & Francis)

[38] Levichev E and Vinokurov N 2010 Undulators and other insertion devices *Rev. Accel. Sci. Technol.* **3** 203–10

[39] Schlueter R D 1995 Wiggler and undulator insertion devices *Synchrotron Radiation Sources —A Primer. Series on Synchrotron Radiation Techniques and Application* ed H Winick (Singapore: World Scientific) pp 377–408

[40] Winick H, Brown G, Halbach K and Harris J 1981 Wiggler and undulator magnets *Phys. Today* **34** 50–63

[41] Alferov D F, Bashmakov Y A and Bessonov E G 1974 *Sov. Phys. Tech. Phys.* **18** 1336

[42] Colson W B 1977 Free electron laser theory *PhD Thesis* Stanford University

[43] Elleaume P 1990 Free electron laser undulators, electron trajectories and spontaneous emission *Laser Handbook* ed W B Colson, C Pellegrini and A Renieri (Amsterdam: North Holland)

[44] Barbini R, Ciocci F, Dattoli G and Giannessi L 1990 Spectral properties of the undulator magnets radiation: analytical and numerical treatment *Nuovo Cimento* **13** 1–65

[45] Dobiasch P, Meystre P and Scully M 1983 Optical wiggler free-electron x-ray laser in the 5 Å region IEEE *J. Quantum Electron.* **19** 1812–20

[46] Ciocci F, Dattoli G and Walsh J E 1985 A short note on the wave-undulator FEL operation *Nucl. Instrum. Methods Phys. Res.* A **237** 401–3

[47] Gea-Banacloche J, Moore G T, Schlicher R R, Scully M O and Walther H 1987 Soft x-ray free-electron laser with a laser undulator IEEE *J. Quantum Electron.* **23** 1558–70

[48] Tang C M, Hafizi B and Ride S K 1993 Thomson backscattered x-rays from an intense laser beam *Nucl. Instrum. Meth.* A **331** 371–8

[49] Bacci A, Ferrario M, Maroli C, Petrillo V and Serafini L 2006 Transverse effects in the production of x rays with a free-electron laser based on an optical undulator *Phys. Rev. ST Accel. Beams* **9** 060704

[50] Yanovsky V 2008 Ultra-high intensity- 300-TW laser at 0.1 Hz repetition rate *Opt. Express* **16** 2109–14

[51] Norby J 2005 *Laser Focus World*

[52] https://http://www.extreme-light-infrastructure.eu

[53] Pantell R H, Soncini G and Puthoff H E 1968 Stimulated photon-electron scattering IEEE *J. Quantum Electron.* **QE-4** 905

[54] Madey J M J 1971 Stimulated emission of bremsstrahlung in a periodic magnetic field *J. Appl. Phys.* **42** 1906–13

[55] Dattoli G and Renieri A 1985 *Laser Handbook* vol IV ed M L Stitch and M S Ball (North-Holland: Amsterdam)

[56] Brown L S 1964 Interaction of intense laser beams with electrons *Phys. Rev.* **133** A705–19

[57] Dattoli G, Di Palma E, Petrillo V, Rau J V, Sabia E, Spassovsky I, Biedron S G, Einstein J and Milton S V 2015 Pathway to a compact SASE FEL device *Nucl. Instrum. Meth.* A **798** 144–51

[58] Dattoli G and Nguyen F 2017 Free electron laser as a tool for fundamental quantum physics *Nucl. Instrum. Methods Phys. Res.* B **402** 336–8

[59] Jackson J D 1998 *Classical Electrodynamics* 3rd edn (New York: Wiley)

[60] Danson C N 2019 Petawatt and exawatt laser worldwide *High Laser Power Laser Sci. Eng.* **VII** 1–54

[61] Shumail M, Bowden G B, Chang C, Neilson J, Tantawi S G and Pellegrini C 2011 Application of the balanced hybrid mode in overmoded corrugated waveguides to short wavelength dynamic undulators *2nd Int. Particle Accelerator Conf. (IPAC) (San Sebastian, Spain)*

[62] Toufexis F and Tantawi S G 2019 Development of a millimeter-period RF undulator *Phys. Rev. Accel. Beams* **22** 120701

[63] Kuzikov S V, Hirshfield J L, Jiang Y, Marshall T C and Vikharev A A 2012 RF undulator for compact x-ray SASE source of variable wavelength *AIP Conf. Proc.* 1507 458–63

[64] Abubakirov E B *et al* 2016 Microwave undulators and electron generators for new-generation free-electron lasers *Radiophys. Quantum Electron.* **58** 755–68

[65] Kuzikov S V, Savilov A V and Vikharev A A 2014 Flying radio frequency undulator *Appl. Phys. Lett.* **105** 033504

[66] Dolgashev V, Tantawi S, Higashi Y and Spataro B 2010 Geometric dependence of radio-frequency breakdown in normal conducting accelerating structures *Appl. Phys. Lett.* **97** 171501

[67] Laurent L, Tantawi S, Dolgashev V, Nantista C, Higashi Y, Aicheler M, Heikkinen S and Wuensch W 2011 Experimental study of RF pulsed heating *Phys. Rev. ST Accel. Beams* **14** 041001

[68] Thompson M C 2008 Breakdown limits on gigavolt-per-meter electron-beam-driven wake-fields in dielectric structures *Phys. Rev. Lett.* **100** 214801

[69] Savilov A V 2010 Compression of complicated RF pulses produced from the super-radiant backward-wave oscillator *Appl. Phys. Lett.* **97** 093501

[70] Rostov V V, Elchaninov A A, Romanchenko I V and Yalandin M I 2012 A coherent two-channel source of cherenkov superradiance pulses *Appl. Phys. Lett.* **100** 224102

[71] Danly B, Bekefi G, Davidson R, Temkin R, Tran T and Wurtele J 1987 Principles of gyrotron powered electromagnetic wigglers for free-electron lasers IEEE *J. Quantum Electron.* **23** 103–16

[72] Tran T, Danly B and Wurtele J 1987 Free-electron lasers with electromagnetic standing wave wigglers IEEE *J. Quantum Electron.* **23** 1578–89

[73] Fliflet A W, Manheimer W M and Fischer R P 1994 Designs for an infrared free-electron laser based on gyrotron-powered electromagnetic wigglers *IEEE Trans. Plasma Sci.* **22** 638–48

[74] Fliflet A W, Manheimer W M and Sprangle P A 1997 Low-voltage infrared free-electron lasers based on gyrotron-powered RF wigglers IEEE *J. Quantum Electron.* **33** 669–76

[75] Granatstein V L, Fliflet A W, Manheimer W M and Gover A 1995 Design of a far infrared fel with an electromagnetic wiggler *Nucl. Instrum. Methods Phys. Res.* B **358** ABS24–5

[76] Granatstein V L, Carmel Y and Gover A 1984 Demonstration of a free electron laser with an electromagnetic wiggler *Proc. of SPIE 0453, Free-Electron Generators of Coherent Radiation* vol 453 344–9

[77] Freund H P and Antonsen T M 2018 Electromagnetic-wave wigglers *Principles of Free Electron Lasers* (Cham: Springer) pp 406–20

Appendices

Appendix A

The dyadic calculus

In the main body of the chapter we have exploited the calculus of dyad [1, 2]. The underlying formalism is easy to learn and tailor suited for many applications, including electromagnetism, crystallography, engineering etc.

We recall that a vector can be defined using either

(a) Cartesian notation

$$\vec{a} \equiv (a_x, a_y, a_z). \tag{A.1}$$

(b) Row/column notation

$$\underset{\sim}{a} \equiv \begin{pmatrix} a_x \\ a_y \\ a_z \end{pmatrix}. \tag{A.2}$$

It is evident that

$$\underset{\sim}{a} \equiv \vec{a}, \tag{A.3}$$

and that the transpose vector, in the notation b), reads

$$\underset{\sim}{a}^T \equiv (a_x \quad a_y \quad a_z). \tag{A.4}$$

Within the same context, we define the dot product

$$\vec{a} \cdot \vec{b} \equiv \underset{\sim}{a}^T \underset{\sim}{b} = (a_x \quad a_y \quad a_z) \begin{pmatrix} b_x \\ b_y \\ b_z \end{pmatrix} = a_x b_x + a_y b_y + a_z b_z. \tag{A.5}$$

and the inner product

doi:10.1088/978-0-7503-2464-9ch8

$$\vec{a} \otimes \vec{b} = \underline{a}\underline{b}^T = \begin{pmatrix} a_x \\ a_y \\ a_z \end{pmatrix} \begin{pmatrix} b_x & b_y & b_z \end{pmatrix}. \tag{A.6}$$

The use of the column vector product properties, allows us to cast (A.6) in the form

$$\hat{c} = \underline{a}\underline{b}^T = \begin{pmatrix} a_x b_x & a_x b_y & a_x b_z \\ a_y b_x & a_y b_y & a_y b_z \\ a_z b_x & a_z b_y & a_z b_z \end{pmatrix}, \tag{A.7}$$

namely a matrix, or a rank 2 tensor[1], whose geometrical interpretation is reported in figure A.1.

A particularly important role is played by the unit matrix, which can be defined in terms of the $\vec{i}, \vec{j}, \vec{k}$ base vectors, according to the identity

$$\begin{aligned} \hat{I} &= \hat{i} + \hat{j} + \hat{k} \\ &= \begin{pmatrix} 1 \\ 0 \\ 0 \end{pmatrix} (1 \ \ 0 \ \ 0) + \begin{pmatrix} 0 \\ 1 \\ 0 \end{pmatrix} (0 \ \ 1 \ \ 0) + \begin{pmatrix} 0 \\ 0 \\ 1 \end{pmatrix} (0 \ \ 0 \ \ 1) \\ &= \begin{pmatrix} 1 & 0 & 0 \\ 0 & 1 & 0 \\ 0 & 0 & 1 \end{pmatrix}, \end{aligned} \tag{A.8}$$

and it is also easily stated that

$$\begin{aligned} \hat{I} \cdot \underline{i} &= \left(\hat{i} + \hat{j} + \hat{k}\right) \cdot \underline{i} = \\ &= \begin{pmatrix} 1 \\ 0 \\ 0 \end{pmatrix} \left((1 \ \ 0 \ \ 0) \begin{pmatrix} 1 \\ 0 \\ 0 \end{pmatrix} \right) + \begin{pmatrix} 0 \\ 1 \\ 0 \end{pmatrix} \left((0 \ \ 1 \ \ 0) \begin{pmatrix} 1 \\ 0 \\ 0 \end{pmatrix} \right) + \begin{pmatrix} 0 \\ 0 \\ 1 \end{pmatrix} \left((0 \ \ 0 \ \ 1) \begin{pmatrix} 1 \\ 0 \\ 0 \end{pmatrix} \right). \end{aligned} \tag{A.9}$$

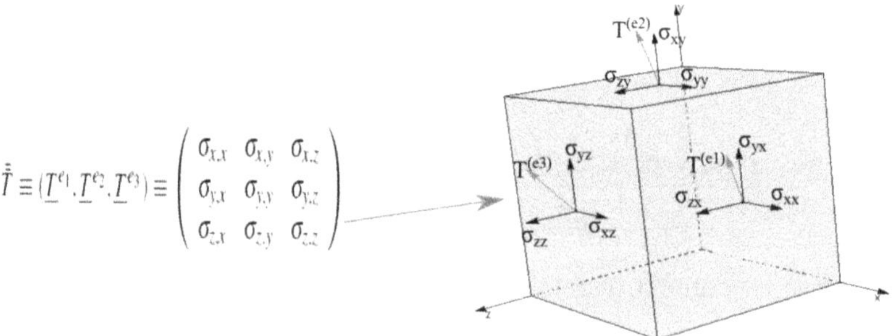

Figure A.1. Geometrical representation of a rank 2 tensor.

[1] We recall that a rank 0 tensor is a scalar, a rank 1 a vector, a rank 2 a matrix. In the following, we omit rank 2, when referring to tensor and explicitly mention if confusion may arise.

With the dot products, in the square brackets of the last two terms, vanishing, we end up with

$$\hat{I} \cdot \underline{i} = \underline{i} \equiv \vec{i}, \tag{A.10}$$

and eventually

$$\begin{aligned} \hat{I} \cdot \underline{j} &= \underline{j} \equiv \vec{j}, \\ \hat{I} \cdot \underline{k} &= \underline{k} \equiv \vec{k}. \end{aligned} \tag{A.11}$$

As a result of the previous identities, we find that the dot product of the unit tensor and a column vector reduces to

$$\hat{I} \cdot \underline{a} = (\hat{i} + \hat{j} + \hat{k}) \cdot \underline{a} = \underline{a} \equiv \vec{a}. \tag{A.12}$$

The operative definition of the vector product, in terms of the outlined formalism, is achieved by introducing the operators

$$\vec{i} \times \equiv \begin{pmatrix} 0 & 0 & 0 \\ 0 & 0 & -1 \\ 0 & 1 & 0 \end{pmatrix}, \vec{j} \times \equiv \begin{pmatrix} 0 & 0 & 1 \\ 0 & 0 & 0 \\ -1 & 0 & 0 \end{pmatrix}, \vec{k} \times \equiv \begin{pmatrix} 0 & -1 & 0 \\ 1 & 0 & 0 \\ 0 & 0 & 0 \end{pmatrix}, \tag{A.13}$$

exploited to check the properties

$$\begin{aligned} \vec{i} \times \vec{j} &= \begin{pmatrix} 0 & 0 & 0 \\ 0 & 0 & -1 \\ 0 & 1 & 0 \end{pmatrix} \begin{pmatrix} 0 \\ 1 \\ 0 \end{pmatrix} = \begin{pmatrix} 0 \\ 0 \\ 1 \end{pmatrix} = \vec{k}, \\ \vec{i} \times \vec{k} &= \begin{pmatrix} 0 & 0 & 0 \\ 0 & 0 & -1 \\ 0 & 1 & 0 \end{pmatrix} \begin{pmatrix} 0 \\ 0 \\ 1 \end{pmatrix} = -\begin{pmatrix} 0 \\ 1 \\ 0 \end{pmatrix} = -\vec{j}, \\ \vec{k} \times \vec{j} &\equiv \begin{pmatrix} 0 & -1 & 0 \\ 1 & 0 & 0 \\ 0 & 0 & 0 \end{pmatrix} \begin{pmatrix} 0 \\ 1 \\ 0 \end{pmatrix} = -\begin{pmatrix} 1 \\ 0 \\ 0 \end{pmatrix} = -\vec{i}. \end{aligned} \tag{A.14}$$

Their use, applied to the cross product between two vectors, allows us to recover the familiar definition

$$\vec{a} \times \vec{b} \equiv \hat{a} \begin{pmatrix} b_x \\ b_y \\ b_z \end{pmatrix} = \begin{pmatrix} -a_z b_y + a_y b_z \\ a_z b_x - a_x b_z \\ -a_y b_x + a_x b_y \end{pmatrix}, \tag{A.15}$$

where

$$\hat{a} = \begin{pmatrix} 0 & -a_z & a_y \\ a_z & 0 & -a_x \\ -a_y & a_x & 0 \end{pmatrix}. \tag{A.16}$$

The tensor $\hat{d}$ is anti-symmetric because $\hat{d}^T = -\hat{d}$.

More in general, it should be noted that if

$$\hat{c} = \underline{m}\underline{n}^T, \tag{A.17}$$

then

$$\hat{c}^T = \underline{n}\underline{m}^T, \tag{A.18}$$

and the tensor

$$\hat{t}_- = \bar{\bar{c}} - \bar{\bar{c}}^T \tag{A.19}$$

is anti-symmetric.

Let us now apply the operational definition in equation (A.13) to establish the vector product of a vector and the identity tensor.

The use of the operational definitions (A.13) and (A.14) allows the derivation of the identities

$$\begin{aligned} \vec{i} \times \hat{I} &\equiv \underline{k}\underline{j}^T - \underline{j}\underline{k}^T = \begin{pmatrix} 0 & 0 & 0 \\ 0 & 0 & -1 \\ 0 & 1 & 0 \end{pmatrix}, \\ \vec{j} \times \hat{I} &\equiv -\underline{k}\underline{i}^T + \underline{i}\underline{k}^T = \begin{pmatrix} 0 & 0 & 1 \\ 0 & 0 & 0 \\ -1 & 0 & 0 \end{pmatrix}, \\ \vec{k} \times \hat{I} &\equiv \underline{j}\underline{i}^T - \underline{i}\underline{j}^T = \begin{pmatrix} 0 & -1 & 0 \\ 1 & 0 & 0 \\ 0 & 0 & 0 \end{pmatrix}, \end{aligned} \tag{A.20}$$

whose geometrical meaning will be discussed later in this appendix.

Let us define the unit vector

$$\begin{aligned} \vec{u} &= \frac{\vec{r}}{|\vec{r}\,|}, \\ \vec{r} &= \left(a\vec{i} + b\vec{j} + c\vec{k}\right), \end{aligned} \tag{A.21}$$

and the associated matrix

$$\hat{u} = \frac{1}{|\vec{r}\,|^2}\begin{pmatrix} a^2 & a\,b & a\,c \\ ab & b^2 & bc \\ ac & bc & c^2 \end{pmatrix}. \tag{A.22}$$

We accordingly define the 'geometrical operators':

(a) Reflection tensor

$$\hat{R}_r = \hat{I} - 2\,\hat{u} \tag{A.23}$$

whose action on a given vector $\vec{p}$ is interpreted as a reflection with respect to a plane perpendicular to the vector $\vec{u}$. It is indeed easily checked that

$$\hat{R}_r \cdot \vec{p} = \vec{p} - 2\hat{u}|\vec{p}|\cos(\theta), \tag{A.24}$$

and the associated vector operations are shown in figure A.2.

(b) Rotation tensor

$$\hat{R}_\vartheta = \hat{u} + \sin(\vartheta)(\vec{u} \times \hat{I}) + \cos\vartheta(\hat{I} - \hat{u}), \tag{A.25}$$

representing a matrix which, once applied to a vector $\vec{x}$, yields the so called Rodrigues rotation

$$\hat{R}_\vartheta \vec{x} = \vec{u}(\vec{u} \cdot \vec{x}) + \sin(\vartheta)(\vec{u} \times \vec{x}) + \cos\vartheta(\vec{u} - \vec{u}\,(\vec{u} \cdot \vec{x})). \tag{A.26}$$

Let us now come back to the geometrical role of the operators in equations (A.14) and (A.19) and note that [3]

$$\begin{aligned}
e^{\alpha\vec{i}\times} &\equiv e^{\alpha\begin{pmatrix}0&0&0\\0&0&-1\\0&1&0\end{pmatrix}} = \begin{pmatrix}1&0&0\\0&\cos(\alpha)&-\sin(\alpha)\\0&\sin(\alpha)&\cos(\alpha)\end{pmatrix},\\
e^{\beta\vec{j}\times} &\equiv e^{\beta\begin{pmatrix}0&0&1\\0&0&0\\-1&0&0\end{pmatrix}} = \begin{pmatrix}\cos(\beta)&0&\sin(\beta)\\0&1&0\\-\sin(\beta)&0&\cos(\beta)\end{pmatrix},\\
e^{\gamma\vec{k}\times} &\equiv e^{\gamma\begin{pmatrix}0&-1&0\\1&0&0\\0&0&0\end{pmatrix}} = \begin{pmatrix}\cos(\gamma)&-\sin(\gamma)&0\\\sin(\gamma)&\cos(\alpha)&0\\0&0&1\end{pmatrix}.
\end{aligned} \tag{A.27}$$

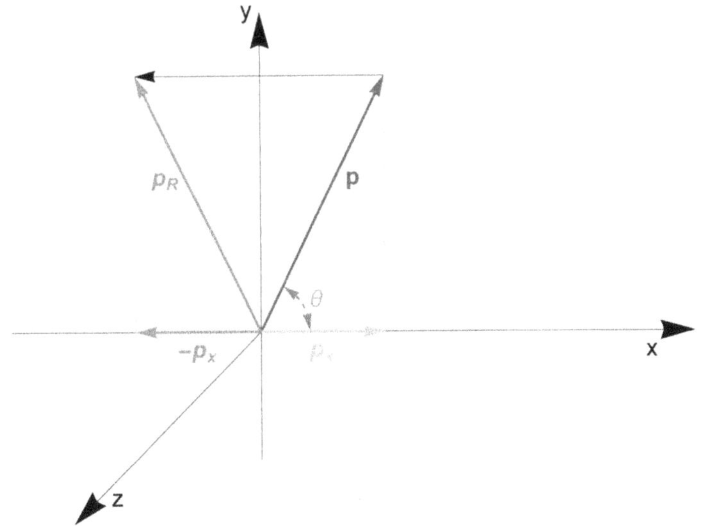

Figure A.2. Geometrical representation of the reflection tensor operator.

The product of the three exponentials realizes the tridimensional matrix rotation and each exponential realizes the rotation in terms of Euler angles.

We have so far not mentioned dyad or dyadic, we will now reconcile the previous definitions with the more sound discussion available in the scientific literature.

We first note that 3 × 3 matrix can always be rewritten as

$$\begin{pmatrix} a_{x,x} & a_{x,y} & a_{x,z} \\ a_{y,x} & a_{y,y} & a_{y,z} \\ a_{z,x} & a_{z,y} & a_{z,z} \end{pmatrix} = a_{x,x}\hat{\imath} + a_{i,i}\hat{\jmath} + a_{z,z}\hat{k} + (a_{x,y}\hat{\lambda} + a_{y,x}\hat{\lambda}^T) + (a_{x,z}\hat{\mu} + a_{z,x}\hat{\mu}^T) + (a_{y,z}\hat{\nu} + a_{z,y}\hat{\nu}^T), \tag{A.28}$$

where

$$\hat{\lambda} = \underline{i}\underline{j}^T, \ \hat{\mu} = \underline{i}\underline{k}^T, \ \hat{\nu} = \underline{j}\underline{k}^T, \tag{A.29}$$

which can eventually be rearranged as

$$\begin{aligned} \bar{\bar{a}} &= \underline{i}\underline{a}_x^T + \underline{j}\underline{a}_y^T + \underline{k}\underline{a}_z^T, \\ \underline{a}_x^T &= a_{x,x}\underline{i}^T + a_{x,y}\underline{j}^T + a_{z,z}\underline{k}^T, \ldots. \end{aligned} \tag{A.30}$$

The matrix (A.29) written in the form (A.30) is called ***dyadic*** written in a '***nonion***' form, the forms we have denoted by $\hat{c} = \underline{m}\underline{n}^T$ denoted by **m n** defines a ***dyad***.

An example of application the bold vector notation is given below

$$\begin{aligned} \hat{I} \times \vec{b} &= (\mathbf{ii} + \mathbf{jj} + \mathbf{kk}) \times (b_x\mathbf{i} + b_y\mathbf{j} + b_z\mathbf{k}) \\ &= b_x(\mathbf{kj} - \mathbf{jk}) + b_y(\mathbf{ik} - \mathbf{ki}) + b_z(\mathbf{ji} - \mathbf{ij}), \\ \vec{b} \times \hat{I} &= (b_x\mathbf{i} + b_y\mathbf{j} + b_z\mathbf{k}) \times (\mathbf{ii} + \mathbf{jj} + \mathbf{kk}) = \hat{I} \times \vec{b}, \end{aligned} \tag{A.31}$$

which have been derived by using the distributive property and noting that, e.g. $\mathbf{ii} \times \mathbf{j} = \mathbf{i}(\mathbf{i} \times \mathbf{j}) = \mathbf{ik}$.

The result in equation (A.31) may sound unexpected, because we might have foreseen a non-commutative product. This point can be clarified by making reference to the geometrical interpretation of a tensor of rank 2; we cannot go deeper in this topic and direct the reader to the specialized literature.

For completeness sake, we check already established identities

$$\begin{aligned} \hat{\imath} &\equiv \mathbf{ii} \equiv \underline{i}\underline{i}^T, \\ \vec{j} \times \hat{\imath} &\equiv (\mathbf{j} \times \mathbf{i})\mathbf{i} \equiv -\underline{k}\ \underline{i}^T, \\ \hat{\imath} \times \vec{j} &\equiv \mathbf{i}(\mathbf{i} \times \mathbf{j}) \equiv \underline{i}\underline{k}^T. \end{aligned} \tag{A.32}$$

We can therefore apply the formalism exposed in the appendix to derive various identities often found in applications, we note, therefore, that the relationship $(\mathbf{bc} - \mathbf{cb}) \cdot \mathbf{d}$ is an alternative way of expressing the triple vector product; we get indeed

$$(\mathbf{bc} - \mathbf{cb}) \cdot \mathbf{d} = \vec{b}(\vec{c} \cdot \vec{d}) - \vec{c}(\vec{b} \cdot \vec{d}) = (\vec{c} \times \vec{b}) \times \vec{d}. \tag{A.33}$$

Before concluding, we would like to comment about the link between dyadic and Dirac formalism. To this aim we recall that in quantum mechanics, vectors describe a state and are expressed in terms of the so called **ket** and **bra** formalism.

To this aim we recall that a **ket** vector is expressed in terms of a proper orthogonal basis as

$$|v\rangle = v^i e_i = |e_i\rangle\langle e^i|v\rangle = \sum_i |e_i\rangle\langle e_i|v\rangle, \tag{A.34}$$

Where, with the notation $\langle v|$, we denote the **bra** counterpart of the **ket** vector. The basis vectors satisfy the conditions

$$\begin{aligned} \langle e_i|e_j\rangle &= \delta_{ij} \text{ orthonormality}, \\ \sum_i |e_i\rangle\langle e_i| &= \mathbb{1} \text{ completeness}. \end{aligned} \tag{A.35}$$

It is therefore evident that the **bra** is the transpose of a **ket**. If we identify the basis with that of the ordinary C^3 space, we find

$$e_1 = \begin{pmatrix} 1 \\ 0 \\ 0 \end{pmatrix}, \quad e_2 = \begin{pmatrix} 0 \\ 1 \\ 0 \end{pmatrix}, \quad e_3 = \begin{pmatrix} 0 \\ 0 \\ 1 \end{pmatrix}, \tag{A.36}$$

and

$$\begin{aligned} \langle e_1|e_1\rangle &= (1 \quad 0 \quad 0) \cdot \begin{pmatrix} 1 \\ 0 \\ 0 \end{pmatrix} = 1, \\ |e_1\rangle\langle e_1| &= \begin{pmatrix} 1 \\ 0 \\ 0 \end{pmatrix} \cdot (1 \quad 0 \quad 0) = \begin{pmatrix} 1 & 0 & 0 \\ 0 & 0 & 0 \\ 0 & 0 & 0 \end{pmatrix}, \end{aligned} \tag{A.37}$$

the correspondence is accordingly naively established.

References

[1] Lindell I V 2004 Advanced Field Theory Course for Graduate Students http://www.ismolindell.com/publications/monographs/pdf/Aftis.pdf

[2] Kolecki J C 2002 *An Introduction to Tensors for Students of Physics and Engineering* (National Aeronautics and Space Administration) http://www.ismolindell.com/publications/monographs/pdf/Aftis.pdf

[3] Dattoli G, Mezi L and Migliorati M 2003 Evolution operators and Euler angles *Nuovo Cimento Soc. Ital. Fis. B* **118** 493–8

IOP Publishing

G Dattoli, E Di Palma, S P Sabchevski and I P Spassovsky

Appendix B

Hot plasma and dielectric tensor

This appendix is devoted to the derivation of equation (3.185), to this aim we premise a few mathematical details, regarding the modified Bessel of the first kind and the properties of the plasma dispersion function in equation (3.121).

Modified Bessel functions of the first kind

(a) Series expansion

$$I_n(x) = \sum_{r=0}^{\infty} \frac{\left(\frac{x}{2}\right)^{n+2\,r}}{r!\,(n+r)!}. \tag{B.1}$$

(b) Generating function

$$\sum_{n=-\infty}^{+\infty} t^n I_n(x) = e^{\frac{x}{2}(t+t^{-1})}. \tag{B.2}$$

(c) Particular cases

$$\begin{gathered} \sum_{n=-\infty}^{+\infty} I_n(x) = e^x,\ \sum\nolimits_{n=-\infty}^{+\infty} I_n'(x) = e^x, \\ \sum_{l=\infty-}^{+\infty} l\ I_l(x) = 0. \end{gathered} \tag{B.3}$$

Where the prime ′ denotes first derivative with respect to the argument.

doi:10.1088/978-0-7503-2464-9ch9

(d) Recurrences

$$I_n'(x) = \frac{1}{2}[I_{n-1}(x) + I_{n+1}(x)],$$
$$2\frac{n}{x}I_n(x) = I_{n-1}(x) - I_{n+1}(x). \tag{B.4}$$

(e) Reflection properties

$$I_n(-x) = (-1)^n I_n(x),$$
$$I_{-n}(x) = I_n(x). \tag{B.5}$$

(f) Small argument expansion

$$I_0(x) \cong 1 + \frac{x^2}{4},$$
$$I_1(x) \cong \frac{x}{2}. \tag{B.6}$$

Plasma dispersion function

$$Z(\zeta) = \frac{1}{\sqrt{\pi}}\int_{-\infty}^{+\infty} \frac{e^{-\beta^2}}{\beta - \zeta}d\beta, \tag{B.7}$$

where β is a complex quantity.

The use of the Laplace transform identity

$$A^{-1} = \int_0^{\infty} e^{-s\,A}ds, \tag{B.8}$$

yields

$$Z(\zeta) = \frac{1}{\sqrt{\pi}}\int_0^{\infty} ds\, e^{s\,\zeta}\int_{-\infty}^{+\infty} e^{-\beta^2 - s\,\beta}d\beta = \int_0^{\infty} ds\, e^{s\,\zeta + \frac{s^2}{4}}. \tag{B.9}$$

Performing the change of variable $s = 2i\sigma$ the last integral in the previous equation reads

$$Z(\zeta) = 2i\int_0^{\infty} d\sigma\, e^{2i\,\sigma\zeta - \sigma^2} = 2ie^{-\zeta^2}\int_0^{\infty} d\sigma\, e^{-(i\zeta - \sigma)^2} = 2ie^{-\zeta^2}\int_{-\infty}^{i\,\zeta} d\lambda\, e^{-\lambda^2}, \tag{B.10}$$

where the error function is defined as

$$\mathrm{erf}(x) = \frac{2}{\sqrt{\pi}}\int_0^{x} e^{-t^2}dt, \tag{B.11}$$

we can write

$$Z(\zeta) = 2ie^{-\zeta^2}\left(\int_{-\infty}^{0} d\lambda\, e^{-\lambda^2} + \frac{\sqrt{\pi}}{2}\mathrm{erf}(i\zeta)\right) \\ = i\,\sqrt{\pi}\, e^{-\zeta^2}(1 + \mathrm{erf}(i\zeta)). \tag{B.12}$$

We can perform a different approximation [1]

(a) Series expansion

$$Z(\zeta) = ie^{-\zeta^2}\left[1 + i\zeta\sum_{r=0}^{\infty}\frac{\zeta^{2r}\sqrt{\pi}}{(2r+1)\,r!}\right], \tag{B.13}$$

which can be cast in the form

$$Z(\zeta) = ie^{-\zeta^2} - \zeta G(\zeta), \tag{B.14}$$

where

$$G(\zeta) = \sum_{p=0}^{\infty}(-1)^p\frac{b(p)}{p!}\zeta^{2p}, \tag{B.15}$$

with

$$b(p) = \sum_{s=0}^{p}\frac{(-1)^s}{2s+1}\binom{p}{s}. \tag{B.16}$$

The behavior of the Gaussian-like function, defined by the infinite series in the second of equation (B.15), is shown in figure B.1 Along with an ordinary Gaussian for comparison.

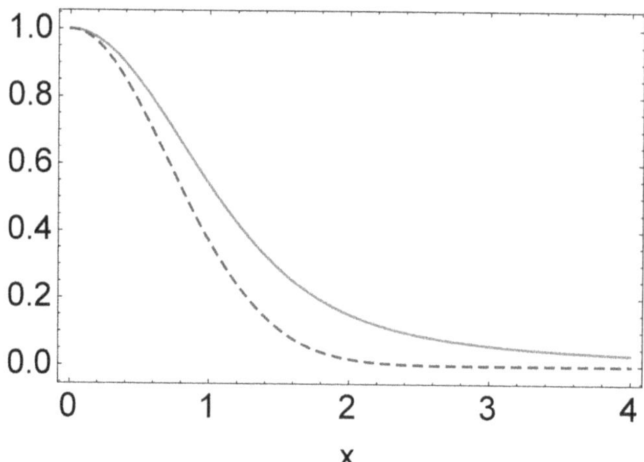

Figure B.1. $G(\zeta)$ versus ζ (continuous line), ordinary Gaussian (dot line).

We have dwelt on the term Gaussian-like because, in umbral form, it is formally equivalent to a Gaussian, we can write indeed [2]

$$G(\zeta) = e^{-\hat{b}\zeta^2}, \tag{B.17}$$

where ζ is an umbral operator defined in such a way that

$$\hat{b}^p = b(p). \tag{B.18}$$

Such an umbral restyling, far from being an academic exercise, offers significant advantages, which will not be discussed here.

(b) Asymptotic expansion

$$\operatorname{erfc}(\lambda)|_{\lambda\gg 1} \cong \frac{2\, e^{-\lambda^2}}{\sqrt{\pi}}\left(\frac{1}{2\lambda} - \frac{1}{4\,\lambda^3} + \ldots\right), \tag{B.19}$$

which for the plasma dispersion function yields

$$Z(\zeta) \cong -\frac{1}{\zeta}. \tag{B.20}$$

This regarding the specific form of the plasma dispersion function, yields (for $l \neq 0$)

$$Z(\zeta_{s,l}) = -\frac{1}{\zeta_{s,l}} + O\left(\frac{1}{\zeta_{s,l}^3}\right) = \frac{k_{||}v_{t,s}}{l\Omega_{c,s}}\frac{1}{1-\omega/l\Omega_s} + O\left(\frac{1}{\zeta_{s,l}^3}\right) \approx \frac{k_{||}v_{t,s}}{l\Omega_{c,s}} + \frac{\omega k_{||}v_{t,s}}{l^2\Omega_{c,s}^2}. \tag{B.21}$$

The case of the compressional Alfvén wave propagating in the direction of the magnetic field ($|k_{||}v^e_{\mathrm{th}||}| \simeq \omega \ll \Omega_{c,i}$) and the previous identities allow the derivation of the approximated forms of the entries of equation (3.119) [3].

On account of equation (B.21), of the assumption of bs small and of equations (B.3) we find

$$\begin{aligned}\varepsilon_{xx} &= 1 + \sum_s \sum_{l=-\infty}^{\infty} \frac{\Omega_{p,s}^2}{\omega|k_{||}|v_{t,s}} \frac{l^2 I_l(b_s)}{b_s}\exp(-b_s)Z(\zeta_{s,l}) \approx \\ &\approx 1 + \sum_s \sum_{l=1}^{\infty} 2\frac{\Omega_{p,s}^2}{\cancel{\omega|k_{||}|v_{t,s}}}(1-b_s) \\ &\quad \left[\underbrace{\frac{l^{\cancel{2}} I_l(b_s)}{b_s}\frac{\cancel{\omega|k_{||}|v_{t,s}}}{\cancel{l}\Omega_{c,s}^2}}_{=0} + \frac{\cancel{l^2} I_l(b_s)}{b_s}\frac{\cancel{\omega|k_{||}|v_{t,s}}}{\cancel{l^2}\Omega_{c,s}^2}\right] \approx \\ &\approx 1 + \sum_s \frac{\Omega_{p,s}^2}{\Omega_{c,s}^2} = \varepsilon_{\perp}.\end{aligned} \tag{B.22}$$

The same holds for the other diagonal elements with the inclusion of equation (B.13)

$$
\begin{aligned}
\varepsilon_{yy} &= 1 + \sum_s \sum_{l=-\infty}^{\infty} \frac{\Omega_{p,s}^2}{\omega|k_{||}|v_{t,s}} \left[\frac{l^2 I_l(b_s)}{b_s} + 2b_s(I_l(b_s) - I_l'(b_s)) \right] \exp(-b_s) Z(\zeta_{s,l}) \approx \\
&\approx 1 + \sum_s \frac{\Omega_{p,s}^2}{\Omega_{c,s}^2} + 2b_s(I_0(b_s) - I_0'(b_s)) \frac{\Omega_{p,s}^2}{\omega|k_{||}|v_{t,s}} (1 - b_s) Z(\zeta_{s,0}) \approx \\
&\approx 1 + \sum_s \frac{\Omega_{p,s}^2}{\Omega_{c,s}^2} + 2b_s \frac{\Omega_{p,s}^2}{\omega|k_{||}|v_{t,s}} i\sqrt{\pi} \exp(-\zeta_{s,0}^2) = \varepsilon_\perp + i\Delta\varepsilon_\perp,
\end{aligned}
\tag{B.23}
$$

$$
\begin{aligned}
\varepsilon_{zz} &= 1 + \sum_s \sum_{l=-\infty}^{\infty} 2 \frac{\Omega_{p,s}^2}{\omega|k_{||}|v_{t,s}} I_l(b_s) \exp(-b_s) \zeta_{s,l}[1 + \zeta_{s,l} Z(\zeta_{s,l})] \approx \\
&\approx \sum_s \cancel{2} \frac{\Omega_{p,s}^2}{\omega|k_{||}|v_{t,s}} I_0(b_s)(1 - b_s) \zeta_{s,0} \left[-\frac{1}{\cancel{2}} \frac{dZ(\zeta_{s,0})}{d\zeta_{s,0}} \right] \approx \\
&\approx \sum_s \frac{\Omega_{p,s}^2}{\omega|k_{||}|v_{t,s}} \zeta_{s,0} \left[2\zeta_{s,0} i\sqrt{\pi} \exp(-\zeta_{s,\,0}^2) - Re(Z') \right] \approx \\
&\approx \sum_s - \left(\frac{\Omega_{p,s}}{|k_{||}|v_{t,s}} \right)^2 Re(Z') + \frac{2\Omega_{p,s}^2}{\omega|k_{||}|v_{t,s}} \zeta_{s,\,0}^2 i\sqrt{\pi} \exp(-\zeta_{s,\,0}^2) = \varepsilon_{||} + i\Delta\varepsilon_{||}.
\end{aligned}
\tag{B.24}
$$

The same procedure applies to non-diagonal terms, which by also including the use of equations (B.4) eventually yields

$$
\begin{aligned}
\varepsilon_{yz} &= - \sum_s \sum_{l=-\infty}^{\infty} \frac{\Omega_{p,s}^2}{\omega|k_{||}|v_{t,s}} \frac{Z_s}{|Z_s|} \sqrt{2b_s} [I_l'(b_s) - I_l(b_s)] \exp(-b_s)[1 + \zeta_{s,l} Z(\zeta_{s,l})] \approx \\
&\approx - \sum_s \frac{\Omega_{p,s}^2}{\omega|k_{||}|\cancel{v_{t,s}}} \frac{Z_s}{|Z_s|} \frac{k_\perp \cancel{v_{t,s}}}{\Omega_{c,s}} [I_0'(b_s) - I_0(b_s)](1 - b_s) \left[-\frac{1}{2} \frac{dZ(\zeta_{s,0})}{d\zeta_{s,0}} \right] \approx \\
&\approx \sum_s \frac{k_\perp}{\omega|k_{||}|} \frac{Z_s}{|Z_s|} \frac{\Omega_{p,s}^2}{\Omega_{c,s}} I_0(b_s) \left[\zeta_{s,0} i\sqrt{\pi} \exp(-\zeta_{s,\,0}^2) - \frac{1}{2} Re(Z') \right] \\
&= \sum_s - \frac{1}{2} \frac{n_\perp}{n_{||}} \frac{Z_s}{|Z_s|} \frac{\Omega_{p,s}^2}{\omega\Omega_{c,s}} Re(Z') + \frac{Z_s}{|Z_s|} \frac{n_\perp}{n_{||}} \frac{\omega}{\Omega_{c,s}} \frac{\Omega_{p,s}^2}{\omega|k_{||}|v_{t,s}} i\sqrt{\pi} \exp(-\zeta_{s,\,0}^2) \\
&= \varepsilon_m + i\Delta\varepsilon_m,
\end{aligned}
\tag{B.25}
$$

$$
\begin{aligned}
\varepsilon_{xy} &= \sum_s \sum_{l=-\infty}^{\infty} \frac{\Omega_{p,s}^2}{\omega|k_{||}|v_{t,s}} l[I_l'(b_s) - I_l(b_s)] \exp(-b_s) Z(\zeta_{s,l}) \approx \\
&\approx \sum_s \sum_{l=-\infty}^{\infty} \frac{\Omega_{p,s}^2}{\omega|k_{||}|v_{t,s}} \cancel{l}[I_l'(b_s) - I_l(b_s)](1 - b_s) \frac{|k_{||}|v_{t,s}}{\cancel{l}\Omega_s} \approx 0,
\end{aligned}
\tag{B.26}
$$

$$\varepsilon_{xz} = \sum_{s}\sum_{l=-\infty}^{\infty} \frac{\Omega_{p,s}^{2}}{\omega|k_{\|}|v_{t,s}} \frac{Z_s}{|Z_s|}\sqrt{\frac{2}{b_s}}\, lI_l(b_s)\exp(-b_s)[1 + \zeta_{s,l}Z(\zeta_{s,l})] \approx 0. \tag{B.27}$$

References

[1] Stix T H and Nierenberg W A 1962 *The Theory of Plasma Wave* (New York: McGraw-Hill Book Company)

[2] Babusci D, Dattoli G, Licciardi S and Sabia E 2020 *Mathematical Methods For Physicists* (Singapore: World Scientific)

[3] Stix T H 1992 *Waves in Plasmas* (New York: Springer)

IOP Publishing

G Dattoli, E Di Palma, S P Sabchevski and I P Spassovsky

Appendix C

Free electron laser small signal high gain equations

The high gain integral equation, we have discussed in the main body of chapter 4 (see equation 4.48), can be can be reduced to a third order ordinary differential equation. The explicit reduction procedure has been detailed in [1, 2], where it has been shown that the small signal FEL dynamics is ruled by (where the apices denote derivatives with respect to τ)

$$a''' - 2i\nu_0 a'' + \nu_0^2 a = \pi\, g_0, \tag{C.1}$$

whose initial conditions specify different operating regimes. We start by assuming that

$$\begin{aligned} a(0) &= a_0, \\ a'(0) &= 0, \\ a''(0) &= 0. \end{aligned} \tag{C.2}$$

The full solution requires a length procedure involving the search of the roots of a third degree algebraic equation and can be cast in the Fang–Torre form

$$\begin{aligned} a(\tau,\ \nu_0) = \frac{a_0}{3\,(\nu_0 + p + q)}\, e^{-\frac{2}{3}\, i\nu\tau} \Bigg\{ \Big(-\nu_0 + p + q \Big)\, e^{-\frac{i}{3}\,(p+q)\,\tau} \\ + 2\Big(2\,\nu_0 + p + q \Big)\, e^{\frac{i}{6}\,(p+q)\,\tau} \\ \left[\cosh\left(\frac{\sqrt{3}}{6}(p - q)\tau \right) + i\frac{\sqrt{3}\,\nu_0}{p - q} \sinh\left(\frac{\sqrt{3}}{6}(p - q)\tau \right) \right] \Bigg\} \end{aligned} \tag{C.3}$$

$$p = \left[\frac{1}{2}\left(r + \sqrt{d} \right) \right]^{\frac{1}{3}}, \quad q = \left[\frac{1}{2}\left(r - \sqrt{d} \right) \right]^{\frac{1}{3}}$$

$$r = 27\,\pi\, g_0 - 2\,\nu_0^3, \quad d = 27\,\pi\, g_0[27\,\pi\, g_0 - 4\,\nu_0^3].$$

doi:10.1088/978-0-7503-2464-9ch10

Which is easily reconciled with the high gain variables by replacing

$$p\,\tau = \sqrt{3}\left(\sqrt[3]{1 + \frac{\sqrt{1 - 4\,\tilde{\nu}_0^3}}{1 - 2\,\tilde{\nu}_0^3}}\right)\tilde{z},\ q\,\tau = \sqrt{3}\left(\sqrt[3]{1 - \frac{\sqrt{1 - 4\,\tilde{\nu}_0^3}}{1 - 2\,\tilde{\nu}_0^3}}\right)\tilde{z}, \tag{C.4}$$

and the dimensionless seed amplitude intensity with

$$I_0 = \frac{\tilde{I}_s}{8\,\pi^2}\,|a_0|^2,\ \tilde{I}_s\left[\frac{MW}{cm^2}\right] = 6.931\,2 \cdot 10^2 \frac{(4\,\pi\,\gamma\,\rho)^4}{\left[\lambda_u[cm]\,K\,f_b(\xi)\right]^2}. \tag{C.5}$$

According to the initial conditions (C.2) the laser field grows from an initial coherent seed. The process described by equation (C.3) is essentially that of a high gain amplifier. On the other side in SASE devices the field growth is triggered by the spontaneous emission noise, determining a kind of bunching on the electron beam itself. The evolution of the dimensionless field amplitude along the longitudinal coordinate is, in this case, provided by

$$\begin{aligned}
a(\tau) &= \frac{2\pi g_0 b_1}{\Delta} e^{-i\frac{2\nu_0}{3}\tau}\left\{\mu e^{-\frac{i}{3}(p+q)\tau} - \left(\chi_- \cosh\left[\frac{\sqrt{3}}{6}\left(p - q\right)\tau\right] + \right.\right.\\
&\quad \left.\left. + \chi_+ \sinh\left[\frac{\sqrt{3}}{6}\left(p - q\right)\tau\right] e^{\frac{i}{6}(p+q)\tau}\right)\right\},\\
\Delta &= \frac{\sqrt{3}}{9}\left(p^3 - q^2\right),\\
\mu &= i\frac{\sqrt{3}}{9}[\nu_0 - (p + q)]\left(p - q\right),\\
\chi &= \frac{1}{6}\left[\nu_0 + \frac{1}{2}\left(p + q\right) - i\frac{\sqrt{3}}{2}\left(p - q\right)\right]\left[p + q + i\frac{\sqrt{3}}{2}\left(p - q\right)\right],\\
\chi_\pm &= \chi \pm \chi^*,
\end{aligned} \tag{C.6}$$

with b_1 being the initial bunching at the fundamental harmonic. The solution given in equation (C.6) is found by solving equation (C.1) with the conditions

$$a(0) = \nu_0,\quad a'(0) = -2\pi g_0 b_1,\quad a''(0) = 2\pi\nu_0 g_0 b_1. \tag{C.7}$$

In figure C.1 we report the evolution of the equation (C.6) for two different initial conditions.

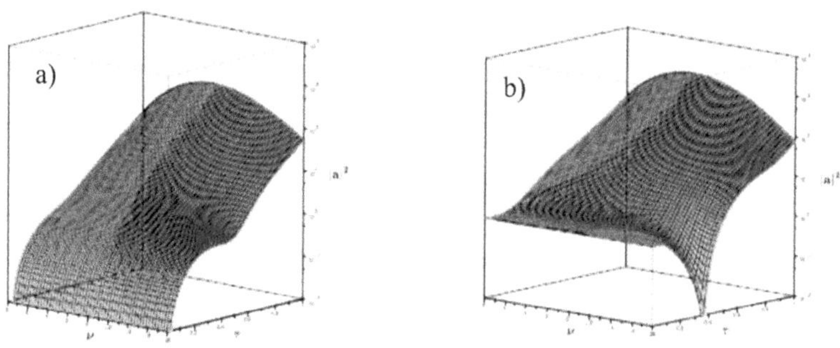

Figure C.1. (a) $a(0) = 0$, $b_1 = 10^{-3}$, $g_0 = 50$. (b) $a(0) = 1$, $b_1 = 0$, $g_0 = 50$.

References

[1] Ciocci F, Dattoli G, Torre A and Renieri A 2000 *Insertion Devices for Synchrotron Radiation and Free Electron Laser* (Series on Synchrotron Radiation Techniques and Applications vol 6) (Singapore: World Scientific)

[2] Dattoli G, Renieri A and Torre A 1993 *Lectures on the Theory of Free Electron Laser and Related Topics* (Singapore: World Scientific) https://doi.org/10.1142/1334

Lightning Source UK Ltd.
Milton Keynes UK
UKHW031822230921
391074UK00004B/249